SCIENTIFIC EXPLANATION

A STUDY OF THE FUNCTION OF THEORY, PROBABILITY AND LAW IN SCIENCE

BY

RICHARD BEVAN BRAITHWAITE

FELLOW OF KING'S COLLEGE, CAMBRIDGE
KNIGHTBRIDGE PROFESSOR OF MORAL PHILOSOPHY
IN THE UNIVERSITY OF CAMBRIDGE

BASED UPON THE
TARNER LECTURES
1946

CAMBRIDGE
AT THE UNIVERSITY PRESS
1955

PUBLISHED BY
THE SYNDICS OF THE CAMBRIDGE UNIVERSITY PRESS
London Office: Bentley House, N.W. 1
American Branch: New York

Agents for Canada, India, and Pakistan: Macmillan

First edition 1953
Reprinted 1955
American printing 1956

First printed in Great Britain at the University Press, Cambridge
Reprinted by offset-litho
MANUFACTURED IN THE UNITED STATES

TO THE MEMORY OF
JOHN MAYNARD KEYNES

CONTENTS

PREFACE

This book contains the substance of the course of lectures which I delivered in 1946 as Tarner Lecturer of Trinity College, Cambridge. In preparing the lectures for publication I have throughout developed, elaborated and (I hope) improved their line of argument.

My primary purpose in this book is to examine the logical features common to all the sciences. It is almost a platitude to say that every science proceeds, more or less explicitly, by thinking of general hypotheses, of greater or less generality, from which particular consequences are deduced which can be tested by observation and experiment. But the implications of this view are by no means platitudinous; and this book aims to show how these implications will throw light upon many features of scientific procedure which appear mysterious and will resolve many of the difficulties which philosophers have found in these procedures.

For science, as it advances, does not rest content with establishing simple generalizations from observable facts: it tries to explain these lowest-level generalizations by deducing them from more general hypotheses at a higher level. Such an organization of a science into a hierarchical deductive system requires the use of subtle deductive techniques, which are provided by pure mathematics. As the hierarchy of hypotheses of increasing generality rises, the concepts with which the hypotheses are concerned cease to be properties of things which are directly observable, and instead become 'theoretical' concepts—atoms, electrons, fields of force, genes, unconscious mental processes—which are connected to the observable facts by complicated logical relationships. The first four chapters of this book are devoted to explaining the parts played by mathematical reasoning and by theoretical concepts and 'models' in the organization of a scientific theory.

There has been too much mystery-mongering about the 'abstract mathematical entities' involved in an elaborate theoretical system like that of contemporary physics. It is true that the mathematics employed in such a system is difficult, and can only be understood after long training. But to understand the way in which the difficult mathematics is used requires no mathematical ability; what is needed is a preparedness to think explicitly about modes of thinking, and about the way in which scientific language and symbolism is used in the expression of a scientific theory. I have therefore been at pains to construct the simplest possible example of a scientific theory—too simple to occur, as it stands, in any science—in order to show both exactly why theoretical concepts find their way into scientific theory and exactly how mathematico-logical deductions are involved. This simple example, in which only the non-numerical algebra of classes is used, enables me also to give a precise meaning to the notion of a 'model' for a scientific theory and to discuss the proper (and the improper) use of such models.

Probability statements, their meaning and the role they play in science, form the theme of the next three chapters. A philosopher of science, viewing recent developments of physics from 'above the battle', will take the most revolutionary change to be the fundamental place given by quantum mechanics to irreducibly statistical ('indeterministic') laws. He must therefore try to give an account of how such probability statements work within scientific deductive systems. In these chapters the peculiarity of scientific probability statements is found to lie in the fact that they are only provisionally, but never definitively, refutable by the sets of observations which test them. The 'looseness of fit' of such statements is, however, itself an exact notion; and sufficient of probability mathematics is developed (here again I have been at pains to use the least possible amount of mathematical apparatus) to elucidate the meaning of probability as used in science. The seventh chapter treats of the principles which should govern the choice between two alternative statistical hypotheses; the new angle from which the problem is viewed is due to the recent work of Abraham Wald.

The justification of induction and the status of natural and of causal laws, problems which have teased philosophers since the time of Hume, are discussed in the next two chapters. Scientists themselves have never been much worried by these problems; and this fact is highly relevant to the way in which a philosopher should treat them, though it does not (as is sometimes thought) excuse a philosopher from having to consider them at all. Here, as elsewhere in the book, I have attempted the difficult task of saying something positive, and which is as definite as I could make it, instead of criticizing rival philosophical views. I am not at all satisfied with what I have said about induction; but I believe that it is worth saying as a useful contribution to further discussion. The final two chapters connect together the various ways in which explanations are provided by science.

If I were asked what there is that is original in this book, I should mention specifically the accounts of probability as used within a science (Chapters V and VI), of the relation between scientific theories and 'models' for them (Chapter IV), and of teleological explanation (Chapter X). But I should hasten to add that I should not have written a book in the form of a general work on the philosophy of science had I not believed that my way of exposition was novel and valuable. I have tried to bridge the gap between the awareness of all educated persons of the general way in which science proceeds (by the hypothetico-deductive method) and their vagueness or puzzlement about such scientific fundamentals as the function of mathematics in science, the nature of theoretical concepts and the principles of statistical inference by examining in great detail simple examples specially constructed to display, without irrelevances, the logically important features. Though these examples may not be, strictly speaking, a contribution to 'learning and research', they are, I believe, a definite contribution to 'education'. Conversely, the omission from this book of many topics which might properly have found a place (e.g. the nature and function of measurement in science) is explained by the fact that I felt that I had nothing fresh to say about them.

My first acknowledgment must be to my pupils and others who have disagreed and argued with me during the past quarter-century, and have thus forced me to improve both the matter and the manner of my thinking upon the philosophy of science. My obligations to particular philosophers, both those who have thought along the same and those who have thought along different lines, are too numerous to be separately mentioned; but it is clear to me that I should not be philosophizing in the way I do had it not been my good fortune to have sat at the feet, in Cambridge, of G. E. Moore and of Ludwig Wittgenstein.

But there are two specific acknowledgments that must be made. My thinking about the role of theoretical concepts and of the philosophy of science in general was re-oriented when I had the melancholy privilege of editing F. P. Ramsey's brilliantly suggestive papers after his death in 1930. Nine years earlier it was the publication of J. M. Keynes's *Treatise on Probability* in the summer of 1921 which attracted both Ramsey and myself (Ramsey at the end of his first year, I at the end of my second year, as undergraduates) to the problems of probability and induction. There is little reference to Keynes's work in this book, and what there is is critical, since I cannot now accept his view of probability. But it was the influence of Keynes—of his friendship and encouragement no less than of his exciting book—which caused me to think seriously about the logic of science. When I was giving the Tarner Lectures, in the Lent Term of 1946, I asked, and obtained, his leave to dedicate to him the book that would come out of the lectures. The dedication can now only be to his memory.

I am very grateful to the Master and Fellows of Trinity College for honouring me by the invitation to give the ninth course of Tarner Lectures. The Aristotelian Society have kindly permitted me to incorporate in Chapter X of this book my Presidential Address to the Society in 1946, already published in their *Proceedings*. I am greatly indebted to F. J. Anscombe for reading critically the three chapters concerned with probability (once in typescript, once in proof), for making very helpful comments and for saving me from one straight mathematical mistake. Harold Jeffreys and Gilbert Ryle, who read the typescript, have made suggestions from which I have profited. Finally I must express my gratitude to my wife, Margaret Masterman, for her constant encouragement during the long-drawn-out writing and rewriting of this book.

R. B. BRAITHWAITE

Cambridge
31 *October* 1951⎫
23 *August* 1952⎭

The reprinting of this book gives me the opportunity of correcting a number of minor errors, and also a defect in my original definition of *impure deductive system* (p. 34) which was pointed out to me by C. G. Hempel. I am most grateful to all those who have helped me in this way.

R. B. B.

4 *November* 1954

CHAPTER I

INTRODUCTORY

WHAT IS SCIENCE?

In this book the word "science" will be taken to include all the natural sciences, physical and biological, and also such parts of psychology and of the social sciences (anthropology, sociology, economics) as are concerned with an empirical subject-matter. It will exclude all philosophy which is not 'general science', all history which is concerned merely with the occurrence of particular historical events, and the disciplines of pure mathematics and symbolic logic which are not (except, perhaps, in a very peculiar sense) about empirical facts at all. This sense of the word "science" corresponds pretty closely with the most frequent modern use of the word (whose first public use was perhaps in the title of the British Association for the Advancement of Science, founded in 1831); it is synonymous with "natural science" if man is included within nature.

The function of a science, in this sense of the word, is to establish general laws covering the behaviour of the empirical events or objects with which the science in question is concerned, and thereby to enable us to connect together our knowledge of the separately known events, and to make reliable predictions of events as yet unknown. This function of establishing general laws is common to all the natural sciences; it is characteristic also of those parts of psychology and of the social sciences which would ordinarily be called scientific as opposed to philosophical. If the science is in a highly developed stage, as in physics, the laws which have been established will form a hierarchy in which many special laws appear as logical consequences of a small number of highly general laws expressed in a very sophisticated manner; if the science is in an early stage of development—what is sometimes called its 'natural-history' stage —the laws may be merely the generalizations involved in classifying things into various classes. But to classify a whale as a mammal is to assert the generalization that all infant whales are provided with milk by their mothers, and this proposition is a general law, although of limited scope. It enables us to predict that the next whale we meet will be a mammal, and it singles out an important feature in which whales differ from fishes.

To emphasize the establishment of general laws as the essential function of a science is not to overlook the fact that in many sciences the questions to which the scientist attaches most importance are historical questions about the causes of particular events rather than questions directly about general laws. Biologists ask for the origin of life upon the earth, astronomers for the origin of the solar system. But the statement that some particular event is the effect of a set of circumstances involves the assertion of a general law; to ask for the cause of an event is always to ask for a general law which applies to the particular event. Though we may be more interested in the application than in the law in itself, yet we need to establish the law in order to know what law it is which we have to apply.

The fundamental concept for science is thus that of scientific law, and the fundamental aim of a science is the establishment of such laws. In order to understand the way in which a science works, and the way in which it provides explanations of the facts which it investigates, it is necessary to understand the nature of scientific laws, and what it is to establish them.

It is self-contradictory to speak of a 'false scientific law'. Since we shall be concerned not with the truth of actual scientific laws, but with the nature of propositions which, if they are true, are scientific laws, we shall call such propositions "scientific hypotheses". A scientific hypothesis is a general proposition about all the things of a certain sort. It is an empirical proposition in the sense that it is testable by experience; experience is relevant to the question as to whether or not the hypothesis is true, i.e. as to whether or not it is a scientific law.

THE EMPIRICAL BASIS OF SCIENCE

What is meant by "experience" here? There are, broad and large, two alternative meanings that may be given to experience in such a context—a wider and a narrower meaning.

The wider meaning is that in which experience covers all the facts which in ordinary language would be said to be observed facts—facts about material objects such as chairs and tables, facts

about physical events such as flashes of lightning, as well as facts
about sensations or other experiences and their objects or contents.
The narrower meaning is that in which the word covers only this
last class of facts—facts which are the data of immediate, direct,
indubitable, incorrigible knowledge (to quote the adjectives used by
philosophers). Philosophers who take the phenomenalist view that
knowledge about material objects and events is in some way or other
analysable in terms of knowledge about sensations or their objects
(sense-data) will tend to use the narrower meaning of experience;
while those who take a realist view that such knowledge is not so
analysable, and that material objects are logically independent of
sensation, will tend to use the term experience with the wider
meaning.

Now a book, like this book, whose subject is the philosophy of
science need not concern itself with the very delicate questions
involved in the discussion of the philosophy of perception or of
'our knowledge of the external world'. These questions, in one
form or another, have engaged the attention of the greatest philosophers since the time of Socrates; and there is no agreement
among the best contemporary philosophers upon how these questions should be answered, nor, indeed, upon how exactly the
questions should be formulated. The phenomenalists have not been
able to give analyses of propositions about material objects which
have convinced their opponents,* and the realists have not been
able fully to satisfy the demand, made by all philosophers in the
empiricist tradition, that immediate experience must be regarded
as the final court of appeal. The controversy in its modern form
has become deeply involved with questions as to the nature of
communication and of the use of language; some realists question
whether statements about immediate experience which by its
nature is private to the experiencer and incommunicable to another
are not more analogous to interjections than to publicly communic-
able propositions, while phenomenalists cannot see how a public
language about material objects can be rooted in fact except by
reference to the incorrigible knowledge provided by immediate
experience. But the whole discussion is irrelevant to the questions

* My own attempt at a quasi-phenomenalist analysis will be found in *Proceedings of the Aristotelian Society*, n.s., vol. 38 (1937-8), pp. 269ff.; and in
Erkenntnis, vol. 7 (1938), pp. 281ff.

with which this book is concerned. What we are concerned with here is the nature of scientific laws and how they are related to observed facts. These observed facts are usually facts about the behaviour of material objects—the motion of a body, the mark on the scale to which the needle of a measuring instrument points, the explosion of an atomic bomb; what we are interested in is the relation of facts such as these to the scientific laws which cover them. We are not in this context interested in the question as to whether or not these facts are immediately experienced, and, if they are not, what is their relation to the sense-data or other entities which are immediately experienced. Thus for the purpose of discussing the philosophy of science we can take experience in the wider sense as embracing all facts that would ordinarily be said to be observed; and we can put aside the question as to whether there is or is not another more fundamental sense of experience—immediate experience—in terms of which experience in the wider sense must ultimately be analysed.

A philosophical realist and a phenomenalist can perfectly well agree upon the analysis of a law of mechanics in terms of the observable motions of material bodies. They will disagree as to whether or not these observable events themselves require analysis in terms of something epistemologically more primitive. But for the phenomenalist the two stages in his analysis are distinct; and the realist and he can agree to discuss his first stage while agreeing to differ as to whether or not there is a second.

Problems in philosophy are so interconnected that the opportunity of separating them should never be neglected when it occurs. In this book, therefore, the experience to which scientific laws are referred will be taken as including all observable facts, where observation is taken in its wide, everyday meaning and with no philosophical limitation. And no reference will be made outside this chapter to the philosophy of perception or to the nature of our knowledge of the external world. Neither a realist nor a phenomenalist epistemology will be presupposed.

Since this policy of separating off the philosophy of science from the philosophy of perception is not one that has been followed by most philosophers of science, who have felt it incumbent upon them to discuss fundamental epistemological questions and to settle on either a realist or a phenomenalist position before passing on to their special problems, it is desirable to defend it against the most important criticisms that can be made against it.

(1) It might be that we have no right to suppose that we have any empirical knowledge at all until we have performed the philosophical task of analysing such knowledge. The answer to this criticism is that the function of a philosopher is not to doubt the propositions accepted by common sense, but to provide analyses or elucidations of them. Were I to say that I did not know that there is a pencil in my hand at this moment, I should be saying something which I knew to be false; and the falsity of my statement would not be mitigated were I to give as a reason for it that I did not know whether the proposition was about objects which existed independently of my experiences or whether it was ultimately about my experiences. This thesis as to the proper function of a philosopher, expounded explicitly in recent times by G. E. Moore, is now so widely accepted (at least the negative part of it—that it is futile to doubt, or to pretend to doubt, common-sense propositions) that it need not be elaborated here.

(2) Many philosophers of the phenomenalist persuasion have thought that the problems involved in knowledge of scientific laws are essentially the same as those involved in perceptual knowledge. An electron, they would say, is a logical construction out of ordinary material objects in the same way as that in which these objects are logical constructions out of sense-data; in each case the philosophical problem is to show how propositions about higher-order entities are to be analysed in terms of propositions about lower-order entities; and, since the two problems are essentially the same, it is foolish to isolate one and to consider it separately. Now I do not want to dispute that consideration of the scientific-law problem may throw a great deal of light upon the perception problem; indeed, I am convinced that the things that will be said in Chapter III about the way in which an electron can be an empirical concept without being explicitly definable in terms of experience are highly relevant to the problem with which the phenomenalist is confronted. But there is one feature in the perception problem which is lacking in the scientific-law problem, a feature which raises very subtle and difficult philosophical puzzles which are not involved in the consideration of scientific laws. This feature is the privacy of the sensations or sense-data or directly

known facts which the phenomenalist wishes to take as his ultimate data. The problem of perception includes considering how I pass from knowledge given by my own sensational experiences—a knowledge which is, for the phenomenalist, essentially private—to the public knowledge of the material objects of common sense. Thus a philosopher of perception has either to refute or to come to terms with solipsism, and has to explain the passage from privacy (or what appears to be privacy) to publicity. But a philosopher of science concerned with the relationship of public scientific laws to public facts of observation need not be concerned at all with the problem of how the privacy of immediate experience is transcended. He has therefore an easier task than the philosopher of perception; and he would be misguided to jeopardize his chance of success in solving the difficult problems which concern him by mixing them with the even more difficult problems of a quite different sort that confront the philosopher of perception.

(3) My critic may, however, retort that, though what I have just said was sufficient to justify this policy for a philosopher of science before 1905, the advent of relativity theory and quantum mechanics has changed all that. The characteristic of the new physics, we are frequently told, is that it builds its structure solely on observables, and that by discarding concepts dear to common sense—absolute simultaneity, material substance, deterministic laws—it has been able to construct fundamental theories which synthesize experience better than do the theories of classical physics. Herbert Dingle* goes so far as to maintain that this much-heralded feature of the new physics has indeed been characteristic of the whole of physics since Galileo, though it is only recently that it has been explicitly recognized. The contention is that post-Einsteinian (or, for Dingle, post-Galilean) physics is not based upon common-sense knowledge at all; its data are events such as pointer readings, not permanent material substances; the scientific construction of electrons, protons, etc., out of these directly observable events does not proceed by way of material objects at all. The logical construction of scientific objects like electrons out of what is directly experienced by-passes the logical construction of ordinary material objects out of the same experience; the two logical constructions are parallel constructions, each serving a different purpose. Physics does not continue the

* *Through Science to Philosophy* (Oxford, 1937), Chapters IV, V.

work of synthesis beyond the stage at which common sense has stopped; on the contrary, its synthesis proceeds along another route and makes use of features in experience which common sense, rightly for its own purpose, ignores.

Now were this contention correct, it would, of course, be impossible to treat the observations which form the basis of scientific laws as being observations (as common sense understands the word) of material objects and their properties. It would be necessary to go down to ground bottom, and to develop a phenomenalist theory of science basing it upon directly experienced objects only. But I see no reason to accept the contention—even about contemporary physics. When quantum physicists say that "science is concerned only with observable things", the contexts make it clear that they are speaking not of sense-data but of observations in the ordinary laboratory sense; they are contrasting laboratory observables with non-observables such as the wave-functions which they use in their theories; all they are maintaining is that science is an empirical study. The 'observations' to which they refer need not even be observations in the ordinary sense, which presuppose a basis of immediate experience; they may be traces on a photographic film or holes punched in a tape by an electron counter. So far as the science is concerned, the 'observing' may all be done by machines, the records of these machines being subsequently inspected and interpreted. Failure to appreciate this point has made many philosophers, and some scientists, believe that modern developments have made physics more subjective and more dependent upon the idiosyncrasies of the human observer than it used to be. But the observer whose motion is relevant to facts of simultaneity, according to relativity theory, is not a disembodied mind sensing sense-data (minds cannot move); it is either a human body or a recording instrument such as a camera. And the observations which disturb the positions or velocities of objects which he is trying to observe, according to quantum mechanics, are not mental acts, but processes in lenses or spectroscopes. There is nothing subjective about Heisenberg's Uncertainty Principle: the impossibility of measuring precisely two correlated physical quantities is not a mental impossibility, but is a consequence of the fundamental statistical laws of quantum mechanics, according to which there is an inverse relationship between the precisions with which these correlated quantities can be measured by any apparatus obeying these fundamental laws.*

The policy to be adopted in this book of ignoring the problems of the philosophy of perception seems therefore to be free from serious objection. A consequence of this policy is that this book will not profess to deal with every philosophical problem that arises in considering the nature of science. It would be ridiculous to expect this; the set of problems with which this book is concerned are those which arise in considering the nature and the verification by observation of scientific theories and laws, and these topics are surely enough for one book.

In this book experience, observation,† and cognate terms will be used in the widest sense to cover observed facts about material objects or events in them as well as directly known facts about the contents or objects of immediate experience. It may be said that it is unnecessary to include this second class of facts at all, if all the empirical data of science are facts about material objects or events in them. But this would be to exclude psychology from the status of a science, unless, indeed, the data of psychology be limited, as behaviourists would wish, to public observable facts about the behaviour of human bodies. Since I do not wish to limit psychology from the beginning in this way, nor so to limit sociology and economics should these also wish to use facts of immediate experience as data, such facts will be included among the observable facts which can be taken to be the empirical basis of a science. In using experience and observation in this way, I am following common usage; I speak of observing that I have toothache as well as of observing that I have a pen in my hand, and both my toothache and my pen are experienced by me, though the latter experience will not be immediate if phenomenalism or a representative theory is the correct philosophy of perception.

The publicity of its data is therefore not used (as many writers would wish to use it) as the hall-mark of a science. The physical

---

* Quantum mechanics is frequently alleged to be subjective in a different way, in that it predicts probabilities instead of certainties. But, as will be shown in Chapter VI, the probabilities with which it is concerned are to be explained in terms of statistical frequencies, and a statistical frequency of 50 % is no more subjective than one of 100 %.

† Observation will also sometimes be used to include facts about recording instruments (photographic films, Geiger counters, etc.) which are interpreted as corresponding to observable facts.

and the ordinary biological sciences use only public data open to observation by all, but the psychological and social sciences may also use private data of immediate experience. These private data are in certain cases deducible from general hypotheses in the same way as that in which public data are so deducible; it is this hypothetico-deductive method applied to empirical material which is the essential feature of a science; and if psychology or economics can produce empirically testable hypotheses, *ipso facto* they are sciences, whether or not the consequences of the hypotheses are private and incommunicable.*

SCIENTIFIC LAWS

We can now turn to the general laws the establishment of which, on the basis of experience, is the function of a science. What is a scientific law?

The one thing upon which everyone agrees is that it always includes a generalization, i.e. a proposition asserting a universal connexion between properties. It always includes a proposition stating that every event or thing of a certain sort either has a certain property or stands in certain relations to other events, or things, having certain properties. The generalization may assert a concomitance of properties in the same thing or event, that everything having the property A also has the property B; e.g. that every specimen of sugar is soluble in water. Or it may assert that, of every two events or things of which the first has the property A and stands in the relation R to the second, the second has the property B; e.g. that in every case of every pair of free billiard balls of which the first is moving and strikes the second, the second will move. Or it may make more complicated but similar assertions about three or four or more things. The relationship between the things may be a relationship holding between simultaneous events in the things, or it may hold between events in the same thing or in two or more things which are not simultaneous. The questions raised by the temporal relationships will be touched upon in Chapter IX. All these types of generalization may be brought under concomitance generalizations—that everything which is A

* This, of course, is not all there is to say upon the problems which under the titles of 'solipsism' or 'other minds' have agitated philosophers. But it is enough for my purpose in this book.

9

is B—by allowing A and B to be sufficiently complex properties. So if we confine our attention for the present to concomitances, our remarks will apply also to generalizations which are of more complicated type.

While there is general agreement that a scientific law includes a generalization, there is no agreement as to whether or not it includes anything else. The apodeictic language in which we frequently express scientific laws (e.g. "Mixtures of hydrogen and oxygen *must* explode if an electric spark occurs in the mixture") tends to make us think that they assert something over and above the mere constant conjunction which would be asserted in the corresponding generalization; and philosophers have propounded views according to which the extra something is a logical relation analogous to that holding between premisses and conclusion in deductive inference, or a relation of activity analogous to that occurring in volition, or a quite specific relation not analogous to anything else. But in the two hundred years since Hume preached the doctrine that there is not this extra something (except a psychological fact about the association of ideas or beliefs in the mind of the person believing the law), the main reason that has made many philosophers develop other views is their sense of the inadequacy of a constant conjunction theory along Hume's lines. So there would seem to be no need to consider views which make scientific laws more than statements of constant conjunction if an adequate constant conjunction view can be maintained.

Such a view will be developed in this book. Scientific laws will be taken as asserting no more (and no less) than the *de facto* generalizations which they include; the law that every hydrogen atom consists of one proton together with one electron will be interpreted as meaning that, as a matter of fact, every hydrogen atom, past, present and future, has this constitution. But the way in which we use such scientific generalizations will be found to be more complicated than the way which presented itself to Hume in the eighteenth century. For Hume the question was how to account for induction by simple enumeration; how, that is, we can pass from knowledge of instances to a rational belief in the generalization under which these instances fall. To do this he made use of a propensity of the mind to associate together ideas corresponding to things which had previously been found associated in experience.

But an adequate theory of science to-day must explain how we come to make use of sophisticated generalizations (such as that about the proton-electron constitution of the hydrogen atom) which we certainly have not derived by simple enumeration of instances. To explain this the generalization must be considered, not in isolation, but in reference to the place which it occupies within a scientific system; and induction must be considered, not primarily as induction by simple enumeration, but as the method by which we establish hypotheses within scientific systems. The combination of the constant-conjunction view that scientific laws are only generalizations with a doctrine of the function of such generalizations within scientific systems puts the constant-conjunction view in a new light. It answers the complaint that the constant-conjunction view underestimates the place of reason in science; by stressing the importance of the logical relationships between generalizations at different levels within the system, it goes some way to satisfy those who hanker after logically necessary scientific laws.

My justification for supposing at the outset that scientific laws are no more than generalizations is that the adequacy of this view can only be shown in the development of a theory of science which includes it. The constant-conjunction view of scientific laws has been held by most writers on the philosophy of science since the time of Mach; what has hindered other philosophers from accepting it is that, in its usual formulations, it is open to damaging criticism. It is my aim to escape these criticisms by incorporating it within a wider view of the function of generalizations in science.

Another reason for starting with the constant-conjunction view is that, according to it, scientific laws are logically weaker propositions than they would be on any alternative view of their nature. On any other view a scientific law, while including a generalization, states something more than the generalization. Thus the assumption that a scientific law states nothing beyond a generalization is the most modest assumption that can be made. This modesty is of great importance in considering the problem of induction. It is difficult enough to justify our belief in scientific laws when they are regarded simply as generalizations; the task becomes more difficult if we are required to justify belief in propositions which are more than generalizations. This fact does not seem to me to have been sufficiently recognized by many of those who criticize the constant-conjunction view for affording no sound basis for induction.

Scientific hypotheses, which, if true, are scientific laws, will then, for the purpose of my exposition, be taken as equivalent to generalizations of unrestricted range in space and time of greater or less degrees of complexity and generality. Other views of the nature of scientific laws will be touched on only incidentally.

THE STRUCTURE OF A SCIENTIFIC SYSTEM

A scientific system consists of a set of hypotheses which form a deductive system; that is, which is arranged in such a way that from some of the hypotheses as premisses all the other hypotheses logically follow. The propositions in a deductive system may be considered as being arranged in an order of levels, the hypotheses at the highest level being those which occur only as premisses in the system, those at the lowest level being those which occur only as conclusions in the system, and those at intermediate levels being those which occur as conclusions of deductions from higher-level hypotheses and which serve as premisses for deductions to lower-level hypotheses.

Let us consider as an example a fairly simple deductive system with hypotheses on three levels. This example has been selected principally because it illustrates excellently the points that need to be made, and partly because the construction and the establishment of a similar system by Galileo marks a turning-point in the history of science.

The system has one highest-level hypothesis:

I. Every body near the earth freely falling towards the Earth falls with an acceleration of 32 feet per second per second.

From this hypothesis there follows, by simple principles of the integral calculus,* the hypothesis:

II. Every body starting from rest and freely falling towards the Earth falls $16t^2$ feet in t seconds, whatever number t may be.

From II there follows in accordance with the logical principle (the *applicative principle*) permitting the application of a generalization to its instances, the infinite set of hypotheses:

---

* Hypothesis I can be expressed by the differential equation $\mathrm{d}^2s/\mathrm{d}t^2 = 32$, whose solution, under the conditions that $s = 0$ and $\mathrm{d}s/\mathrm{d}t = 0$ when $t = 0$, is $s = 16t^2$.

III*a*. Every body starting from rest and freely falling for 1 second towards the Earth falls a distance of 16 feet.

III*b*. Every body starting from rest and freely falling for 2 seconds towards the Earth falls a distance of 64 feet.

And so on.

In this deductive system the hypotheses at the second and third levels (II, III*a*, III*b*, etc.) follow from the one highest-level hypothesis (I); those at the third level (III*a*, III*b*, etc.) also follow from the one at the second level (II).

The hypotheses in this deductive system are empirical general propositions with diminishing generality. The empirical testing of the deductive system is effected by testing the lowest-level hypotheses in the system. The confirmation or refutation of these is the criterion by which the truth of all the hypotheses in the system is tested. The establishment of a system as a set of true propositions depends upon the establishment of its lowest-level hypotheses.

The lowest-level hypothesis III*a* is tested by applying it to a particular case. A body is allowed to fall freely for 1 second and the distance it falls measured.* If it is found that it falls 16 feet, the hypothesis is confirmed; if it is found that it falls more, or less, than 16 feet, the hypothesis is refuted.

It is convenient to treat the logic of this procedure as consisting of two steps. A case is either observed, or experimentally produced, of a body falling for 1 second. The following proposition is then empirically known:

e_1. This body freely falls for 1 second towards the Earth, starting from rest.

The general hypothesis III*a* is then applied to this case by first deducing from III*a* the proposition:

III*a'*. It is only the case that this body, starting from rest, freely falls for 1 second towards the Earth if it falls a distance of 16 feet.

From this application of the general hypothesis, together with the proposition e_1, there is deduced:

f_1. This body falls 16 feet.

---

* The system tested by Galileo was, in fact, more complicated than our example. Galileo was unable to measure times of fall of bodies accurately enough to test III*a*; what he tested empirically was a more elaborate system in which the lowest-level hypotheses were propositions about descents of bodies rolling down grooves in inclined planes.

The testing of a scientific hypothesis thus consists in deducing from it a proposition of the form «e_1 only if f_1». Then there follows from the conjunction of e_1 with this proposition the third proposition f_1, whose truth or falsity is observed.

If f_1, the logical consequence of e_1 and IIIa', is observed to be true, the hypothesis IIIa is ordinarily said to be confirmed. The piece of evidence f_1, conjoined with e_1 (which conjunction will be called an *instance* of the hypothesis), is said to support the hypothesis. But it is clear that this one piece of evidence is insufficient to prove the hypothesis. It would only do so if the hypothesis were a logical consequence of the conjunction of f_1 with e_1. This, of course, is not the case. It is perfectly possible for the hypothesis to hold in this one instance, but to be false in some other instance, and consequently false as a general proposition. And, indeed, this is the case however many times the hypothesis is confirmed. However many conjunctions f_1 with e_1, f_2 with e_2, etc., have been examined and found to confirm the hypothesis, there will still be unexamined cases in which the hypothesis might be false without contradicting any of the observed facts.* Thus the empirical evidence of its instances never proves the hypothesis: in suitable cases we may say that it *establishes* the hypothesis, meaning by this that the evidence makes it reasonable to accept the hypothesis; but it never *proves* the hypothesis in the sense that the hypothesis is a logical consequence of the evidence.

The situation is different if f_1 is observed to be false. For the conjunction of not-f_1 with e_1 is logically incompatible with the hypothesis being true; the falsity of the hypothesis is a logical consequence of the conjunction of not-f_1 with e_1. Calling this conjunction a *contrary instance* of the hypothesis, we may say that a hypothesis is proved to be false, or refuted, by one known contrary instance.

This asymmetry of confirmation and refutation is a consequence of the fact that all the hypotheses of a science are general proposi-

* Unless, of course, the hypothesis has only a limited number of instances, and not only have all these instances been examined, but it is known that there are no unexamined instances. Generalizations with only a limited number of instances, which can be proved from a knowledge of these instances by what logicians have called *perfect induction*, present no logical problem; since they are of little interest in science, they will not be considered further. All scientific hypotheses will be taken to be generalizations with an unlimited number of instances.

tions of the form « Every A is B ». Propositions of the form « Some A's are B's » (existential propositions),* which are the contradictories of general propositions, have the reverse asymmetry; they can be proved by one instance, but no number of contrary instances will suffice to disprove them.

It has been said that there is no greater tragedy than the murder of a beautiful scientific hypothesis by one discordant instance. As will be seen, it is usually possible to save any particular higher-level hypothesis from this fate by choosing instead to sacrifice some other higher-level hypothesis which is essential to the deduction. But the fact that in principle a scientific hypothesis (or a conjunction of scientific hypotheses, if more than one are required for the deduction of the observable facts) can be conclusively disproved by observation, although it can never be conclusively proved, sharply distinguishes the question of the refutation of a scientific theory from that of its establishment. The former question is a simple matter of deductive logic, if the system of hypotheses is taken as a whole;† the latter question involves the justification of inductive inference, a problem which has worried philosophers since the time of Hume and to which Chapter VIII of this book will be devoted.

So far as confirmation is concerned the relation between lowest-level hypotheses and a hypothesis on the next higher level is similar to that between instances of a lowest-level hypothesis and the hypothesis itself. In our examples, the third-level hypotheses III a, III b, etc., are special cases of the second-level hypothesis II; each of them can be seen to follow from II in accordance with the applicative principle, but II is not a logical consequence of any finite number of hypotheses such as III a, III b, etc. The formula $s = 16t^2$ can hold for any finite number of values of s and t without being true in general. This is most easily seen by representing the lowest-level hypotheses as points on a graph (fig. 1). The second-level hypothesis is represented by a curve passing through these points. Any number of curves beside the parabola $s = 16t^2$ can be

* In the terminology of traditional logic my existential propositions would be called "particular propositions" and my general propositions "universal propositions", the term "general proposition" being used to cover both types.

† So simple that Karl Popper takes the possibility of falsification by experience as the criterion for a system of hypotheses being an empirical scientific system (*Logik der Forschung* (Vienna, 1935), § 6).

drawn to pass through any finite number of points. Thus a refutation of III*a* serves to refute II, but a proof of III*a* does not prove II. Nevertheless, evidence in favour of III*a* is evidence in favour of II; and if the evidence is good enough for us to regard III*a*, III*b*, etc., as established, it may also be good enough for us to regard II also as established.

A similar relationship holds between II and I as holds between III*a* and II. The difference is that, whereas the method of deducing

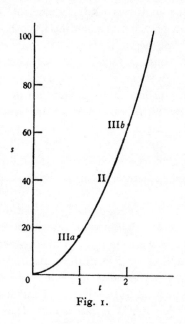

Fig. 1.

III*a* from II is merely the logical principle involved in implying a general proposition to a special case of it (the applicative principle), and this is implicit in the use of any general proposition, the deduction of II from I is made by using methods of the integral calculus, either explicitly by using known theorems of the calculus, or implicitly by constructing a special geometrical proof, as Galileo, ignorant of the calculus, had to do. But this difference is irrelevant to the general nature of the procedure. II is just as much a logical consequence of I as III*a* is of II, although to know the former relationship requires a knowledge of mathematics which must be specially learnt, whereas the latter relationship we learnt when we were taught how to use the word "every".*

The general characteristic of a deductive system is that the logical strength of the hypotheses increases the higher their level. Sometimes, although each of the hypotheses at a certain level is weaker than the one hypothesis at the next higher level from which they are all deducible, yet the conjunction of them all is equivalent to that hypothesis. This will happen when there are a limited number of special cases of the higher-level hypothesis, each of which is asserted by one of the lower-level hypotheses. It should be noted that this situation never arises in the relation of a lowest-level hypothesis to its instances, since no hypothesis is a mere enumerative generalization of a finite set of instances.

There are other important points about scientific deductive systems which are illustrated by our example. Since observed instances of IIIa are evidence for II as well as for IIIa, they are *indirect* evidence† for all the logical consequences of II, e.g. IIIb. Thus a hypothesis in a deductive system not at the highest level is empirically supported not only by observation of its instances, or of instances of hypotheses lying below it in the system, but also by observations of instances of other hypotheses in the system. The evidence for a scientific hypothesis is thus frequently much stronger than the *direct* evidence for it of its instances, or of instances of a hypothesis which logically follows from it; it also includes, as indirect evidence, the direct evidence for any higher-level hypothesis from which it logically follows. My reasons for believing that all men are mortal are not confined to knowledge that a great number of men have died; they include also knowledge that a great number of animals have died, which knowledge supports the wider generalization that all animals are mortal. One of the main purposes in organizing scientific hypotheses into a deductive system is in order that the direct evidence for each lowest-level hypothesis may become indirect evidence for all the other

* See below, pp. 82 ff.

† Observed facts will be said to be *indirect evidence* for a hypothesis p if they are direct evidence for a hypothesis q (or for a set of hypotheses q_1, q_2, etc.) from which p logically follows. A corollary of this definition is that if the observed facts are direct evidence for a set of hypotheses q_1, q_2, ..., q_n, they are indirect evidence for any one of these hypotheses, since each logically follows from the set.

17

lowest-level hypotheses; although no amount of empirical evidence suffices to prove any of the hypotheses in the system, yet any piece of empirical evidence for any part of the system helps towards establishing the whole of the system.

But there is one important point about most scientific deductive systems which our simple example does not illustrate. In our example only one higher-level hypothesis is used as a premiss in deductions to lower-level hypotheses. II logically follows from I alone, III a, III b, etc., from II alone. In most scientific deductive systems, however, each deduction requires more than one premiss. For instance, the example of Galileo's deductive system was incorporated by Newton into a larger deductive system in which I ceased to be the highest-level hypothesis, but was instead presented as deducible from the conjunction of two higher-level hypotheses, one being that called (in the plural) Newton's Laws of Motion and the other being his Law of Universal Gravitation.* Consequently most of the deductive systems used in science are not of the simply branching type exemplified by Galileo's system, but are systems in which there are a number of higher-level hypotheses all of which are required to serve as premisses in one or other of the deductions in the system. Of course if we were to extend our meaning of scientific hypothesis to include conjunctions of generalizations, we could lump all the premisses together into one conjunctive hypothesis, and have only one highest-level hypothesis in every system. Thus any scientific system could be treated as a simply branching system. But to combine disparate hypotheses into one conjunctive hypothesis would confuse thought. And it would have the added disadvantage that we should have to admit as scientific hypotheses propositions which were not themselves generalizations.

Since the consequences of any set of hypotheses are also consequences of any set of hypotheses which includes this set, the highest-level hypotheses could always be increased by adding any hypothesis to them. But to do this would be to make the observed facts evidence for a set of hypotheses which included one which played no part in their deduction from the set, and would thus make them indirect evidence for the supernumerary hypothesis and

---

* I am not stating the hypotheses explicitly, because they make use of the terms *force* and *mass* which are theoretical concepts of the sort to be discussed in Chapter III.

for its consequences. Since the supernumerary hypothesis might be any generalization whatever, this would have the undesirable result that any observable fact would be indirect evidence for any generalization whatever.* To avoid this result we stipulate that each of the highest-level hypotheses in a system must be necessary for the deduction of the lower-level hypotheses in the system; none must be included which play no part in the system. Similarly, we must treat as two systems, and not conflate into one, two systems whose sets of highest-level hypotheses have no hypothesis in common.

The fact that most scientific deductive systems employ more than one highest-level hypothesis has an important bearing upon the empirical testing of these hypotheses. As has been shown, one contrary instance is sufficient to refute a generalization, and the refutation of this generalization (a lowest-level hypothesis) will be sufficient to refute a higher-level hypothesis from which it logically follows. But suppose, as is frequently the case, that we are considering a deductive system in which there is no one higher-level hypothesis from which this lowest-level hypothesis follows, but instead the system is such that this follows from two or more higher-level hypotheses. Then what will be refuted by the refutation of the lowest-level hypothesis will be the conjunction of these two or more higher-level hypotheses; what will be a logical consequence of the falsity of the lowest-level hypothesis will be that at least one of the higher-level hypotheses is false.

Thus in the case of almost all scientific hypotheses, except the straightforward generalizations of observable facts which serve as the lowest-level hypotheses in the deductive system, complete refutation is no more possible than is complete proof. What experience can tell us is that there is something wrong somewhere in the system; but we can make our choice as to which part of the system we consider to be at fault. In almost every system it is

---

* C. G. Hempel (*Mind*, n.s., vol. 54 (1945); p. 104) raises a point similar to this as an objection to defining the 'confirmation' of a hypothesis in terms of what can be deduced from the hypothesis in conjunction with observable propositions. I do not profess to give a precise definition of 'confirmation'; but I believe that the limitation of hypotheses to be themselves general propositions, and the exclusion of supernumerary hypotheses from a scientific system, will avoid the difficulties pointed out by Hempel. There is an elaborate discussion of the whole subject in Rudolf Carnap, *Logical Foundations of Probability* (Chicago, 1950), §§ 87 f.

possible to maintain any one hypothesis in the face of apparently contrary evidence at the expense of modifying the others. Ptolemy was able to save the geocentric hypothesis by supposing that the planets moved in complicated orbits round the earth. But at some time a point is reached at which the modifications in a system required to save a hypothesis become more unplausible than the rejection of the hypothesis; and then the hypothesis is rejected.

The scientific deductive system which physics has gradually built up by incorporating the original deductive systems of Galileo and his contemporaries has developed by the rejection of hypotheses when the system which included them led to the prediction of observable results which were found not to be observed. But exactly which hypothesis was to be rejected at each point was a matter for the 'hunch' of the physicist. Generally speaking, a hypothesis was not rejected until an alternative hypothesis was available to take its place. Long before Einstein propounded his theory of gravitation it was known that Newton's theory could not account by itself for the observed motion of Mercury's perihelion. But Newton's theory was not dethroned until Einstein's theory was available to take its place. The process of refuting a scientific hypothesis is thus more complicated than it appears to be at first sight.

There is no hard and fast line at the point at which the common-sense synthesis of experience becomes a scientific ordering in a scientific system. Just as in tracing back common-sense thought either in the individual or in the race there is no point at which there were no generalizations believed, so in the history of a science there is rarely any one historical date at which it is possible to say that the first hypothesis was adumbrated. The history of a science is the history of the development of scientific systems from those containing so few generalizations, and these so flimsily established that one might well hesitate to call them systems at all, into imposing structures with a hierarchy of hypotheses. This development takes place by the establishment of some of the original hypotheses, by the replacement of others by better established hypotheses, and by the construction of higher-level hypotheses under which the lower-level hypotheses can be subsumed. The problems raised by this development are of many different kinds. There are historical problems, both as to what causes the individual scientist to discover a new idea, and as to what causes the general acceptance of scientific ideas. The solution of these historical problems involves the individual psychology of thinking and the sociology of thought. None of these questions are our business here. What we are concerned with are the straight logical problems of the internal structure of scientific systems and of the roles played in such systems by the formal truths of logic and mathematics, and also the problems of inductive logic or epistemology concerned with the grounds for the reasonableness or otherwise of accepting well-established scientific systems. The next three chapters will be devoted to the former of these sets of problems.

3

CHAPTER II

SCIENTIFIC DEDUCTIVE SYSTEMS
AND THEIR REPRESENTATIONS

A scientific theory is a deductive system in which observable consequences logically follow from the conjunction of observed facts with the set of the fundamental hypotheses of the system. A study of the nature of a scientific theory is thus a study of the nature of the deductive system used in the theory; and this chapter will be devoted to the internal logic of scientific deductive systems.

Every deductive system consists of a set of propositions (to be called the *initial propositions*) from which all the other propositions (to be called the *deduced propositions*) follow according to logical principles. Some of these propositions follow immediately from the set of initial propositions, others follow immediately from propositions which follow immediately from the initial propositions, others follow from these propositions, and so on. Every proposition of the system follows either immediately or mediately from the set of initial propositions. Every deduced proposition in a deductive system occurs at the end of a chain of deductive steps which starts with the set of initial propositions. The chain which leads to any particular proposition may be short or long, but it is always of a finite length, so that the proposition is reached after a limited number of steps of immediate deduction.

The natural way to represent a deductive system is to write the sentences, formulae, or other symbolic expressions, which express the propositions in order so that the spatial relationship of the sentences on the page corresponds to the logical relationship between the propositions of the system. Thus the initial propositions will be written down at the top of the page, below them a proposition which is an immediate consequence of them, below it a proposition which is an immediate consequence either of the initial propositions alone or of the first deduced proposition alone or of the conjunction of these, and so on. Thus every sentence on the page will express a proposition which is an immediate consequence of propositions expressed by sentences which precede it on the page; and a chain of deductive steps in the deductive system will be represented by a series of sentences on the page. This method of representing a deductive system is as old as Euclid; various examples of it will be found later in this chapter.

This natural method of representing a deductive system has a characteristic which has only become completely clear during the last century. It has been found that, by a suitable choice of the symbolic language in which the propositions of the system are expressed, to the fact that one proposition is an immediate consequence of other propositions there can be made to correspond the fact that the sentence expressing that proposition can be derived from the sentences expressing the other propositions by some simple manipulation of the symbols in the sentences. For example, in the deductive system of elementary mathematics the expression of the arithmetical proposition

$$\text{``}(9-2).(9+2)=9^2-2^2\text{''}$$

may be obtained from the expression of the algebraic identity

$$\text{``}(x-y).(x+y)=x^2-y^2\text{''}$$

by substituting the symbol "9" for the symbol "x" and the symbol "2" for the symbol "y". If the language to express the deductive system be chosen in a suitable way, it is possible, given a set of sentences, to write down a series of sentences each of which is obtained by a symbolic manipulation of some of the preceding sentences without thinking of the meanings of the sentences, but so that the series of sentences will nevertheless represent a deductive chain in the deductive system.

A representation of a deductive system in such a way that to each principle of deduction there corresponds a rule of symbolic manipulation will be called a *calculus*. The use of a calculus to represent a deductive system has the enormous practical advantage that it enables deductions to be effected merely by symbolic manipulation, and the correctness of these deductions can be checked automatically merely by inspecting the relationships between the symbols; it is for this reason that the Indian invention of Arabic numerals was such a landmark in the history of civilization. But the use of a calculus also ensures that thinking should be completely explicit, since no principle of deduction can be used in the deductive system which is not represented by a rule of

SCIENTIFIC DEDUCTIVE SYSTEMS

symbolic manipulation in the calculus. Various eighteenth-century mathematicians produced mathematical proofs which have subsequently been found to contain fallacies; their pseudo-proofs implicitly used invalid deductive principles whose invalidity became obvious when nineteenth-century mathematicians tried to represent them in an adequate calculus.

A philosopher thinking about deduction will realize that this correspondence between a deductive system and a calculus representing it is no accident. Whatever may be the case with less precise forms of thought, deductive thinking is not independent of the possibility of its expression. That syllogistic deduction can be expressed in the way Aristotle expressed it, or that deduction from algebraic identities can be expressed by making substitutions for variables, tells us something about these modes of deduction. Indeed, some philosophers of the logical positivist school maintained in the nineteen-thirties that knowledge of a calculus representing a pure deductive system (i.e. one containing no contingent propositions)* gave complete knowledge of that system, so that there was nothing to formal logic or mathematics except the relationships of symbols to one another, the principles of deduction being no more than the rules of manipulation of a symbolic calculus. This extreme view has few advocates to-day; it has been recognized that the whole truth about the nature of mathematical reasoning cannot be told without considering its 'semantics' as well as its 'syntax'. But semantics is concerned with a calculus in relation to a deductive system which it expresses; and it appears to be impossible to philosophize in any way about formal logic or mathematics without explicitly considering modes of expression. All profitable writing on the foundations of mathematics at the present time involves the construction and discussion of suitable calculi.†

* By a *contingent* proposition is meant one to whose truth or falsity experience is relevant. I shall use the term "contingent" (or "logically contingent") rather than "empirical" to characterize such propositions because some noncontingent propositions (e.g. If there are two chairs in one room and three chairs in another, there are altogether five chairs in both rooms) can be said to be 'empirical' in the sense that they are *about* empirical objects—though in a 'vacuous' sense of "about".

† *Note on books.* Introductions to the modern doctrine of calculi as games with symbols and of deductive systems as interpretations of calculi are to be found in two separately published parts of the *International Encyclopaedia of Unified Science*: Rudolf Carnap, *Foundations of Logic and Mathematics*, [= vol. 1, no. 3] (Chicago, 1939) and J. H. Woodger, *The Technique of Theory Construction*,

We are here concerned, not with the philosophy of mathematics but with the philosophy of science in which contingent propositions enter. In this field it will be less readily agreed that it is essential explicitly to consider the language in which scientific propositions are expressed. But there are two excellent reasons which have induced me to imitate the philosophers of mathematics, and to approach consideration of scientific deductive systems by way of consideration of calculi representing them. The first reason is that we shall need to consider the role played by mathematics in a scientific deductive system, and this can best be shown by considering first a calculus representing the system. The second reason is one which will be developed in Chapter III, where it will be found that the function of theoretical terms, such as electrons or wave-functions, which occur in an elaborate scientific theory, cannot properly be appreciated without an explicit examination of the way in which words for them are used in theoretical treatises. I shall therefore start by considering the nature of a calculus in abstraction from its relation to a deductive system—never forgetting, I hope, that such consideration is only a means to the end of understanding the nature of a deductive system used in science.

A calculus, considered in isolation from any interpretation of it, is a game for one player played with marks on paper in the following way. The set-up of the game is a sheet of paper ruled in horizontal lines, with certain marks or series of marks already written along the first few lines. Each of these marks or series of marks will be called a *formula*, a formula being the mark or series of marks written along one line. The formulae which the player finds already written on the sheet before he starts playing will be called the *initial formulae*. The rules of play are instructions to the player as to how he may write down new formulae on the sheet of paper, each of the new formulae being derived from some or all of the preceding formulae in accordance with the rules of play. If we call writing down a new formula *making a move*, playing the game will consist in making a series of moves in accordance with the

[= vol. 2, no. 5] (Chicago, 1939). A more thorough exposition will be found in Carnap, *Introduction to Semantics* (Cambridge, Mass., 1942).

Carnap's earlier *Logical Syntax of Language* (London, 1937) (original German edition, Vienna, 1934) attempts to subsume the whole of philosophy under the study of calculi: in his later books Carnap agrees that this must be supplemented by the 'semantics' due to Tarski and other Polish logicians.

rules of play. Such a game, played with a specified set-up (the initial formulae) and specified rules of play, is a calculus.

The rules of play of a calculus may be such that after a limited number of moves have been made, the rules allow no further moves and the game comes to an end. This is always the case with the rules of play of a game in the ordinary sense of the word, in which it is always arranged that there shall be some rule (or combination of rules) which ensure that the game shall come to an end after a finite number of moves. Most of the calculi with which we shall be concerned will not be of this limited sort; each will be such that another move can be made at any stage of the game. We may, and shall, stop playing the game when we have derived the particular formulae we desire; but we always could continue the game and derive more formulae.

The rules of play may be such that at each stage of the game they allow only one move to be made. If only one move is permitted, and the player were to make a different move, he would be disobeying the rules of the game. He can, of course, at any stage, decide to cease playing the game (and can start playing a different game with a different set of rules); but while he is playing the game, he has no freedom of choice. Games in the ordinary sense like this (e.g. Beggar-my-neighbour) are boring to all except young children. But calculus-games of this sort are of the greatest practical importance, since they can be played automatically without thought, and consequently they can be taught to children by rote, and also machines can be constructed to play them. All calculating machines, from the simplest abacus to the newest 'electronic brain', operate within no-choice calculi. The importance of these machines is due to the fact that large parts of mathematics, and in particular the mathematical methods used for producing numerical solutions of closer and closer degrees of approximation, can be represented by no-choice calculi.

The calculi with which we shall be concerned as representing scientific deductive systems are rarely of this no-choice type. More usually the rules are such that at every stage of the game some choice is possible. The choice of move will not be unrestricted, since the move made will have to obey the rules of the game. But within the restrictions imposed by the rules, the player is free to make whatever move he likes. Here his free creative activity has scope.

Indeed, unless he makes a move purely at random (and, of course, a machine can be constructed to make a random move at any stage in the game at which a choice is required), he will have to make the move with some purpose in mind. In playing calculus-games the player's purpose is usually that of obtaining a particular formula; and he will therefore at each stage choose a move which he hopes will be a stage towards the desired result. If he does not see one which will lead directly towards it, he will choose one which will improve his position (as in the 'middle game' at chess) by giving him a powerful formula which may be useful to him later in attaining his desired end. Anyone who has ever tried to construct a proof of a mathematical theorem or to solve a mathematical problem will appreciate the mixture of insight and trial-and-error that is involved. But what determines the choice of move is a matter for psychology, not for logic, which is concerned only with whether or not the move is in accordance with the rules. A bad chess-player is playing chess just as much as a good one; and an incompetent and fumbling mathematician obeys the rules of the mathematical calculus just as much as a mathematical genius. All that is relevant to our purpose is to emphasize that to describe the process of deductive thinking in terms of playing a game with a calculus is not to belittle the intelligence and creative activity employed therein.

THE FIRST CALCULUS AND ITS INTERPRETATION

This chapter will expound in detail two related examples of calculi and of their interpretation as deductive systems. The examples have been constructed to be as simple as possible, while at the same time displaying the points that need to be made with regard to the function of logically necessary truths in scientific reasoning. The deductive systems represented by the calculi are of great importance in various branches of science; and considering them and their relations to the calculi representing them will assist us in considering the problem with which we shall be concerned in the next chapter—that of the status of the theoretical terms used in an advanced science.

Both the calculi with which we shall be concerned make use of marks and series of marks of a special kind. These will be called *elements* and defined as follows. A *primitive element* is defined as

a small Greek letter. A *non-primitive element* is defined as a series
of marks obtained by adjoining any two elements, which may be
primitive or non-primitive, the same or distinct, and enclosing the
adjoined couple in a pair of curved brackets. Thus α, β, γ, etc.,
are primitive elements; $(\alpha\beta)$, $(\alpha\alpha)$, $((\alpha\beta)\gamma)$, $((\alpha\beta)((\gamma\gamma)(\delta\alpha)))$, etc.,
are non-primitive elements. An element can be distinguished from
other marks or series of marks which occur in the calculus by the
fact that an element is either a single small Greek letter or is a series
of marks of which the first and last marks are a twin pair of curved
brackets, i.e. a pair which consists of a left-hand bracket paired
with a right-hand bracket lying to the right of it and which is such
that the brackets (if any) lying between the pair all form twin pairs
of brackets. Thus by inspecting the brackets we can tell that, in the
last example of a non-primitive element given, the part $((\gamma\gamma)(\delta\alpha))$
is itself an element, but the part $(\alpha\beta)((\gamma\gamma)$ is not.

The formulae of both calculi are defined as being series of
marks consisting of two elements with a double-arrow $\leftrightarrow$ between
them. There is one and only one double-arrow in each formula;
the element which is the entire expression to the left of the
double-arrow will be called the *left element* (abbreviated to L),
that to the right the *right element* (abbreviated to R), of the
formula.

The rules of play of the First Calculus are such as to permit of
five, and only five, sorts of moves for deriving a new formula from
formulae already written down (the existing formulae).

I. The first sort of move permitted is to write down a new
formula which copies exactly one of the existing formulae, except
that, for an element at one or more places in it which occurs as
either the left element or the right element of some existing
formula, there is substituted the right element or the left element
respectively of this formula. Thus, if $(\alpha\beta)\leftrightarrow((\lambda\mu)\beta)$ and $\beta\leftrightarrow(\mu\nu)$
are already written down, we may write down the new formula
$(\alpha\beta)\leftrightarrow((\lambda\mu)(\mu\nu))$ obtained by substituting for the β occurring on
the right-hand side of the first formula the right element of the
second formula.

II. The second sort of move is to construct a new formula by
writing as its left element any element whatever which either is
identical with or contains an element which occurs as either the
left or the right element in some existing formula, and by writing

as the right element of the new formula this same element with the substitution in it of the right element or the left element respectively of the existing formula (adding, of course, a double-arrow between the two elements to complete the new formula). Thus if $\alpha \leftrightarrow (\lambda\mu)$ is already written down, we may construct the new formula $(\alpha\beta) \leftrightarrow ((\lambda\mu)\beta)$, of which the right element is the same as the left element except that $(\lambda\mu)$ is substituted for α.

Both these types of move make a substitution of the right or the left element for the left or the right element of an existing formula; rule I makes this substitution in a complete existing formula, rule II constructs a new formula of which the right element is identical with the left element but for this substitution having been made.

III. The third sort of move permitted is to write down a new formula obtained by substituting for any element at one place in an existing formula the element obtained by writing this element twice and enclosing the adjoined couple in a pair of curved brackets, or by making the reverse substitution at one place in a formula where there is an element consisting of a couple of identical elements enclosed in a pair of curved brackets. Thus if $(\alpha\beta) \leftrightarrow (\nu(\lambda\mu))$ is already written down, we may write down the new formula $(\alpha\beta) \leftrightarrow (\nu((\lambda\lambda)\mu))$; and if $(\alpha\beta) \leftrightarrow ((\lambda(\mu\mu))\nu)$ is already written down, we may write $(\alpha\beta) \leftrightarrow ((\lambda\mu)\nu)$.

IV. The fourth sort of move permitted is to write down a new formula obtained by substituting for any adjoined couple of elements at one place in an existing formula the same two elements but in reversed order. Thus if $(\alpha\beta) \leftrightarrow ((\lambda\mu)\nu)$ is already written down, we may write $(\alpha\beta) \leftrightarrow (\nu(\lambda\mu))$.

V. The fifth sort of move is to write down a new formula obtained by making the following substitution in an existing formula at one place where there is an adjoined couple of elements, one of which is itself an adjoined couple of elements enclosed in a pair of curved brackets. Move this pair of brackets either to the left or to the right so that the pair encloses the element not originally enclosed together with that element of the couple originally enclosed nearest to this element. Thus if $(\alpha\beta) \leftrightarrow ((\lambda\mu)(\mu\nu))$ is already written down, we may move the pair of brackets enclosing $(\mu\nu)$ to the left and write down

$$(\alpha\beta) \leftrightarrow (((\lambda\mu)\mu)\nu);$$

and if $(\alpha\beta)\leftrightarrow(((\lambda\mu)\mu)\nu)$ is already written down, we may move the pair of brackets enclosing $(\lambda\mu)$ to the right and write down

$$(\alpha\beta)\leftrightarrow((\lambda(\mu\mu))\nu).$$

The initial formulae of the First Calculus are the three formulae

$$\alpha\leftrightarrow(\lambda\mu), \quad \beta\leftrightarrow(\mu\nu), \quad \gamma\leftrightarrow(\nu\lambda).$$

Every derived formula in the calculus will be obtained from some one or two of the initial formulae and previously derived formula by making a move of one of the five specified sorts.

Table I gives a chain of fourteen moves leading to a formula which we shall require in our later discussion. For convenience the three initial formulae are printed first and numbered [1], [2], [3] (their order is irrelevant); and the method by which each of the derived formulae is obtained is shown.

TABLE I

[1] $\alpha\leftrightarrow(\lambda\mu)$

[2] $\beta\leftrightarrow(\mu\nu)$

[3] $\gamma\leftrightarrow(\nu\lambda)$

.   .   .   .

[4] $(\alpha\beta)\leftrightarrow((\lambda\mu)\,\beta)$ by II from [1], substituting R for L of [1] in $(\alpha\beta)$

[5] $(\alpha\beta)\leftrightarrow((\lambda\mu)\,(\mu\nu))$ by I from [4] and [2], substituting R for L of [2] in [4]

[6] $(\alpha\beta)\leftrightarrow(((\lambda\mu)\,\mu)\,\nu)$ by V from [5], substituting $(((\lambda\mu)\,\mu)\,\nu)$ for $((\lambda\mu)\,(\mu\nu))$

[7] $(\alpha\beta)\leftrightarrow((\lambda(\mu\mu))\,\nu)$ by V from [6], substituting $(\lambda(\mu\mu))$ for $((\lambda\mu)\,\mu)$

[8] $(\alpha\beta)\leftrightarrow((\lambda\mu)\,\nu)$ by III from [7], substituting μ for $(\mu\mu)$

[9] $(\alpha\beta)\leftrightarrow(\nu(\lambda\mu))$ by IV from [8], substituting $(\nu(\lambda\mu))$ for $((\lambda\mu)\,\nu)$

[10] $(\alpha\beta)\leftrightarrow(\nu((\lambda\lambda)\,\mu))$ by III from [9], substituting $(\lambda\lambda)$ for λ

[11] $(\alpha\beta)\leftrightarrow(\nu(\lambda(\lambda\mu)))$ by V from [10], substituting $(\lambda(\lambda\mu))$ for $((\lambda\lambda)\,\mu)$

[12] $(\alpha\beta)\leftrightarrow((\nu\lambda)\,(\lambda\mu))$ by V from [11], substituting $((\nu\lambda)\,(\lambda\mu))$ for $(\nu(\lambda(\lambda\mu)))$

[13] $(\alpha\beta)\leftrightarrow(\gamma(\lambda\mu))$ by I from [12] and [3], substituting L for R of [3] in [12]

[14] $(\alpha\beta)\leftrightarrow(\gamma\alpha)$ by I from [13] and [1], substituting L for R of [1] in [13]

[15] $(\alpha\beta)\leftrightarrow((\gamma\gamma)\,\alpha)$ by III from [14], substituting $(\gamma\gamma)$ for γ

[16] $(\alpha\beta)\leftrightarrow(\gamma(\gamma\alpha))$ by V from [15], substituting $(\gamma(\gamma\alpha))$ for $((\gamma\gamma)\,\alpha)$

[17] $(\alpha\beta)\leftrightarrow(\gamma(\alpha\beta))$ by I from [16] and [14], substituting L for R of [14] in [16]

Similar chains of moves, starting by making substitutions in $(\beta\gamma)$ and in $(\gamma\alpha)$, will lead to

[31] $(\beta\gamma)\leftrightarrow(\alpha(\beta\gamma))$

and to

[45] $(\gamma\alpha)\leftrightarrow(\beta(\gamma\alpha))$

respectively.

This calculus can be given many different interpretations. The interpretation with which we shall be most concerned is the following one. Take each primitive element (each small Greek letter) to stand for a class of things having a certain empirical property. Thus α may be taken as standing for the class of apricots, or for the class of aluminium atoms, or for the class of people who are astigmatic, or for any other class specified as being the class of those things which have a particular empirical property. Since it is irrelevant what this particular empirical property is it will be called "A"; and α will therefore be taken as standing for the class of things having the property A, or (for short) the class of A's. Similarly, β will be taken as standing for the class of B's, γ for the class of C's, λ for the class of L's, μ for the class of M's, and ν for the class of N's, where B, C, L, M, N are all empirical properties. The element formed by the adjunction of two elements, together with the enclosure of the adjoined couple in a pair of curved brackets, will be taken to stand for the class of those things which are members both of the class represented by the first element of the adjoined couple and of the class represented by the second element of this couple. This class is what is called the *intersection* or the *meet* or the *logical product* of the two classes. Thus $(\alpha\beta)$ will stand for the class of things which are both members of the class designated by α and members of the class designated by β, i.e. the class of things which have both the property A and the property B. If A is an abbreviation for the property of being astigmatic, and B an abbreviation for the property of being British, $(\alpha\beta)$ will stand for the class of astigmatic Britons. Similarly, $(\alpha\alpha)$ will stand for the class of things which are members both of the class designated by α and of that designated by α, $(\gamma(\alpha\beta))$ for the class of things which are members both of the class designated by γ and of that represented by $(\alpha\beta)$, i.e. for the class of things which are members both of the class designated by γ and of the class whose members are members both of the class designated by α and of the class designated by β. Every non-primitive element represents the class whose members are members of both the classes represented by the elements in the adjoined couple which, together with the enclosing brackets, make up the non-primitive element; if either one of these adjoined elements is itself non-primitive, this element also represents a class whose members are members of both of two classes. Thus every non-primitive element represents a class built up out of classes designated by the primitive elements by forming classes whose members are the members in common of two classes. Every element, primitive or non-primitive, represents a class of things having an empirical property.

The formulae of the calculus will be interpreted by taking the double-arrow $\leftrightarrow$ to stand for the relation of class-identity. Thus $\alpha \leftrightarrow (\lambda\mu)$ is to be interpreted as representing the proposition that the class of A's is the same class as the class of things which are both L and M. Since every formula consists of two elements connected by a double-arrow, and since it always makes sense to say of any pair of classes with which we shall be concerned that they are identical, every formula of the calculus represents a proposition. The proposition represented by the formula $\alpha \leftrightarrow (\lambda\mu)$ may be expressed in various ways. In class language it is expressed by saying that the class of A's is identical with the class of things which are both L and M; it can also be expressed without using class language as the conjunction of the general proposition that everything which is A is also L and M with the general proposition that everything which is both L and M is also A. Similarly, the formula $(\alpha\beta) \leftrightarrow (\gamma(\alpha\beta))$ represents the proposition which may be expressed by saying that the class of things which are both A and B is identical with the class of things which are both C and both A and B. Using the notion of inclusion this proposition may be expressed by saying that the class of things which are both A and B is included in the class of C's; and it may also be expressed, without using class language, as the single general proposition that everything which is both A and B is also C.

On this interpretation the rules of play of the calculus represent logical principles of deduction. Rules I and II represent logical properties of class-identity; if two classes are identical, rule II expresses the truth that any class built up out of one of these classes is identical with the class built up in the same way out of the other, and rule I expresses the truth that from any statement about one of these classes expressible by a formula of the calculus there follows the corresponding statement about the other class. Rules III, IV and V represent logical properties of the operation of class-intersection. Rule III expresses the truth that, from any expressible statement about a class, there follows the corresponding statement about the intersection of that class with itself, and vice versa. Rule IV expresses the truth that, from any expressible statement about the intersection of one class with another, there follows the corresponding statement about the intersection of the latter class with the former. Rule V expresses the truth that, from any expressible statement about the intersection of one class with the intersection of a second class with a third, there follows the corresponding statement about the intersection with the third class of the intersection of the first class with the second, and vice versa.

Thus each of the rules of play of the calculus represents a deductive principle according to which one proposition may be deduced from another, or, in the case of rule I, from two others. Since any proposition which is deducible from one or more propositions which are themselves deducible from a set of propositions is also deducible from that set of propositions, every proposition represented by a formula of the calculus is deducible from the propositions represented by the initial formulae. So a chain of moves in the calculus starting with these initial formulae represents a chain of deductions from the corresponding initial propositions.

The series of formulae given in Table I represents a deductive chain in which, from the initial propositions

[1] Everything which is A is both L and M, and vice versa;

[2] Everything which is B is both M and N, and vice versa;

[3] Everything which is C is both N and L, and vice versa;

fourteen deductive steps lead to

[17] Everything which is both A and B is also C.

Another fourteen deductive steps will lead to

[31] Everything which is both B and C is also A,

and another fourteen steps to

[45] Everything which is both C and A is also B.

The First Calculus, and the chain of forty-two formulae derived from the three initial formulae, is thus interpreted as representing a deductive system (S_1) in which forty-two propositions are deduced from three initial propositions. The deductive principles according to which they are deduced form part of the logic of classes, the first two being concerned with the formal properties of class-identity, the last three with the formal properties of class-intersection. The validity of these deductive principles is a matter of logic. The six empirical properties A, B, C, L, M, N will be taken as being logically independent, so that none of the three initial propositions connecting them are logically necessary. A deductive system, such as S_1, in which all the initial propositions are contingent will be called an *impure deductive system*. It will be contrasted with a deductive system all of whose initial propositions are logically necessary, which will be called a *pure deductive system*. Since a contingent proposition can only be deduced from a set of propositions which include at least one contingent proposition, all the propositions in a pure deductive system are logically necessary. A deductive system in which some but not all of the initial propositions are contingent will be called a *mixed deductive system*; and a system which is either impure or mixed (i.e. a system which is not a pure one) will be called an *applied deductive system.**

* There are also deductive systems (e.g. *reductio ad absurdum* proofs) in which some or all of the initial propositions are logically impossible. Such deductive systems will not be considered in this book.

It is important to notice that the adjectives "impure", "pure", "mixed", and "applied" are applicable to deductive systems but not to the calculi representing them. The same calculus can usually be interpreted in different ways either as a pure, or as an impure, or mixed, deductive system. The First Calculus can be interpreted perfectly well as a pure deductive system in the following way. Take each primitive element (each small Greek letter) to stand for a positive integer, and take the element formed by the adjunction of two elements, together with the enclosure of the adjoined couple in a pair of brackets, to stand for the highest common factor of the integers represented by the two adjoined elements. Take the double-arrow sign to stand for the relation of identity between numbers. Then the first two rules of play express truths about this relation of identity, and the last three rules express arithmetical truths about highest common factors. If α, β, γ, λ, μ, ν are taken as standing for 6, 10, 8, 24, 30, 40 respectively, the three initial propositions which are the interpretations of the three initial formulae are

[1]   6 is the highest common factor of 24 and 30,

[2]   10 is the highest common factor of 30 and 40,

[3]   8 is the highest common factor of 40 and 24.

Then Table I interpreted in this way will represent a chain of propositions in which each step produces a new proposition deducible from the initial set. For example, eleven steps will bring us to

[14]   The highest common factor of 6 and 10 is the highest common factor of 8 and 6.

Since the initial propositions are logically necessary, being arithmetical truisms, all the propositions in the deductive system are also logically necessary, and the system is a pure deductive system.

The special problems connected with pure deductive systems fall within the purview of the logic of mathematics rather than that of the logic of science, and hence will not concern us directly. But mixed deductive systems, in which some of the initial propositions are logically necessary while others are logically contingent, are of the greatest importance in an advanced science such as physics, and it will be necessary to consider them in some detail.

The reason why mixed deductive systems are used in an advanced science is that such a science makes use of so much mathematical apparatus that it is convenient to use this apparatus *explicitly* by including propositions of pure mathematics within the deductive system instead of using this apparatus *implicitly* for providing deductive principles according to which the deductions are made. To explain this a second calculus will be constructed.

THE SECOND CALCULUS AND ITS INTERPRETATION

In the impure deductive system, S_1, which is our interpretation of the First Calculus, all the mathematical apparatus is used in the form of deductive principles, the initial propositions being all contingent propositions. But it is possible to include the impure deductive system within a mixed deductive system, S_2, containing both logically necessary and contingent initial propositions, so that the contingent propositions will be exactly the same, and will be deduced in exactly the same order, as in the impure deductive system S_1. The advantage of thinking by means of the mixed system S_2 instead of by means of the impure system S_1 will be that S_2 will make use of fewer deductive principles, since a large part of the mathematical apparatus required will appear in the form of propositions of the pure part of the system. This pure part will require additional initial propositions (since mathematico-logical propositions cannot be deduced from contingent propositions alone), but these logically necessary propositions will take the place of some of the deductive principles which will no longer be necessary.

The calculus (call it the Second Calculus) representing this mixed system S_2 will therefore differ from the First Calculus in containing more initial formulae, but it will make its moves according to fewer rules of play. Treated as a game, it will be an easier game in the sense that, instead of five, only two rules of play need be borne in mind in playing it.

The first rule is the same as rule I of the First Calculus. It permits of substitutions in any existing formula of the left element for the right element (or vice versa) of any existing formula.

The second rule is used in making moves from formulae containing one or more of a special set of primitive elements to be called *variables*. Only four variables are required in the use we shall make of the Second Calculus or of similar calculi, and ξ, η, ζ, ω will be reserved to serve as variables. Rule II then states that we may always write down a new formula by substituting for one or more variables at every place in an existing formula which contains them any element or elements whatever, provided that, if any variable for which a substitution is to be made occurs at several places in the original formula, the same element must be substituted for it at every place at which it occurs. Thus if $(\xi\eta) \leftrightarrow (\eta\xi)$ is already written down, we may write down the new formula $((\lambda\mu)\nu) \leftrightarrow (\nu(\lambda\mu))$ obtained by substituting $(\lambda\mu)$ for ξ in each of the two places in which ξ occurs, and ν for η in each of the two places in which η occurs in the original formula. But the rule will not permit us to write down $((\lambda\mu)\nu) \leftrightarrow (\nu\xi)$, where $(\lambda\mu)$ is substituted for ξ at only one of the two places at which ξ occurs.

The moves permitted by both these rules are substitutions in complete formulae already written down. The first rule permits the substitution for any element at any one or more places in a formula in which it occurs of an element which occurs as either the left element or the right element of a formula in which the element in question occurs as the right or the left element respectively. The second rule permits the substitution for any one or more variables at every place in a formula in which they occur of any element or elements whatever, provided the same substitution is made at every place in the formula at which the same variable occurs.* The two types of substitution will be called *double-arrow-substitution* and *variable-substitution* respectively.

Besides the three initial formulae of the First Calculus, the Second Calculus employs the three further initial formulae

$$(\xi\xi) \leftrightarrow \xi, \quad (\xi\eta) \leftrightarrow (\eta\xi), \quad (\xi(\eta\zeta)) \leftrightarrow ((\xi\eta)\,\zeta).$$

In these three new initial formulae the only primitive elements which occur are variables.

Table II shows a chain of twenty-five moves in the Second Calculus (formulae [7]–[31]), according to which the fourteen formulae derived in the First Calculus (formulae [4]–[17] of

---

* The variable-substitution rule can, if desired, be taken in a simpler form permitting of substitution for only one variable at a time. Successive steps will then be necessary to make substitutions for more than one variable. The extended form of the variable-substitution rule given serves to reduce the number of steps by telescoping into one the several steps that would be required were the simpler form of the rule employed.

Table I) are successively derived. For convenience the six initial formulae ([1]–[6]) have been printed at the head, and the formula (or formulae) from which each derived formula is obtained is indicated. The first derived formula [7] is obtained by a double-arrow-substitution in formula [4] of the right element for the left element in [4] itself; all the other derived formulae in the left-hand column are obtained by variable-substitutions, and all the derived formulae in the right-hand column by double-arrow-substitutions.

Similar chains of moves can be constructed which will lead to $(\beta\gamma) \leftrightarrow (\alpha(\beta\gamma))$ and to $(\gamma\alpha) \leftrightarrow (\beta(\gamma\alpha))$.

The reason why the formulae have been arranged in two columns is that those printed in the left-hand column have the important characteristic that none of them are derived from any of the formulae printed in the right-hand column. The left-hand column contains the three new initial formulae, which may be distinguished from the other three initial formulae by the fact that all the primitive elements occurring in the new initial formulae are variables, together with formulae derivable, immediately or mediately, from these new initial formulae by means of the two rules of play. Thus the formulae in the left-hand column are formulae of a calculus whose initial formulae are

$$(\xi\xi) \leftrightarrow \xi, \quad (\xi\eta) \leftrightarrow (\eta\xi), \quad (\xi(\eta\zeta)) \leftrightarrow ((\xi\eta)\,\zeta),$$

and whose rules of play are those of the Second Calculus (i.e. the double-arrow-substitution rule and the variable-substitution rule). A calculus with a variable-substitution rule whose initial formulae contain only variables as primitive elements will be called an *algebraic calculus*.* The formulae in the left-hand column thus belong to an algebraic calculus which is an independent part (to be called the *algebraic part*) of the Second Calculus. The essential distinction between the Second and the First Calculus is that the former has an algebraic part, whereas the latter, which does not use variables, has no algebraic part. The formulae which appear on the right-hand side of Table II are exactly the same, and are derived in the same order, as those which appear in the First

---

* This name has been chosen because the use of symbols as variables with a suitable variable-substitution rule is the distinguishing characteristic of algebra in the ordinary school-book sense of the word, and because an algebraic calculus in my sense corresponds pretty closely with what contemporary mathematicians call "an algebra".

TABLE II

[4]	$(\xi\xi)\leftrightarrow\xi$		[1]	$\alpha\leftrightarrow(\lambda\mu)$
[5]	$(\xi\eta)\leftrightarrow(\eta\xi)$		[2]	$\beta\leftrightarrow(\mu\nu)$
[6]	$(\xi(\eta\zeta))\leftrightarrow((\xi\eta)\ \zeta)$		[3]	$\gamma\leftrightarrow(\nu\lambda)$

.   .   .   .

[7] $\xi\leftrightarrow\xi$
            from [4] and [4]

[8] $(\alpha\beta)\leftrightarrow(\alpha\beta)$ from [7]

[9] $(\alpha\beta)\leftrightarrow((\lambda\mu)\ \beta)$
            from [8] and [1]

[10] $(\alpha\beta)\leftrightarrow((\lambda\mu)\ (\mu\nu))$
            from [9] and [2]

[11] $((\lambda\mu)\ (\mu\nu))\leftrightarrow(((\lambda\mu)\ \mu)\ \nu)$
            from [6]

[12] $(\alpha\beta)\leftrightarrow(((\lambda\mu)\ \mu)\ \nu)$
            from [10] and [11]

[13] $(\lambda(\mu\mu))\leftrightarrow((\lambda\mu)\ \mu)$
            from [6]

[14] $(\alpha\beta)\leftrightarrow((\lambda(\mu\mu))\ \nu)$
            from [12] and [13]

[15] $(\mu\mu)\leftrightarrow\mu$ from [4]

[16] $(\alpha\beta)\leftrightarrow((\lambda\mu)\ \nu)$
            from [14] and [15]

[17] $((\lambda\mu)\ \nu)\leftrightarrow(\nu(\lambda\mu))$ from [5]

[18] $(\alpha\beta)\leftrightarrow(\nu(\lambda\mu))$
            from [16] and [17]

[19] $(\lambda\lambda)\leftrightarrow\lambda$ from [4]

[20] $(\alpha\beta)\leftrightarrow(\nu((\lambda\lambda)\mu))$
            from [18] and [19]

[21] $(\lambda(\lambda\mu))\leftrightarrow((\lambda\lambda)\ \mu)$
            from [6]

[22] $(\alpha\beta)\leftrightarrow(\nu(\lambda(\lambda\mu)))$
            from [20] and [21]

[23] $(\nu(\lambda(\lambda\mu)))\leftrightarrow((\nu\lambda)\ (\lambda\mu))$
            from [6]

[24] $(\alpha\beta)\leftrightarrow((\nu\lambda)\ (\lambda\mu))$
            from [22] and [23]

[25] $(\alpha\beta)\leftrightarrow(\gamma(\lambda\mu))$
            from [24] and [3]

[26] $(\alpha\beta)\leftrightarrow(\gamma\alpha)$
            from [25] and [1]

[27] $(\gamma\gamma)\leftrightarrow\gamma$ from [4]

[28] $(\alpha\beta)\leftrightarrow((\gamma\gamma)\ \alpha)$
            from [26] and [27]

[29] $(\gamma(\gamma\alpha))\leftrightarrow((\gamma\gamma)\ \alpha)$
            from [6]

[30] $(\alpha\beta)\leftrightarrow(\gamma(\gamma\alpha))$
            from [28] and [29]

[31] $(\alpha\beta)\leftrightarrow(\gamma(\alpha\beta))$
            from [30] and [26]

Calculus chain printed as Table I. In the First Calculus these formulae are derived, immediately or mediately, from three variable-free initial formulae; in the Second Calculus, on the other hand, their derivation makes use of the algebraic part of the Second Calculus.

The mixed deductive system S_2 will be obtained by interpreting those elements and formulae of the Second Calculus in which variables do not occur in exactly the same way as that in which the elements and formulae of the First Calculus were interpreted to yield the impure deductive system S_1. As regards the variables and elements, and the formulae containing them, no direct interpretation will be given of the variables ξ, η, ζ themselves, nor of elements, such as $(\xi(\eta\zeta))$, which contain one or more of them, and no direct interpretation will be given of the double-arrow sign as it occurs in formulae containing one or more variables. Instead, an interpretation is given of a complete formula containing one variable in the following way: The formula is to be understood as expressing that statement about every class which it would be stating about a particular class designated by χ were χ to be substituted in the formula in the place of the variable wherever the variable occurs. Thus the initial formula $(\xi\xi)\leftrightarrow\xi$ is to be interpreted as expressing the proposition that the intersection of every class with itself is the same as the class itself, since $(\chi\chi)\leftrightarrow\chi$ would be interpreted as expressing the proposition that the intersection of the class designated by χ with itself is the same class as itself. Similarly, $((\lambda\mu)\,\eta)\leftrightarrow(\eta(\lambda\mu))$ is to express the proposition that it is true of every class that the intersection with this class of the intersection of the class of L's with the class of M's is the same as the intersection of this class with the intersection of the class of L's with the class of M's. If there are two or three variables in the formula, it is to be interpreted as expressing that statement about every pair or triad of classes which it would be stating about a particular pair of classes designated by χ, θ, or a particular triad of classes designated by χ, θ, ϕ, were each of these letters to be substituted in the formula in the place of the variables wherever the variables occur. Thus the initial formula $(\xi\eta)\leftrightarrow(\eta\xi)$ is to be understood as expressing the proposition that it is true of every pair of classes that the intersection of the first of these with the second is the same class as is the intersection of the second with the first.

On this interpretation the rules of play represent logical principles of deduction. The double-arrow-substitution rule expresses, as before, the truth that, from any statement about a class expressible by a formula in the calculus, there follows the corresponding statement about an identical class. The variable-substitution rule expresses the truth that, from any expressible statement about every class, there follows the corresponding statement about any particular class. This is a special case of the deductive principle involved in all uses of general propositions which states that what is true of everything of a certain logical category applies to any instance of that logical category. As will be mentioned in the next chapter, the notion of general proposition is intimately connected with the use of this *applicative principle* (as W. E. Johnson called it).

With this interpretation the series of formulae of Table II represent a chain of deductions from the six initial propositions in the mixed deductive system S_2. The propositions expressed by the formulae appearing in the right-hand column are contingent propositions; they are the same propositions as occurred in the chain of deduction of the impure deductive system S_1 represented by the formulae of Table I, and they appear in the same order. The right-hand column represents the impure part of the mixed deductive system S_2. The propositions expressed by formulae appearing in the left-hand column, formulae belonging to the algebraic part of the calculus, are all logically necessary propositions. Those represented by formulae containing one or more variables are general logical truths about the properties and relations of classes; those represented by formulae containing no variable are the applications of such general truths to particular cases. For example, $\xi \leftrightarrow \xi$ represents the general truth that every class is identical with itself; $(\alpha\beta) \leftrightarrow (\alpha\beta)$ represents the application of this general truth to the case of the class of things which are both A and B, i.e. the proposition that this particular class is identical with itself. Both the general truths and their particular applications are logically necessary propositions; the algebraic part of the Second Calculus, therefore, represents the pure part of the mixed deductive system S_2. It represents the mathematical apparatus required for deduction of the propositions of the impure part. Since the algebraic part of the Second Calculus is independent of the rest, all the propositions in the pure part of S_2 follow from propositions also in the pure part; consequently the pure part is independent of the impure part in a way in which this latter is not independent of the former. The pure part forms a deductive system—a pure deductive system—on its own. It is represented in the Second Calculus by the algebraic calculus which forms the algebraic part of this calculus.

But, it may well be asked, what is the advantage gained by using the mixed deductive system S_2 represented by the Second Calculus in preference to the impure deductive system S_1 represented by the First Calculus? What is the virtue in reducing the number of deductive principles required at the expense of increasing the number of initial propositions, and of greatly increasing the number of steps required to deduce the propositions which we want? To deduce the proposition expressed by the formula $(\alpha\beta) \leftrightarrow (\gamma(\alpha\beta))$ requires twenty-five steps in system S_2, but only fourteen in system S_1.

The answer to this criticism is twofold. In the first place it should be noticed that the increased number of steps in a deduction in system S_2 is compensated by the greater simplicity of the deductive steps. The two deductive principles used in system S_2 are so simple that they are unlikely to be used incorrectly. In making moves in the Second Calculus all that has to be done is to copy formulae already written down making appropriate simple substitutions. But in making moves in the First Calculus it is necessary as well to construct new formulae, to double elements in formulae, to reverse the order of elements and to move pairs of brackets to the right or to the left. Neither the First nor the Second Calculus are no-choice calculi; so intelligence is required to select the appropriate move at each stage. But checking the correctness of the moves is, of course, automatic; and this is much simpler in the case of the Second Calculus, since all that has to be checked is that the move has been made according to one or other of two simple rules.

The second advantage obtained by using the deductive system S_2 represented by the Second Calculus is that the use of this system enables us to construct deductions of contingent propositions in which almost all the deductive steps fall within the pure part of the deductive system represented by the algebraic part of the Second Calculus. This point can best be expounded in connexion with an example; so a series of formulae in the Second Calculus will be given which represent a chain of deductions in the mixed deductive system S_2 leading to the formula [26] of Table II by a route in which the first and major part lies wholly within the pure part of the mixed deductive system. As in Table II all the formulae on the left-hand side belong to the algebraic part of the Second Calculus.

If we compare the chain of deduction represented by Table III with that represented by the series of formulae [1]–[26] of Table II both chains using the same initial propositions and the same deductive principles, we notice, besides the minor fact that the former chain contains four steps less than does the latter chain, the important fact that the first thirteen of the sixteen steps in the former chain are all deductions within the pure part of the mixed deductive system represented by the algebraic part of the calculus. The importance of this fact lies in the way it opens for a division of labour. Theorems in the pure deductive system can be deduced by pure mathematicians, operating an algebraic calculus, without any concern whatever for the application of these theorems to any empirical field; they can then be applied by the scientist to enable him to deduce consequences of his scientific hypotheses. The scientist can therefore leave all the heavy deductive work to the mathematician; all he need do is to add the two or three steps at the end which are required to apply the mathematician's labours to the purpose he wishes.* And he can arrange with the mathematician that he need use only one deductive principle in making these two or three steps, so that the possibility of his making a mathematical mistake can be reduced to the minimum.

This possibility of division of labour, so that all the really difficult deductions can be left to the professional mathematician, is the reason why mixed deductive systems such as the system S_2 are nearly always preferred by scientists to impure deductive systems such as the system S_1. From the point of view of the philosophical logician use of an impure system displays the

---

* To derive the formula $(\beta\gamma) \leftrightarrow (\alpha\beta)$ according to the method of Table III only four more steps are necessary, since we can start with formula [18]. According to the method of Table II we cannot start later than formula [7], and nineteen more steps will be necessary.

Table III

[4] $(\xi\xi)\leftrightarrow\xi$

[5] $(\xi\eta)\leftrightarrow(\eta\xi)$

[6] $(\xi(\eta\zeta))\leftrightarrow((\xi\eta)\,\zeta)$

· · · · ·

[7] $((\xi\eta)\,(\eta\zeta))\leftrightarrow(((\xi\eta)\,\eta)\,\zeta)$
from [6]

[8] $(\xi(\eta\eta))\leftrightarrow((\xi\eta)\,\eta)$
from [6]

[9] $((\xi\eta)\,(\eta\zeta))\leftrightarrow((\xi(\eta\eta))\,\zeta)$
from [7] and [8]

[10] $(\eta\eta)\leftrightarrow\eta$
from [4]

[11] $((\xi\eta)\,(\eta\zeta))\leftrightarrow((\xi\eta)\,\zeta)$
from [9] and [10]

[12] $((\xi\eta)\,\zeta)\leftrightarrow(\zeta(\xi\eta))$
from [5]

[13] $((\xi\eta)\,(\eta\zeta))\leftrightarrow(\zeta(\xi\eta))$
from [11] and [12]

[14] $((\xi\eta)\,(\eta\zeta))\leftrightarrow(\zeta((\xi\xi)\,\eta))$
from [13] and [4]

[15] $(\xi(\xi\eta))\leftrightarrow((\xi\xi)\,\eta)$
from [6]

[16] $((\xi\eta)\,(\eta\zeta))\leftrightarrow(\zeta(\xi(\xi\eta)))$
from [14] and [15]

[17] $(\zeta(\xi(\xi\eta)))\leftrightarrow((\zeta\xi)\,(\xi\eta))$
from [6]

[18] $((\xi\eta)\,(\eta\zeta))\leftrightarrow((\zeta\xi)\,(\xi\eta))$
from [16] and [17]

[19] $((\lambda\mu)\,(\mu\nu))\leftrightarrow((\nu\lambda)\,(\lambda\mu))$
from [18]

[1] $\alpha\leftrightarrow(\lambda\mu)$

[2] $\beta\leftrightarrow(\mu\nu)$

[3] $\gamma\leftrightarrow(\nu\lambda)$

· · · · ·

[20] $(\alpha(\mu\nu))\leftrightarrow((\nu\lambda)\,\alpha)$
from [19] and [1]

[21] $(\alpha\beta)\leftrightarrow((\nu\lambda)\,\alpha)$
from [20] and [2]

[22] $(\alpha\beta)\leftrightarrow(\gamma\alpha)$
from [21] and [3]

44

essential philosophical features better than the use of a mixed one, for in the former all initial propositions are contingent, the whole of the logic lying in the deductive principles employed. Any mixed deductive system can be transformed into an equivalent impure deductive system by using in the latter additional deductive principles corresponding to the logically necessary initial propositions of the mixed system; and the philosophical logician, concerned primarily with distinguishing between necessary and contingent propositions, may prefer a scientific system to be represented as deducible from a set of contingent propositions only.

Indeed, a chain of deductions leading to a particular contingent proposition in a mixed system by a chain in which all the earlier part consists of logically necessary propositions (e.g. the propositions represented by [4]–[19] in Table III) can be transformed into a very short chain of deductions in an impure system if one is prepared to use a sufficiently complicated deductive principle. For example, if we are prepared to use as a deductive principle the truth that from any statement (expressible in a calculus like the First Calculus) about the intersection of the intersection of one class with a second class with the intersection of the second class with a third class, there follows the corresponding statement about the intersection of the intersection of the third class with the first class with the intersection of the first class with the second class, we can then employ a deductive system having this deductive principle in addition to the first two deductive principles and the three initial propositions of the system S_1. In this new deductive system the proposition represented by the formula $(\alpha\beta) \leftrightarrow (\gamma\alpha)$ can be reached in only five steps:

$(\alpha\beta) \leftrightarrow ((\lambda\mu)\,\beta)$	by II of the First Calculus,
$(\alpha\beta) \leftrightarrow ((\lambda\mu)\,(\mu\nu))$	by I of the First Calculus,
$(\alpha\beta) \leftrightarrow ((\nu\lambda)\,(\lambda\mu))$	by the calculus-rule representing the new deductive principle,
$(\alpha\beta) \leftrightarrow (\gamma(\lambda\mu))$	by I of the First Calculus,
$(\alpha\beta) \leftrightarrow (\gamma\alpha)$	by I of the First Calculus.

But such a special deductive principle as this new one is not intuitively obvious. Nor will it enable more than a very limited class of propositions to be deduced. It is impossible, for instance, to deduce in this new deductive system the proposition expressed by $(\alpha\beta) \leftrightarrow (\gamma(\alpha\beta))$.

The use of deductions in a mixed deductive system in which the earlier propositions are all propositions of a pure system goes a long way to satisfy the philosophical logician's demand for a sharp division between the logically necessary and the logically contingent. Moreover, all the propositions which occur in the pure part of such a deduction, except the last one, are not only logically necessary propositions but are also completely general; they are thus propositions of formal logic or of pure mathematics. In the pure part of the deduction represented by Table III the propositions expressed by formulae [4]–[18] are propositions about every class, or about every pair of classes, or about every triad of classes; no one of them refers to any particular class in a way in which it does not also refer to every other particular class. They are all propositions belonging to the pure logic of classes. This complete generality is represented in the calculus by the fact that no primitive elements other than variables occur in any of the formulae [4]–[18]. It is only when formula [19] is obtained from formula [18] by using the variable-substitution rule that a formula appears in which there occur primitive elements which are not variables; this transition represents the deduction of a proposition about particular classes—the class of L's, the class of M's, the class of N's—from the corresponding completely general proposition about all triads of classes. The proposition corresponding to formula [19] belongs no longer to the pure logic of classes; it is the application of a theorem of this logic to a particular case. Like the preceding fifteen propositions, it is a proposition which is logically necessary, and which is logically necessary by virtue of its form alone, since its logical necessity does not depend upon any special properties of the classes with which it is concerned, but depends only upon its being an application of the general theorem. It may thus, like its fifteen predecessors, be called a *formal* proposition. But it is not completely general, since it is about three particular classes; it is thus a proposition of applied mathematics, not of pure mathematics.

In a deduction in a mixed deductive system like that represented in Table III, three stages can thus be distinguished. First there is the deduction of the completely general theorem required as mathematical apparatus; secondly, there is the application of this completely general theorem to the particular subject-matter involved; thirdly, there is the use of this particular applied logically necessary proposition to serve with contingent propositions for the deduction of contingent propositions. The first stage is formal logic or pure mathematics, the second stage applied logic or applied mathematics, the third stage a deduction falling wholly within the science in question.

The deductive principle used in passing from the first to the second stage is the simple principle that what is true of *every* is true of *any* (the *applicative principle*), a principle which will be seen to involve no more than the way in which general statements are given meaning.* The principle used in deducing the final steps within the science is, in the case represented in Table III, the principle that what is true of a class is true of an identical class, which involves no more than the way in which the word "identical" is used of classes. It is possible to construct a more elaborate deductive system equivalent to system S_2 in which the logic of class-identity would be put in the form of additional initial propositions in the pure part of the system. In this system (which it would take too much space to expound here) the only principle used in deducing the contingent propositions forming the final steps would be the logical principle known as the *detachment principle* or the *modus ponendo ponens*, which states that, from two propositions of the forms respectively p and «Not both p and not-q», the proposition of the form q may be deduced. It is always possible to construct a deduction in a mixed deductive system equivalent to a deduction in any applied deductive system in which the only principle used in deducing the contingent propositions is this detachment principle. Thus the internal logic of the deductions in both the intermediate 'applied-mathematics' stage and in the final 'science' stage is of extreme simplicity, all the difficult deduction occurring in the first 'pure-mathematics' stage.

SEPARATION OF THE MATHEMATICAL APPARATUS

The fact that the propositions of pure mathematics which occur in a scientific deductive system form an independent subsystem is

* See below, pp. 82 ff.

reflected by a division which is made in most serious treatises which expound a science organized as a deductive system of a fair degree of complexity (e.g. physics, genetics using statistical techniques, factor psychology), a division between an introductory mathematical section giving the pure mathematics which cannot be presupposed as common knowledge but which will be required in the rest of the book, and the main part of the book in which this mathematics is 'applied' to the subject in question. Indeed, in modern physics the amount of advanced pure mathematics is so great that it has been found convenient to write separate mathematical treatises for mathematical physicists to keep on their shelves as books of reference; these books, usually entitled "Methods of Mathematical Physics", contain no physics properly speaking, but are compendia of pure mathematics, differing only from other mathematical treatises in that they omit branches of pure mathematics which have as yet no physical application, and that the mathematical theorems are proved in forms less general than would satisfy the aesthetic sense of a Professor of Pure Mathematics. A scientist developing the consequences of a scientific hypothesis can then construct a deductive system by taking over from one of these books the pure mathematics he requires, and implicitly incorporating this as a subsystem of his mixed deductive system. If he is prepared to take the validity of this pure-mathematical subsystem for granted on the authority of the pure mathematicians, he need not follow all the stages in their argument. Indeed, all he need do explicitly is to select from his "Methods of Mathematical Physics" or "Statistical Methods for Biologists" those theorems which he needs directly to apply to his subject-matter. Thus he need only have sufficient knowledge of mathematics to know his way about the mathematical treatises and to understand the mathematical theorems he uses; he need not be a good enough mathematician to appreciate the proofs, still less a great enough mathematician to construct these proofs *ab initio*.

It has been a fortunate fact in the modern history of physical science that the scientist constructing a new theoretical system has nearly always found that the mathematics he required for his system had already been worked out by pure mathematicians for their own amusement. Thus Einstein, in developing general relativity (1915), had Riemann's non-Euclidean geometry (1854) and

Ricci's tensor calculus (1887) ready to hand; and the non-commutative multiplication used in quantum mechanics (1925–7) had been worked out in connexion with Cayley's matrices (1858) and with operational methods for handling differential equations (Boole, 1844). Modern biologists have not been so fortunate, and have had for the most part to work out for themselves the statistical mathematics they required. It is tempting to speculate as to how much progress the Greeks would have made in mechanics had they been acquainted with algebra, or how much more rapidly physics would have advanced in the seventeenth century had the infinitesimal calculus been discovered at its beginning instead of at its end. The moral for statesmen would seem to be that, for proper scientific 'planning', pure mathematicians should be endowed fifty years ahead of scientists.

CHAPTER III

THE STATUS OF THE THEORETICAL TERMS OF A SCIENCE

In the last chapter the function of deductive systems in science was expounded by first explaining the notion of a calculus as a game with marks on paper, and afterwards showing how such a calculus could be interpreted to represent a deductive system in which the conclusions were propositions testable by observation. To this method of exposition it might well be objected that it complicated the issue, that in discussing the logic of science it was unnecessary to introduce the notion of a calculus at all; what was essential was that of a deductive system, and this could be understood perfectly well by considering the logical relationships between its propositions, the formulae by which they could be expressed and the rules by which their calculus-game could be played being of no importance whatever. For example, the deductive system of p. 40 as an interpretation of the Second Calculus could be described by giving its six initial propositions, remarking that three of these were contingent and three logically necessary, and by pointing out that the two principles of deduction used— the applicative principle, and that asserting that whatever is true of a class is true of any class identical with it—would permit certain contingent conclusions to be drawn. Science, it might be said, is concerned with propositions, not with expressions of these propositions. Whether a scientist writes in English or in French, whether he uses geometrical or algebraical notations to express his propositions, is quite irrelevant; and his procedure of deducing observable consequences from general hypotheses should be explained without bringing in any such irrelevancies:

Were it the case that every sentence occurring in a scientific treatise could be first understood by itself and subsequently considered in its logical relationships to the meanings of the other sentences, this criticism would be perfectly valid; and we could discuss the logic of science by considering only the propositions expressed by these sentences and ignoring the sentences themselves. Of course we should have to use sentences (or equivalent symbols) to express these propositions and their relationships; but these sentences would be 'transparent' (to use a metaphor of Bertrand Russell's); we should 'use' them but not 'mention' them (to employ a distinction of W. V. Quine's). But it is far from being the case that every scientific sentence can be understood in itself as having a direct meaning. Consider a comparatively simple example from contemporary physics: "Every hydrogen atom consists of one proton and one electron." Even if it be granted that the phrase "hydrogen atom" has a straightforward meaning which can be understood by itself, it cannot be pretended that the words "electron" and "proton" can be properly understood without reference to the deductive system of physics in which the propositions expressed by means of them occur. The words may have independent meaning in some subjective sense of meaning; I may think of an electron as a minute sphere and of a proton as a minute sphere with a greater mass; but this is not how the words are used in a treatise on physics. There they are used as symbols in a calculus which is to be interpreted as an applied deductive system; they are not understood as having any meaning apart from their place in such a calculus.

What happens in an abstract science is that we use, as in all inference, a calculus which we interpret as a deductive system; but we do not interpret the calculus by attaching meanings to its formulae separately. We give direct meanings to those formulae of the calculus which we take to represent propositions about observable entities; we give indirect meanings to the other formulae as representing propositions in a deductive system in which the observable propositions are conclusions. Thus we do not interpret the calculus all in a piece, as it were; we interpret the final part of it first, and work backwards to the beginning. A zip-fastener is a better simile for the fitting of a deductive system to a calculus than is the measurement of a rod by the simultaneous superposition of its ends on points of a scale. A calculus designed to represent electrical theory will be constructed so that its final formulae express propositions about observable flashes of light or pointer-readings of a measuring instrument; like a zip-fastener, each side will be firmly attached at one end. But its earlier formulae about fields of force, wave-functions, electrons and other *theoretical concepts* (as we shall call them) will be fitted to the deductive system derivatively in virtue of their place in the calculus. And the expressions "electric-field vector", "electron", etc., will be given meaning by virtue of their occurrence in these formulae.

This indirectness of meaning of the theoretical terms of a science immediately poses the philosopher of science with the problem of how these terms are given this indirect meaning. As a philosopher he cannot rest content to say that they occur in 'mathematical equations' and to ignore considering the way in which these equations signify.* Nor can he be satisfied with saying that the theoretical terms are 'mere symbols' given meaning only by their relationships to one another.† For though their interrelationships will doubtless be found to play a large part in determining their meaning, experience must, one way or another, enter into this determination; for otherwise the symbols would not stand for empirical concepts at all, and what we took to be physics would turn out to be pure mathematics. The question is in what way a theoretical concept like an electron is an empirical concept; it cannot be answered by denying that an electron is an empirical concept at all.

An answer to this question, implicit in the writings of many philosophers of science such as Mach and Karl Pearson, was given explicitly by Bertrand Russell in his doctrine of 'logical constructions'.‡ Electrons, on this view, are logical constructions out of the observed events and objects by which their presence can be detected; this is equivalent to saying that the word "electron" can be explicitly defined in terms of such observations. Every sentence containing the word "electron" can, on this view, be

* Sir William Dampier wrote in 1944: "The last trace of the old, hard, massy atom has disappeared, mechanical models of the atom have failed, and the ultimate concepts of physics have, it seems, to be left in the decent obscurity of mathematical equations" (*A Shorter History of Science* (Cambridge, 1944), p. 154). But a philosopher cannot allow obscurity to be decent, especially since the obscurity in question is not that of the pure mathematics concerned (which is abstruse but not obscure) but that of the application of the mathematics.

† As was Sir Arthur Eddington: "I never discover what carbon really is. It remains a symbol....Carbon is a symbol definable only in terms of the other symbols belonging to the cyclic scheme of physics" (*The Nature of the Physical World* (Cambridge, 1928), p. 269).

‡ "The supreme maxim in scientific philosophising is this: *Wherever possible, logical constructions are to be substituted for inferred entities*" (*Mysticism and Logic and other essays* (London, 1918), p. 155).

translated without loss of meaning into a sentence in which there occur only words which denote entities (events, objects, properties) which are directly observable. It may be very difficult to make this translation, but it is always possible to do so. It is the business of a philosopher of science to exhibit how these translations are to be made, and thus to show how the theoretical terms of a science can be explicitly defined by means of observable entities. A philosopher of science, of course, will not be expected to define the particular theoretical terms of any particular science; but he will be expected to give descriptions of how such definitions might be given—descriptions which may be different for different sorts of theoretical terms, or for the theoretical terms of different sciences.

This 'logical construction' view of the status of theoretical terms was criticized by F. P. Ramsey, who constructed an example of a simple applied deductive system to illustrate his criticism.* I have been able to construct even simpler examples, the discussion of which will take up most of the rest of this chapter. Though my examples and many of the things I shall say about how they elucidate the functioning of theories are not derived from Ramsey, the main criticism of the logical construction view—namely, that if the theoretical terms of a theory are logically constructed out of observable entities, the theory will be incapable of being modified to explain new sorts of facts—is Ramsey's, which my exposition serves merely to elaborate.

'FACTOR' THEORIES AS EXAMPLES

The theories which will be used to illustrate this criticism are theories which explain empirical generalizations which associate the occurrence of one property with another by positing unobservable theoretical 'factors' to account for these empirical generalizations. The theories which will be constructed are far too simple to be explanatory theories in any branch of science; but they are the simplest example of theories which would explain complicated phenomena as being due to the combinations of a limited number of atoms or other factors, theories which play a great part in genetics and in the psychology of 'factors of the mind' as well as in chemistry and physics. The deductive apparatus used in my

---

* "Theories" (1929), published posthumously in *The Foundations of Mathematics and other logical essays* (London, 1931), pp. 212 ff.

examples consists only in that of Boolean algebra, a particularly simple kind of mathematics; and the calculi and deductive systems with which we shall be principally concerned are those which have already been developed in the last chapter.

Both the calculi expounded in the last chapter used three initial formulae

$$\alpha \leftrightarrow (\lambda\mu), \quad \beta \leftrightarrow (\mu\nu), \quad \gamma \leftrightarrow (\nu\lambda),$$

and in both of them there appeared as derived formulae

$$(\alpha\beta) \leftrightarrow (\gamma(\alpha\beta)), \quad (\beta\gamma) \leftrightarrow (\alpha(\beta\gamma)), \quad (\gamma\alpha) \leftrightarrow (\beta(\gamma\alpha)).$$

In the two applied deductive systems which were given as the interpretations of the two calculi, meanings were attached to the symbols of the two calculi in such a way that all these formulae were given the same interpretation in each deductive system. For convenience these interpretations are repeated here, the capital letters A, B, C, ... standing for empirical properties:

$\alpha \leftrightarrow (\lambda\mu)$	to represent	Everything which is A is also L and M, and vice versa; $[p_1]$
$\beta \leftrightarrow (\mu\nu)$		Everything which is B is also M and N, and vice versa; $[p_2]$
$\gamma \leftrightarrow (\nu\lambda)$		Everything which is C is also N and L, and vice versa; $[p_3]$
$(\alpha\beta) \leftrightarrow (\gamma(\alpha\beta))$		Everything which is A and B is also C; $[q_1]$
$(\beta\gamma) \leftrightarrow (\alpha(\beta\gamma))$		Everything which is B and C is also A; $[q_2]$
$(\gamma\alpha) \leftrightarrow (\beta(\gamma\alpha))$		Everything which is C and A is also B. $[q_3]$

In both the deductive systems the first three of these propositions are the only contingent initial propositions; in S_1 they are the only initial propositions, in S_2 there are also three logically necessary initial propositions. If we call only contingent propositions *premisses*, the first three of these propositions (p_1, p_2, p_3) are the only premisses required for the deduction of the last three (q_1, q_2, q_3). These three propositions, like all the propositions in S_1 and all those in the impure part of S_2, follow logically from the conjunction of these three premisses. All six propositions are general propositions; in the language of Chapter I q_1, q_2, q_3 are hypotheses at a lower level than p_1, p_2, p_3. Since we shall be concerned in this chapter only with the relation between the p-propositions and the q-propositions, we shall call the former simply *the hypotheses* and the latter *the empirical generalizations*. For the sake of simplicity we shall use the Second Calculus of the last chapter (and a slight modification of it) and the deductive system S_2 (or a slight modification of it) as an interpretation of the calculus.

In the exposition in the last chapter the calculi were interpreted by attaching empirical meanings directly to all the six symbols α, β, γ, λ, μ, ν. But the situation with which we are presented in an advanced science like physics is one in which this is not possible; there, meaning is attached directly only to some of the symbols which are to have empirical meaning, the meaning of the others being given only indirectly.

A 'THREE-FACTOR' THEORY

In order to see how this can be done, let us consider how a calculus can be interpreted without attaching a meaning directly to each of its symbols.

Suppose that there are three observable properties A, B, C about which it has been found by experiment or observation that the two following empirical generalizations hold:

q_1 Everything which is A and B is also C;

q_2 Everything which is B and C is also A.

Let us now try to explain these generalizations by a scientific theory; that is to say, let us try to find hypotheses from which they can be deduced. Try a theory which will explain them by positing three theoretical properties L, M, N which are to be such that the occurrence of each of the observable properties A, B, C is associated with the joint occurrence of two of these three theoretical factors. This is equivalent to proposing the three hypotheses:

p_1 Everything which is A is also L and M, and vice versa;

p_2 Everything which is B is also M and N, and vice versa;

p_3 Everything which is C is also N and L, and vice versa.

These three hypotheses are the premises in either of the deductive systems of the last chapter, and in both these systems the generalizations logically follow from the hypotheses taken together. The theory with these hypotheses will thus serve as an explanation of the empirical generalizations.

The theory, however, not only explains the known; it predicts the yet unknown. For a third empirical generalization follows from the hypotheses, namely,

q_3 Everything which is C and A is also B.

This generalization can be tested by observation or experiment: if it is found to be confirmed, it supports the theory; if it is refuted by experience, the theory will have to be discarded. This empirical generalization might well not have been thought of by the scientist had he not gone through the stage of constructing a theory to account for the empirical generalizations he had already discovered.

The theory makes use of the three factors L, M, N. What is the nature of these factors? If they are directly observable, there is no problem: the symbols λ, μ, ν in a calculus representing the deductive system are taken to stand for the class of L's, the class of M's, the class of N's in the same way as α, β, γ are taken to stand for the class of A's, the class of B's, the class of C's, respectively. But they may not be directly observable; they may be 'supposed' or 'assumed' or 'posited' in order to 'explain' or 'account for' the empirical generalizations that have been discovered to hold of the observable properties A, B, C. What then is the status of L, M, N? Or, to put the matter more precisely, how are the symbols in a calculus representing the theory's deductive system to be given a meaning?

The answer that first suggests itself is that the symbols λ, μ, ν must be *defined* in terms of the meanings of α, β, γ, in which case the factors L, M, N for which these symbols stand will be exhibited as what Russell calls *logical constructions* out of A, B, C.

To examine this answer it is necessary to be clear as to what is meant by definition in this context. Since definition of a symbol (the definiendum) is in terms of the *meaning* of another (the definiens), definition cannot be explained by reference to an uninterpreted calculus, that is, to a calculus considered apart from its interpretation. But in either the First or the Second Calculus, each of which contains a double-arrow-substitution rule, a formula in which there is a primitive element which both occurs by itself as either the left or as the right element and also does not occur in the other element of the formula can be interpreted as giving a definition of this primitive element in terms of the meaning of the element which is on the other side of the formula, provided that this latter element has already been given a meaning. For two symbols are *universally synonymous* if one of the symbols can always be substituted for the other in any sentence in which the latter occurs without changing the meaning of the sentence. Then if we know the meaning of the latter symbol but do not know the meaning of the former, the fact of their universal synonymity will enable us to define the former in terms of the latter. Now the double-arrow-substitution rule enables us to substitute in any formula of the calculus the left element of a formula for the right element, or vice versa. So the left and the right elements in a formula of the calculus are *synonymous within the calculus*; and, if one of these is so far uninterpreted, we may if we wish define it as the meaning of the other element. Of course if both elements have already been given meanings, they can be synonymous within the calculus without being synonymous outside the calculus, and hence without being universally synonymous. If θ and ϕ stand for two classes which are identical (i.e. if they stand for the same class), θ and ϕ would be synonymous within a calculus having a double-arrow-substitution rule if it contained the formula $\theta \leftrightarrow \phi$ representing this identity; but if θ is taken to stand for the class of rational animals and ϕ for the class of featherless bipeds, θ and ϕ are not universally synonymous, since the proposition that I am a rational animal (i.e. that I am a member of the class of rational animals, a proposition not expressible within the calculus) is not the same as the proposition that I am a featherless biped. If, however, a primitive element in the calculus has not been given a meaning, one method of doing so is to use a formula of the calculus in which it occurs only as the left or only as the right element, and to define it as the meaning of the element on the other side of the formula. The double-arrow-substitution rule of the calculus will then ensure synonymity within the calculus of the definiendum and the definiens.

A formula which could be used to provide a definition of a primitive element θ will, for convenience, be called a *definitory formula for θ*. In all the calculi with which we shall be concerned a definitory formula for θ is any formula in which both θ is the complete left (or right) element and also θ does not occur in the right (or in the left) element. A definitory formula for θ cannot be used to define θ unless the element which is the other side of the formula has already been given a meaning; thus a definitory formula is only sometimes a defining formula.

For example, in our First or Second Calculus, if λ, μ, ν be taken to stand for the class of four-sided plane figures, the class of equilateral plane figures and the class of equiangular plane figures respectively, the three initial formulae

$$\alpha \leftrightarrow (\lambda\mu), \quad \beta \leftrightarrow (\mu\nu), \quad \gamma \leftrightarrow (\nu\lambda)$$

can be taken as providing definitions for α, β, γ. α will then be defined as designating the class of rhombuses (plane figures which are both four-sided and equilateral), β the class of regular figures (plane figures which are both equilateral and equiangular), γ the class of rectangles (plane figures which are both equiangular and four-sided). The derived formulae will then express the propositions that all rhombuses which are regular are rectangular, that all regular rectangles are rhombuses, and that all rectangular rhombuses are regular, all of which are logically necessary propositions following logically from the propositions in the deductive system corresponding to the definitions of α, β, γ or of "rhombus", "regular figure", "rectangle".

What is required, however, for it to be possible to treat the theoretical factors as logical constructions out of the observable properties is a reversal of this process.* What are needed are definitions of the symbols λ, μ, ν standing for the theoretical factors in terms of the observable properties corresponding to α, β, γ. We require, therefore, formulae in the calculus in which λ, μ, ν appear only as left (or right) elements. Can such definitory formulae for λ, μ, ν be derived?

In the case either of the First or of the Second Calculus the answer is in the negative. But this negative answer does not completely settle the question. Both these calculi are very simple in the sense that their rules only permit of very few types of move.

---

* Russell puts the requirement in the form: "Physics exhibits sense-data as functions of physical objects, but verification is only possible if physical objects can be exhibited as functions of sense-data. We have therefore to solve the equations giving sense-data in terms of physical objects, so as to make them instead give physical objects in terms of sense-data" (*Mysticism and Logic and other essays*, p. 146).

Indeed, in constructing them I deliberately made them as simple as possible for the purpose of explaining the relations between calculi and deductive systems. So their interpretation as deductive systems applying to empirical classes (the class of A's, etc.) does not in fact include all that there is of the logic of classes. There are propositions about relationships between the empirical classes which are logical consequences of the three premisses but which cannot be expressed in either calculus; and there are logical truths about the relationships between classes which, since they are inexpressible in the Second Calculus, do not occur in the pure part of S_2. A hopeful way of attacking the problem will therefore be to add additional rules to the algebraic part of the Second Calculus so that this algebraic part will represent the whole, and not merely part, of the logic of classes. Then we can see whether formulae permitting the definition of λ, μ, ν will appear in this modified calculus. So let us construct a new calculus with set-up and rules similar to those of the Second Calculus with two modifications. The first is that the definition of a *non-primitive element* is to be extended to include a new type of such element, namely, any element obtained by adding a prime mark $'$ after any element, primitive or non-primitive. Thus α', $(\alpha')'$, $(\alpha\beta)'$, $(\alpha'\beta')'$, etc., are to be elements in the new calculus. The second modification is to substitute the more complicated formula

$$((\xi'\eta')'\,(\xi'\eta)')\leftrightarrow\xi$$

for the initial formula $(\xi\xi)\leftrightarrow\xi$ of the Second Calculus (formula [4] of Tables II and III). (This latter formula will appear in the new calculus as a derived formula.) This new calculus may be interpreted as a deductive system S_3 in exactly the same way as that in which the Second Calculus was interpreted as S_2 if an interpretation is given for an element being primed (i.e. having a prime mark attached to it). The interpretation to be given for a primed element is that it is to stand for the class of things which lack the empirical property which specifies the class represented by the unprimed element, i.e. a primed element is to stand for what is called the *complement* of this latter class. Thus if α stands for the class of things having the property A (for short, the class of A's), α' will stand for the class of things lacking the property A (for short, the class of non-A's).

In this deductive system S_3, represented by the new calculus, all the propositions of the logic of classes find a place. The pure part of S_3 forms a complete pure deductive system of classes; it is represented by the algebraic calculus which is part of the new calculus. Since we shall need to consider other calculi with the same algebraic part, the algebraic calculus whose initial formulae are

$$((\xi'\eta')'\,(\xi'\eta)')\leftrightarrow\xi,\quad (\xi\eta)\leftrightarrow(\eta\xi),\quad (\xi(\eta\zeta))\leftrightarrow((\xi\eta)\,\zeta),$$

and whose rules of play are the double-arrow-substitution rule and the variable-substitution rule will be called the *Huntington Calculus*.* The new calculus representing the system S_3 may therefore be considered as the Huntington Calculus with the addition of three variable-free initial formulae

$$\alpha\leftrightarrow(\lambda\mu),\quad \beta\leftrightarrow(\mu\nu),\quad \gamma\leftrightarrow(\nu\lambda).$$

Let us now ask whether in this new calculus formulae can be derived in which λ, μ, ν stand alone as left (or right) elements. The answer is still in the negative. But since the new calculus represents a deductive system which contains the whole of the logic of classes, we may profitably ask whether it is possible to add to this calculus additional initial formulae of the required sort containing λ, μ, ν without affecting formulae of the calculus in which λ, μ, ν do not occur, i.e. without enabling any more formulae not including λ, μ, ν to be derived in the calculus than would have been derivable without these additional initial formulae. For if we can do this, we can define λ, μ, ν in a way which is not indeed necessitated by their use in the calculus, interpreted as system S_3, but which leaves the rest of the formulae derivable in the calculus undisturbed. So far as the working of the calculus is concerned, the new initial formulae would be *idle*.

Such idle formulae can be found.† They can be expressed in the

---

* The three initial formulae and set-up of this calculus are equivalent to the 'simplified set of postulates' for Boolean algebra published by E. V. Huntington in *Mind*, n.s., vol. 42 (1933), pp. 203 ff. For a proof that this calculus represents a complete deductive system of classes see also E. V. Huntington, *Transactions of the American Mathematical Society*, vol. 35 (1933), pp. 274 ff. and 557 ff. My way of exposition, in terms of calculi as games with symbols, is different from Huntington's traditional mathematical treatment. Huntington developed many sets of postulates for Boolean algebra; and his 1933 set which I am using is quite different from his famous 'first set of postulates' published in 1904.

† In the language of ordinary algebra the problem is that of solving the set of three 'simultaneous equations' $\alpha\leftrightarrow(\lambda\mu)$, $\beta\leftrightarrow(\mu\nu)$, $\gamma\leftrightarrow(\nu\lambda)$ for λ, μ, ν. The

most general form with the aid of elements θ, ϕ, ψ, each of which can be interpreted as standing for any class whatever.* The idle formulae are

$$\lambda \leftrightarrow ((\gamma'\alpha')\,(\beta'(\theta'(\phi\psi))))',$$
$$\mu \leftrightarrow ((\alpha'\beta')\,(\gamma'(\theta(\phi'\psi))))',$$
$$\nu \leftrightarrow ((\beta'\gamma')\,(\alpha'(\theta(\phi\psi'))))'.$$

They can be written in a simpler form if we introduce a new symbol $\cup$ into the Huntington Calculus by agreeing that any pair of elements with this symbol between them and with the whole enclosed in brackets, e.g. $(\alpha\cup\beta)$, $(\alpha\cup(\alpha\cup(\beta\gamma')'))$, is to be an element of the calculus, and by adding the initial formula

$$(\xi\cup\eta)\leftrightarrow(\xi'\eta')',$$

which will serve to define $(\alpha\cup\beta)$ in terms of the meaning of $(\alpha'\beta')'$. Since system S_3 interprets $(\alpha'\beta')'$ as standing for the complement of the intersection of the complement of the class of A's with the complement of the class of B's, which is the same as the class whose members are members either of the class of A's or of the class of B's (or of both), this interpretation can be used to define $(\alpha\cup\beta)$ as standing for this class, called the *union* or the *join* or the *logical sum* of the two classes represented by α and by β. The introduction of the symbol $\cup$ by means of the new initial formula $(\xi\cup\eta)\leftrightarrow(\xi'\eta')'$ has no significance except to abbreviate formulae, and the Huntington Calculus which will be used in the future with this fourth initial formula added will continue to be called simply the Huntington Calculus.

In this calculus the idle formulae can be put in the forms

$$\lambda \leftrightarrow ((\gamma\cup\alpha)\cup(\beta'(\theta'(\phi\psi)))),$$
$$\mu \leftrightarrow ((\alpha\cup\beta)\cup(\gamma'(\theta(\phi'\psi)))),$$
$$\nu \leftrightarrow ((\beta\cup\gamma)\cup(\alpha'(\theta(\phi\psi')))),$$

where θ, ϕ, ψ are any elements whatever.

general solution given here, together with other forms of solution, will be found in A. N. Whitehead, *Universal Algebra* (Cambridge, 1898), pp. 79 f. In his language the derived formulae $(\alpha\beta)\leftrightarrow(\gamma(\alpha\beta))$, $(\beta\gamma)\leftrightarrow(\alpha(\beta\gamma))$, $(\gamma\alpha)\leftrightarrow(\beta(\gamma\alpha))$ form the 'resultant' of 'eliminating' λ, μ, ν from the set of simultaneous equations.

* The elements θ, ϕ, ψ are not variables, since the variable-substitution rule does not apply to them.

The position now is that (1) in the Huntington Calculus with variable-free initial formulae $\alpha \leftrightarrow (\lambda\mu)$, $\beta \leftrightarrow (\mu\nu)$, $\gamma \leftrightarrow (\nu\lambda)$ added, it is possible to derive the formulae

$$(\alpha\beta) \leftrightarrow (\gamma(\alpha\beta)), \quad (\beta\gamma) \leftrightarrow (\alpha(\beta\gamma)), \quad (\gamma\alpha) \leftrightarrow (\beta(\gamma\alpha));$$

(2) in this same calculus it is impossible to derive formulae in which λ or μ or ν stands alone as left (or right) element; (3) in the Huntington Calculus with these variable-free initial formulae added, the addition of the three idle formulae as initial formulae does not enable us to derive any formulae in which only α, β, γ occur as primitive elements which we could not derive without this addition; and (4) in the Huntington Calculus whose added variable-free formulae are the three idle formulae together with the formulae

$$(\alpha\beta) \leftrightarrow (\gamma(\alpha\beta)), \quad (\beta\gamma) \leftrightarrow (\alpha(\beta\gamma)), \quad (\gamma\alpha) \leftrightarrow (\beta(\gamma\alpha)),$$

it is possible to derive the formulae*

$$\alpha \leftrightarrow (\lambda\mu), \quad \beta \leftrightarrow (\mu\nu), \quad \gamma \leftrightarrow (\nu\lambda).$$

If we interpret the various calculi to represent the relationships between the hypotheses of the theory and the empirical generalizations, (3) shows that formulae are available of a sort suitable to allow us to define the theoretical terms by means of the observable properties; (1) and (4) together show that, if such a definition could be effected by the use of these formulae, the hypotheses of the theory would be logically equivalent to the empirical generalizations, since the set of each would be deducible from the other.

A SPATIAL DIAGRAM FOR THE THREE-FACTOR THEORY

It may be helpful to explain the position in a more concrete way by the use of a diagram. Since we are interpreting all the calculi with which we are now concerned as deductive systems concerned with the relationships between classes, and since such relationships can be exemplified in spatial diagrams, the things we wish to say can be said less abstractedly by reference to a suitably constructed diagram.†

---

* The derivation is too long to be more than indicated here. The expansion of $(\lambda\mu)$ by means of the idle formulae can be reduced, by using formulae of the Huntington Calculus, to $(\alpha \cup (\beta\gamma))$. Hence, since $(\beta\gamma) \leftrightarrow (\alpha(\beta\gamma))$,

$$(\lambda\mu) \leftrightarrow (\alpha \cup (\alpha(\beta\gamma)));$$

whence $(\lambda\mu) \leftrightarrow \alpha$.

† No *metrical* properties of the spatial figures concerned are relevant.

Suppose that meaning is attached to the primitive elements of our calculi in the following way by reference to fig. 2:

α is to designate the class of points on or inside the rhombus *afde*.

β is to designate the class of points on or inside the rhombus *bdef*.

γ is to designate the class of points on or inside the rhombus *cefd*.

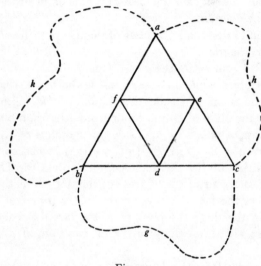

Fig. 2.

Then $(\alpha\beta)$ will represent the class of points which are both on or inside the rhombus *afde* and on or inside the rhombus *bdef*, i.e. the class of points on or inside the triangle *def*.

$(\gamma(\alpha\beta))$ will represent the class of points which are both on or inside the rhombus *cefd* and on or inside the triangle *def*, i.e. the class of points on or inside the triangle *def*.

Similarly $(\beta\gamma)$, $(\alpha(\beta\gamma))$, $(\gamma\alpha)$, $(\beta(\gamma\alpha))$ all represent the class of points on or inside the triangle *def*.

The figure has been drawn in such a way that the overlap of each pair of rhombuses falls within the third rhombus. These relationships between the rhombuses are expressed by the formulae

$$(\alpha\beta)\leftrightarrow(\gamma(\alpha\beta)), \quad (\beta\gamma)\leftrightarrow(\alpha(\beta\gamma)), \quad (\gamma\alpha)\leftrightarrow(\beta(\gamma\alpha)),$$

which correspond to the empirical generalizations which it is our purpose to explain by a scientific theory.

The theory, as applied to the diagram, deduces these relationships from relationships between each of the rhombuses separately and three other spatial figures corresponding to λ, μ, ν. In the diagram as drawn:

λ is taken to designate the class of points on or inside the figure *hafdc* in the diagram,

μ is taken to designate the class of points on or inside the figure *kbdea* in the diagram,

ν is taken to designate the class of points on or inside the figure *gcefb* in the diagram.

Then $(\lambda\mu)$ represents the class of points which are both on or inside the figure *hafdc* and on or inside the figure *kbdea*, i.e. the class of points on or inside the rhombus *afde*. Similarly, $(\mu\nu)$ represents the class of points on or inside the rhombus *bdef*, and $(\nu\lambda)$ the class of points on or inside the rhombus *cefd*.

The formulae $\alpha \leftrightarrow (\lambda\mu)$, $\beta \leftrightarrow (\mu\nu)$, $\gamma \leftrightarrow (\nu\lambda)$, which correspond to the hypotheses of the theory, together express the fact that the overlap of each pair of the figures is a rhombus. The theory therefore deduces the fact that the overlap of each pair of rhombuses falls within the third rhombus from the fact that each rhombus may be regarded as the overlap of two out of three of the figures.

It will be seen from the diagram that, in order that each rhombus should be regarded as the overlap of two figures, all that is required of the parts of the three figures lying outside the triangle *abc* (i.e. the regions bounded by the sides of this triangle and the dotted line) is that no two of these regions should overlap one another. For if, for example, the region *haec* had a common part with the region *kafb*, the points of this common part would be members of the class represented by $(\lambda\mu)$ and would therefore have to be members of the class designated by α, all the points of which lie on or inside the triangle *abc*.

This indeterminateness of the dotted parts of the boundaries of the regions corresponding to λ, μ, ν is reflected in the occurrence in the formulae proposed as possible definitions of λ, μ, ν in terms of α, β, γ of elements θ, ϕ, ψ with an indeterminate interpretation. The class of points of that part of the region corresponding to λ

which lies outside the triangle abc will be represented by $(\alpha'(\beta'(\gamma'(\theta'(\phi\psi)))))$, the similar class for μ by $(\alpha'(\beta'(\gamma'(\theta(\phi'\psi)))))$ and the similar class for ν by $(\alpha'(\beta'(\gamma'(\theta(\phi\psi')))))$. The intersection of the first two of these classes will be represented by $(\theta\theta')$, since in the Huntington Calculus we can derive the formula

$$((\alpha'(\beta'(\gamma'(\theta'(\phi\psi)))))(\alpha'(\beta'(\gamma'(\theta(\phi'\psi))))))\leftrightarrow(\theta\theta').$$

$(\theta\theta')$. represents the class of things which both have and lack the property corresponding to θ. Since there is nothing which both has and lacks this property, this class is empty, i.e. has no members. All empty classes may be regarded as being one and the same class, since $(\xi\xi')\leftrightarrow(\eta\eta')$ is a formula derivable in the Huntington Calculus. This unique class will be called the *null-class*; it will be designated by o (called the *null-element*). The intersection of the other similar pairs of classes will also each be the null-class.

We seem now to have reached an impasse in the attempt to define the symbols λ, μ, ν which are to stand for our theoretical concepts by means of the symbols α, β, γ which are directly interpretable in empirical terms. We have indeed succeeded in finding formulae which enable us to identify the classes designated by λ, μ, ν with classes constructed out of those designated by α, β, γ; but these constructions make use of three undetermined classes, which may be any classes whatever. It is impossible to impose restrictions on these classes by strengthening the initial formulae of the algebraic part of our calculus (in the way in which the Second Calculus was strengthened), for the Huntington Calculus represents the whole of the logic of classes and is thus all the logical apparatus available for treating our subject-matter. Thus the position, in terms of our diagram, is that the figures corresponding to λ, μ, ν are determined by the rhombuses corresponding to α, β, γ only in so far as concerns the parts of these figures lying on or within the triangle abc corresponding to $(\alpha\cup(\beta\cup\gamma))$; outside this triangle the figures may be of any shape or size you wish, provided only that no two of them overlap one another. If the dotted lines in the diagram which are the boundaries of these extra-triangular regions are imagined as being made of elastic string, they can be pulled out in any way we like, provided only that they are kept from crossing one another, without affecting the relationships of the regions within the triangle.

There would seem to be two methods by which these extra-triangular regions might be fixed in a less arbitrary way. We might agree to fix them as being as large as possible. Or we might fix them as being as small as possible. By so doing we should not get rid of the arbitrariness; but there would be some justification for our choice, and thus for our definition of λ, μ, ν.

The method by which one of these extra-triangular regions would be fixed to be as large as possible would be to make it co-extensive with the whole of the plane minus the triangle. It would then be the region corresponding to the element $(\alpha'(\beta'\gamma'))$. But, since no two of the regions can be allowed to overlap, only one of the regions can be made to be identical with this region, and the other two regions will have to vanish, since they will be (as it were) crowded out of the logical space available. If λ is to represent the favoured class, the formulae specifying λ, μ, ν would then be obtained by taking θ to be the null-element o, and both ϕ and ψ to be its complement o'. This will enable us to derive the formulae

$$\lambda \leftrightarrow ((\gamma \cup \alpha) \cup \beta'), \quad \mu \leftrightarrow (\alpha \cup \beta), \quad \nu \leftrightarrow (\beta \cup \gamma).$$

But since λ, μ, ν enter perfectly symmetrically into the formulae

$$\alpha \leftrightarrow (\lambda \mu), \quad \beta \leftrightarrow (\mu \nu), \quad \gamma \leftrightarrow (\nu \lambda)$$

expressing the hypotheses of the theory, there can be no reason for arranging to maximize any one of them rather than any other.

The situation is different, however, if we try to make the extra-triangular regions as small as possible. For they can all be made to vanish, which corresponds to taking θ, ϕ, ψ as each being the null-element o. In this case the regions corresponding to λ, μ, ν would be the trapeziums *afdc*, *bdea*, *cefb* respectively. By taking each of θ, ϕ, ψ to be the null-element, we can derive the formulae

$$\lambda \leftrightarrow (\gamma \cup \alpha), \quad \mu \leftrightarrow (\alpha \cup \beta), \quad \nu \leftrightarrow (\beta \cup \gamma),$$

which are perfectly symmetrical between α, β, γ.

This would seem to be exactly what we want. The symbols λ, μ, ν can be defined by these formulae in terms of the meanings of α, β, γ and of no other empirical concept. And the theoretical concepts designated by λ, μ, ν can be analysed in terms of the observable properties using ordinary English without any symbolic language; λ stands for the class of L's, where L is the property of being either C or A (or both), just as being a doctor is being either a physician or a surgeon or both. There would be no mystery as to the status of L, M, N. These factors would be exhibited as logical constructions of a very simple kind out of the observable properties A, B, C. Moreover, their collective field of application would be exactly the same as that of A, B, C collectively; the class of things which have one or more of the three properties L, M, N would be the same as the class of things having one or more of A, B, C. It is true that there would still be an arbitrary element in the choice of L, M, N, since they might have been chosen to be wider properties without affecting their value in the scientific theory. But their choice as the narrowest properties that would do what was required of them might well be justified by a methodological principle of parsimony: *entia explicantia non sunt amplificanda praeter necessitatem.*

We have seen that if in our calculus we take the three formulae corresponding to the proposed definitions of λ, μ, ν as initial formulae in the place of the formulae

$$\alpha \leftrightarrow (\lambda\mu), \quad \beta \leftrightarrow (\mu\nu), \quad \gamma \leftrightarrow (\nu\lambda),$$

representing the hypotheses of the theory, and add as initial formulae the three formulae

$$(\alpha\beta) \leftrightarrow (\gamma(\alpha\beta)), \quad (\beta\gamma) \leftrightarrow (\alpha(\beta\gamma)), \quad (\gamma\alpha) \leftrightarrow (\beta(\gamma\alpha))$$

representing the empirical generalizations, we can derive in this modified calculus the formulae representing the hypotheses of the theory. This comes to saying that, if we define the theoretical terms λ, μ, ν in this way, the hypotheses of the theory will be logically deducible from the empirical generalizations which they were put forward to explain. Since the empirical generalizations are, of course, logically deducible from the hypotheses, such a definition of the theoretical terms would make the set of hypotheses logically equivalent to the set of the empirical generalizations. The deductive system leading from the hypotheses to the generalizations would be reversible, and would lead equally well from the generalizations to the hypotheses.

This conclusion seems most welcome at first sight. A little reflexion, however, makes it clear that if the theoretical terms are defined in such a way as to make the theory logically equivalent to the facts it explains, the theory becomes merely an alternative way of stating these facts. The hypotheses of the theory become translations of the empirical generalizations rather than, in any important sense, explanations of them; they do not stretch out beyond the limited number of generalizations; not only do they have exactly the same field of application, but they say exactly the same things about this field. A definition of the theoretical terms would thus sacrifice one of our principal objects in constructing a scientific theory, that of being able to extend it in the future, if way opens, to explain facts about new things by incorporating the theory in a more general theory having a wider field of application. This point can be shown excellently by explaining how a 'three-factor' theory can be incorporated in a 'four-factor' theory.

A 'FOUR-FACTOR' THEORY

Suppose that two further observable properties D, E are noticed, about which it is found that the following generalizations hold:

Everything which is A and D is also E,
Everything which is E and A is also D.

Then it is natural to attempt to explain them, not by constructing an entirely new theory, but by modifying the three-factor theory we already have by positing a fourth theoretical property R such that the occurrence of D is associated with the joint occurrence of R with one of the three factors already posited and the occurrence of E with the joint occurrence of R with another of these factors. It is easy to see that if we add to the three hypotheses of the three-factor theory the two hypotheses

Everything which is D is also L and R, and vice versa,
Everything which is E is also M and R, and vice versa,

we can deduce the new empirical generalizations. The addition of one additional factor R to the three factors L, M, N of the original theory—the change from a three-factor to a four-factor theory—thus accounts for the new empirical generalizations most economically, and connects together these new generalizations with the old ones by constructing a theory in which they are all logically deducible.

The deductive system of the four-factor theory may be represented by the Huntington Calculus with the addition of five variable-free initial formulae

$$\alpha \leftrightarrow (\lambda\mu), \quad \beta \leftrightarrow (\mu\nu), \quad \gamma \leftrightarrow (\nu\lambda), \quad \delta \leftrightarrow (\lambda\rho), \quad \epsilon \leftrightarrow (\mu\rho),$$

where $\alpha, \beta, \gamma, \delta, \epsilon$ are to stand for the class of A's, of B's, of C's, of D's, of E's respectively. In this modified calculus the formulae

$$(\alpha\delta) \leftrightarrow (\epsilon(\alpha\delta)), \quad (\delta\epsilon) \leftrightarrow (\alpha(\delta\epsilon)), \quad (\epsilon\alpha) \leftrightarrow (\delta(\epsilon\alpha))$$

are derivable in the same way as the similar formula derived in Table I or in Table II. The first and third of these represent the two empirical generalizations which gave rise to the four-factor theory, the second represents a new empirical generalization

Everything which is D and E is also A

predicted by the theory for empirical confirmation or rejection. And, since $((\mu\nu)(\lambda\rho)) \leftrightarrow ((\nu\lambda)(\mu\rho))$ is a formula derivable in the calculus, we can derive the formula

$$(\beta\delta) \leftrightarrow (\gamma\epsilon)$$

representing the double empirical generalization

Everything which is B and D is also C and E, and vice versa,

which the four-factor theory also presents for empirical test. This consequence of the theory was to me a surprising one; I should not have thought of it had I not been deliberately looking for all the possible conclusions deducible from the five hypotheses of the theory.

The four-factor theory will, however, do more than provide new empirical generalizations about properties already known; it will suggest the possibility of there being a new observable property related to the known ones. For the four factors have been posited to account for the interrelationships of five observable properties by treating each of these as the combination of two out of the four factors. In the calculus which represents the theory the variable-free initial formulae expressing the hypotheses

$$\alpha \leftrightarrow (\lambda\mu), \quad \beta \leftrightarrow (\mu\nu), \quad \gamma \leftrightarrow (\nu\lambda), \quad \delta \leftrightarrow (\lambda\rho), \quad \epsilon \leftrightarrow (\mu\rho)$$

make use of five combinations of the four Greek letters λ, μ, ν, ρ two at a time. If we consider how many possible combinations there are of these four letters two at a time, we find that there is a sixth combination ν with ρ so far unused. Consequently we may modify the calculus representing the theory by adding a sixth variable-free initial formula

$$\kappa \leftrightarrow (\nu\rho).$$

In this calculus the following formulae connecting κ with α, β, γ, δ, ϵ will be derivable:

$$(\beta\epsilon)\leftrightarrow(\kappa(\beta\epsilon)), \quad (\epsilon\kappa)\leftrightarrow(\beta(\epsilon\kappa)), \quad (\kappa\beta)\leftrightarrow(\epsilon(\kappa\beta)),$$

$$(\gamma\kappa)\leftrightarrow(\delta(\gamma\kappa)), \quad (\kappa\delta)\leftrightarrow(\gamma(\kappa\delta)), \quad (\delta\gamma)\leftrightarrow(\kappa(\delta\gamma)),$$

$$(\alpha\kappa)\leftrightarrow(\beta\delta), \quad (\gamma\epsilon)\leftrightarrow(\alpha\kappa).$$

Since we have already succeeded in interpreting α, β, γ, δ, ϵ as standing for classes of observable properties, it would seem profitable to look to see if there is not a sixth observable property K for which κ can be taken to stand and which will have the relationships expressed by these formulae.* If such a property can be found, no new factor will be required to account for it; the four factors of the four-factor theory will be sufficient. Thus the transition from a three-factor to a four-factor theory will not only have connected together known empirical generalizations and have predicted previously unknown ones, but it will also have led to the discovery of a property and of its relationships which had previously escaped notice. The search, which has proved so successful, for chemical atoms having specific nuclear and electronic constitutions, and for chemical molecules having specific atomic constitutions, has been stimulated by the previous construction of theories to explain the constitutions of atoms or of molecules already known, theories which had, as it were, 'gaps' in them asking to be filled by the discovery of the yet unknown atoms or molecules.

It should be noticed that, whereas discovery of observable properties which fill such gaps in a theory is rightly thought to provide a weighty confirmation of the theory (for there will be a number of empirical generalizations which will have to be found true of each such property for it to fill a gap, and each of these generalizations

* These new empirical generalizations are not all logically independent of the empirical generalizations already established. Any one of the three formulae $(\beta\delta)\leftrightarrow(\gamma\epsilon)$, $(\alpha\kappa)\leftrightarrow(\beta\delta)$, $(\gamma\epsilon)\leftrightarrow(\alpha\kappa)$ is derivable from the other two.

will support the theory), failure to discover such properties would not be regarded as weighing against the theory, except perhaps in very special cases. There may easily be excellent reasons, did we know them, either why two of the theoretical factors should not be able to occur together or why, if they do so occur, they should produce no observable effect. Until a property for which κ can be taken to stand is discovered, the symbol κ must be treated as representing a purely theoretical concept on a par with the theoretical concepts represented by λ, μ, ν, ρ; it may be convenient to use a calculus in which it is introduced by the initial formula $\kappa \leftrightarrow (\nu\rho)$ which will serve to define it in terms of the meanings of ν and of ρ, but this calculus will serve no better (and no worse) than the calculus without this initial formula to represent the observed facts. And the four-factor theory, expressed by either of these calculi, will furnish a satisfactory explanation of the generalizations obeyed by the five observable properties by treating them as being each due to the combination of two out of four factors, irrespective of the fact that there is no observable property known which might be regarded as the combination of the sixth pair of these factors.

Now the utilization of a theory which explains one set of generalizations to explain a further set by a slight modification of the theory—by the transition from a three-factor to a four-factor theory—would have been severely restricted if the symbols standing for the three factors had been defined in terms of the observable properties whose relationships the three-factor theory was designed to explain. For, as we have seen, the only plausible way of defining λ, μ, ν would be by adding as new initial formulae to the calculus the formulae

$$\lambda \leftrightarrow (\gamma \cup \alpha), \quad \mu \leftrightarrow (\alpha \cup \beta), \quad \nu \leftrightarrow (\beta \cup \gamma);$$

and were this done there would be no proper logical room, as it were, for the three factors L, M, N to combine with a fourth factor R to yield new observable properties.

The restriction that would be imposed by these definitions can be shown in various ways. The best way, perhaps, is to notice that if the calculus representing the four-factor theory is modified by adding as extra initial formulae the formulae that might be used to define λ, μ, ν, it will be possible to derive formulae connecting

71

α, β, γ, δ, ϵ, κ that are not derivable in the unmodified calculus. Three such formulae* are

$$\delta\leftrightarrow((\alpha\epsilon)\cup(\gamma\kappa)), \quad \epsilon\leftrightarrow((\alpha\delta)\cup(\beta\kappa)), \quad \kappa\leftrightarrow((\beta\epsilon)\cup(\gamma\delta)).$$

The first of these formulae expresses the empirical generalization that the class of D's is the same as the class formed by joining the intersection of the class of A's with the class of E's to the intersection of the class of C's with the class of K's, i.e. that everything which is D is also either both A and E or both C and K (or both), and vice versa. The other two formulae express corresponding empirical generalizations providing identifications for the class of E's and for the class of K's.

Now these empirical generalizations may turn out to be true. But they go beyond the logical consequences of the four-factor theory. What can be deduced from the hypotheses of this theory is that everything which is either both A and E or both C and K (or both) is also D;† but it is impossible to deduce the converse proposition that everything which is D is also either both A and E or both C and K (or both), which would be necessary in order to establish the identity of the classes concerned. Thus the formulae which it is proposed should be added to the calculus to provide definitions of λ, μ, ν, namely

$$\lambda\leftrightarrow(\gamma\cup\alpha), \quad \mu\leftrightarrow(\alpha\cup\beta), \quad \nu\leftrightarrow(\beta\cup\gamma),$$

though they are idle formulae and do no harm if added to a calculus representing the three-factor theory, are by no means idle and harmless if added to a calculus representing the four-factor theory. For this calculus would then cease to represent the four-factor theory as we want it to be; it would, instead, represent a theory

---

* The first of these formulae may be derived in the following way, where in the derivation only formulae which occur in the non-algebraic part of the calculus are explicitly mentioned:

$\delta\leftrightarrow(\lambda(\lambda\rho))$	from	$\delta\leftrightarrow(\lambda\rho)$,
$\delta\leftrightarrow((\gamma\cup\alpha)\delta)$	from	$\lambda\leftrightarrow(\gamma\cup\alpha)$,
$\delta\leftrightarrow((\alpha\delta)\cup(\delta\gamma))$		

$\delta\leftrightarrow((\epsilon(\alpha\delta))\cup(\kappa(\delta\gamma)))$ from $(\alpha\delta)\leftrightarrow(\epsilon(\alpha\delta))$ and $(\delta\gamma)\leftrightarrow(\kappa(\delta\gamma))$,
$\delta\leftrightarrow((\alpha\epsilon)\cup(\gamma\kappa))$ from $(\epsilon\alpha)\leftrightarrow(\delta(\epsilon\alpha))$ and $(\gamma\kappa)\leftrightarrow(\delta(\gamma\kappa))$.

It is only the step to $\delta\leftrightarrow((\gamma\cup\alpha)\delta)$ that makes use of the extra initial formula.

† This proposition is expressed by the formula

$$((\alpha\epsilon)\cup(\gamma\kappa))\leftrightarrow(\delta((\alpha\epsilon)\cup(\gamma\kappa))),$$

which is derivable in a calculus including the Huntington Calculus from the formulae $(\epsilon\alpha)\leftrightarrow(\delta(\epsilon\alpha))$ and $(\gamma\kappa)\leftrightarrow(\delta(\gamma\kappa))$.

with a more stringent set of hypotheses, logical consequences of which would assert not only that the intersections of certain pairs out of the six empirically given classes were included in certain classes, but that each of three of these classes was actually identical with a logical combination of four other classes.

A consequence of the proposed definition of λ, μ, ν is that the classes designated by δ, ϵ, κ would all be restricted to falling within the classes represented by $(\gamma \cup \alpha)$, $(\alpha \cup \beta)$, $(\beta \cup \gamma)$, respectively, and would thus all have to fall within the class represented by $(\alpha \cup (\beta \cup \gamma))$. The field of application of the properties D, E, K would all have to lie within that of A, B, C collectively; the class of things having one or more of the three properties D, E, K would have to be included in the class of things having one or more of the three properties A, B, C. So the new properties D, E, K, whose relationships the four-factor theory was put forward to explain, could only be such as would belong only to things having the properties A, B, C covered by the three-factor theory. The four-factor theory would then account for new properties of instances of the old properties: it would not apply to a class of things about which the three-factor theory had said nothing.

This point may be illustrated by a diagram used in a similar way to that on p. 63. As before, α, β, γ are to designate the classes of points on or inside the rhombuses *afde*, *bdef*, *cefd* respectively (fig. 3). Then, since λ, μ, ν are now to be defined as standing for the joins of these classes of points in pairs, λ, μ, ν are to designate the classes of points on or inside the trapeziums *afdc*, *bdea*, *cefb* respectively. The class of points designated by ρ is to be the class of points on or inside the triangle *lpq*; this triangle has been drawn so that it overlaps (without covering) each of the four triangles *afe*, *bdf*, *ced*, *efd*, and has a part outside all of them. Then if δ, ϵ, κ are to designate the classes of points represented by $(\lambda\rho)$, $(\mu\rho)$, $(\nu\rho)$ respectively,

δ will designate the class of points on or inside the trapezium *fnpq*,

ϵ will designate the class of points on or inside the pentagon *fmoeq*,

κ will designate the class of points on or inside the quadrilateral *fmpe*.

It can easily be verified from the diagram that not only those relationships between the figures hold which are represented by the formulae corresponding to the empirical generalizations which the four-factor theory was propounded to explain, e.g.

$$(\alpha\delta)\leftrightarrow(\epsilon(\alpha\delta)), \quad (\beta\delta)\leftrightarrow(\gamma\epsilon),$$

but that the relationships hold which are represented by the unwanted formulae

$$\delta\leftrightarrow((\alpha\epsilon)\cup(\gamma\kappa)), \quad \epsilon\leftrightarrow((\alpha\delta)\cup(\beta\kappa)), \quad \kappa\leftrightarrow((\beta\epsilon)\cup(\gamma\delta))$$

and which are due to the restriction of the classes designated by λ, μ, ν to being classes of points lying on or inside the triangle *abc*.

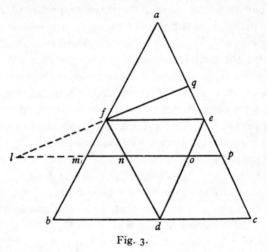

Fig. 3.

It will also be seen that the only way in which the class expressed by $(\delta\cup(\epsilon\cup\kappa))$ could be as extensive as the class expressed by $(\alpha\cup(\beta\cup\gamma))$ would be for the former class to be identical with the latter. In this case the class would be wholly included within the class designated by ρ, and the classes designated by δ, ϵ, κ would become identical with those designated by λ, μ, ν, i.e. with those expressed by $(\gamma\cup\alpha)$, $(\alpha\cup\beta)$, $(\beta\cup\gamma)$.* So the field of application of the properties D, E, K collectively would always be smaller than that of A, B, C collectively unless further restrictive identifications of the empirically given classes were valid.

* From $(\delta\cup(\epsilon\cup\kappa))\leftrightarrow(\alpha\cup(\beta\cup\gamma))$ we can derive $((\lambda\cup(\mu\cup\nu))\,\rho)\leftrightarrow(\lambda\cup(\mu\cup\nu))$ since $((\lambda\cup(\mu\cup\nu))\,\rho)\leftrightarrow(\delta\cup(\epsilon\cup\kappa))$ and $(\lambda\cup(\mu\cup\nu))\leftrightarrow(\alpha\cup(\beta\cup\gamma))$; whence $(\lambda\rho)\leftrightarrow$

If we define λ, μ, ν in terms of observable properties when we construct the three-factor theory, we must define ρ similarly when we introduce it to stand for the fourth factor in the four-factor theory. The similar way of definition, giving the minimum extension to the class designated by ρ, is to add to the calculus we are using the new initial formula*

$$\rho \leftrightarrow (\delta \cup (\epsilon \cup \kappa)).$$

In our diagram this corresponds to eliminating the dotted part of the triangle *lpq* lying outside the triangle *abc*, i.e. by taking ρ to designate only the class of points lying on or inside the quadrilateral *mpqf*. This limitation imposed on the extension of ρ will have no effect on the relationships of regions within the triangle *abc*; the new initial formula will be an idle formula so far as concerns the calculus representing the four-factor theory with λ, μ, ν defined. But if we wish to expand the four-factor theory into a five-factor theory with a fifth factor designated by σ, this formula will cease to be an idle formula in the new calculus required, and there will be the consequence that all the classes corresponding to combinations of two out of these five factors will have to be included in the class expressed by $(\alpha \cup (\beta \cup \gamma))$. So the five-factor theory will only be able to account for further properties of some of the class of things having properties already covered by the four-factor theory. In general, an *n*-factor theory in which each of the *n*-factors has been 'logically constructed' out of observable properties (in the sense that the symbol designating each factor has been defined in terms of observable properties) can only be expanded into an $(n + 1)$-factor theory at the price of its only applying to properties possessed only by things whose properties are already covered by the *n*-factor theory. In our diagram it will only give further relationships of other regions falling within the fundamental triangle *abc*. If the factors of such a theory are to be regarded as logical constructions, the theory, however much it be expanded, will never increase its field of application. It will always be limited to the class of things the relation of three of whose properties the original three-factor theory was propounded to explain.

---

* Since $\delta \leftrightarrow ((\alpha\epsilon) \cup (\gamma\kappa))$ is already derivable in the calculus, we could instead use $\rho \leftrightarrow (\epsilon \cup \kappa)$ as the new initial formula. Similarly we could instead use $\rho \leftrightarrow (\delta \cup \kappa)$ or $\rho \leftrightarrow (\delta \cup \epsilon)$.

This long exposition has been given in order to make it clear exactly what price has to be paid for treating the theoretical terms of a scientific theory as logical constructions out of observable entities. The example taken of a factor theory is one in which the mathematical apparatus used in deducing the empirical generalizations from the hypotheses is of the simplest kind, and consequently the example is one in which it would appear prima facie to be easiest to put this apparatus into reverse. We have seen that even here the theoretical terms can only be defined by means of observable properties on condition that the theory cannot be adapted properly to apply to new situations. Still more will this be the case with a theory using a more complicated mathematical apparatus. Frequently, indeed, the possibility of using a useful piece of mathematical apparatus depends upon the theoretical terms having a wider range than that of the observable properties taken together.*

It is only in theories which are not intended to have any function except that of systematizing empirical generalizations already known that the theoretical terms can harmlessly be explicitly defined. A theory which it is hoped may be expanded in the future to explain more generalizations than it was originally designed to explain must allow more freedom to its theoretical terms than would be given them were they to be logical constructions out of observable entities. A scientific theory which, like all good scientific theories, is capable of growth must be more than an alternative way of describing the generalizations upon which it is based, which is all it would be if its theoretical terms were limited by being explicitly defined.

IMPLICIT DEFINITION OF THEORETICAL TERMS

We can, however, extend the sense of definition if we wish to do so. In explicit definition, which we have so far considered, the possibilities of interpreting a certain symbol occurring in a calculus are reduced to one possibility by the requirement that the symbol should be synonymous (within the calculus) with a symbol or combination of symbols which have already been given an inter-

---

* The Schrödinger wave-functions of quantum mechanics are complex functions whose phase factors cancel out when deductions to empirical generalizations are made from hypotheses containing them.

pretation. But the possibilities of interpreting a certain symbol occurring in a calculus may be reduced without being reduced to only one possibility by the interpretation already given of other symbols occurring in formulae in the calculus. If we wish to stress the resemblance between the reduction of the possibilities of interpreting a symbol to only one possibility and the reduction of these possibilities but not to only one possibility, instead of wishing to stress (as we have so far stressed) the difference between these two sorts of reduction, we shall call the second reduction as well as the first by the name of definition, qualifying the noun by such words as "implicit" or "by postulate". With this extension of the meaning of definition the thesis of this chapter can be expressed by saying that, while the theoretical terms of a scientific theory are *implicitly defined* by their occurrence in initial formulae in a calculus in which there are derived formulae interpreted as empirical generalizations, the theoretical terms cannot be *explicitly defined* by means of the interpretations of the terms in these derived formulae without the theory thereby becoming incapable of growth.

For the case of our three-factor and four-factor theories implicit definitions can be given separately for the theoretical terms. In the three-factor theory and in the four-factor theory there can be derived respectively the formulae*

$$\lambda \leftrightarrow ((\gamma \cup \alpha) \cup \lambda), \quad \lambda \leftrightarrow (((\gamma \cup \alpha) \cup \delta) \cup \lambda),$$

with similar formulae for μ, ν and (in the four-factor theory) ρ. For the three-factor theory with γ and α interpreted as designating the class of C's and the class of A's respectively the formula $\lambda \leftrightarrow ((\gamma \cup \alpha) \cup \lambda)$ has the effect of restricting the interpretation of λ to being a class which includes the class of things which are either C or A (or are both), i.e. which includes the union of the class of C's with the class of A's. Since to say that a class includes the union of one class with another is equivalent to saying that it includes both these classes, this restriction upon the interpretation of λ is the twofold restriction that the class designated by λ must include both the class of C's and the class of A's. Similarly, for the four-factor theory with γ, α, δ designating the class of C's, the class of A's and the class of D's respectively, the restriction upon the

---

* From $(\gamma \alpha) \leftrightarrow ((\nu \lambda)(\lambda \mu))$ we can derive $(\gamma \alpha) \leftrightarrow ((\gamma \alpha) \lambda)$, whence $\lambda \leftrightarrow ((\gamma \cup \alpha) \cup \lambda)$ is derivable, since $\xi \leftrightarrow (\eta \cup \xi)$ is derivable from $\eta \leftrightarrow (\eta \xi)$.

interpretation of λ imposed by the formula $\lambda \leftrightarrow (((\gamma \cup \alpha) \cup \delta) \cup \lambda)$ is the threefold restriction that the class designated by λ must include the class of C's and the class of A's and the class of D's. Clearly the twofold restriction imposed by the three-factor theory upon the interpretation of λ is in no way incompatible with the third restriction—that of having to include the class of D's—to which it will also be subjected were the three-factor theory expanded into the four-factor theory. An implicit definition of theoretical terms is perfectly compatible with a growth of the theory in which they occur.

In the case of our factor theories (e.g. the three-factor theory) it was possible to express the restriction upon λ, μ, ν separately by formulae in which each of these occurred without the other two occurring, and in which each occurred as the sole element on the left-hand side of the formula. In most calculi it is impossible to express the restrictions so simply. Nor do the formulae

$$\lambda \leftrightarrow ((\gamma \cup \alpha) \cup \lambda), \quad \mu \leftrightarrow ((\alpha \cup \beta) \cup \mu), \quad \nu \leftrightarrow ((\beta \cup \gamma) \cup \nu)$$

express all the restrictions upon λ, μ, ν derived from the initial formulae of the three-factor theory

$$\alpha \leftrightarrow (\lambda\mu), \quad \beta \leftrightarrow (\mu\nu), \quad \gamma \leftrightarrow (\nu\lambda),$$

for these also impose the restrictions

$$(\alpha'(\beta'(\gamma'(\lambda\mu)))) \leftrightarrow o, \quad (\alpha'(\beta'(\gamma'(\mu\nu)))) \leftrightarrow o, \quad (\alpha'(\beta'(\gamma'(\nu\lambda)))) \leftrightarrow o.$$

It is therefore usually easiest to express the implicit definitions of the theoretical terms by reciting the initial formulae of the theory's calculus, omitting of course those which are the initial formulae of any algebraic calculus which is a part of the calculus. E.g. for implicitly defining the λ, μ, ν of the three-factor theory, the initial formulae $\alpha \leftrightarrow (\lambda\mu)$, $\beta \leftrightarrow (\mu\nu)$, $\gamma \leftrightarrow (\nu\lambda)$ would be recited. Then if all the initial formulae containing the theoretical terms are given, we shall know that we have before us all the restrictions which are imposed by the fact that formulae derived from the initial formulae are to be interpreted as representing empirical generalizations about observable properties. The implicit empirical definition of the theoretical terms in a scientific deductive system consists in the fitting of the calculus to the system from the bottom upwards. This is done by first fitting derived formulae of the calculus to the empirical generalizations which are the lowest-level hypotheses in the deductive system, and then working backwards so that the formulae containing the theoretical terms are interpreted as representing those higher-level hypotheses from which the lowest-level hypotheses logically follow in the scientific deductive system.

THE 'REALITY' OF THEORETICAL CONCEPTS

At this point in the discussion the metaphysical question that will have been in the minds of many of my readers can no longer be postponed: if the theoretical terms of a science are not explicitly definable, but instead are implicitly defined by the way in which they function in a calculus representing a scientific deductive system, in what consists the 'reality' of the entities which these theoretical terms denote? If the words "theoretical concept" (an expression whose use has so far been avoided as much as possible) are used to denote that for which a theoretical term stands, what is the status of these theoretical concepts?

One way of answering this question which is in essence the answer given by Ramsey is to say that the status of a theoretical concept (e.g. an electron) is given by the following proposition which specifies the status of an electron within the deductive system of contemporary physics: There is a property E (called "being an electron") which is such that certain higher-level propositions about this property E are true, and from these higher-level propositions there follow certain lowest-level propositions which are empirically testable. Nothing is asserted about the 'nature' of this property E; all that is asserted is that the property E exists, i.e. that there are instances of E, namely, electrons.*

In the case of our three-factor theory this will amount to saying that the status of the theoretical concepts L, M, N is that these specify classes—the class of L's, the class of M's, the class of N's respectively—which are such that the class of A's is identical with the intersection of the class of L's with the class of M's, the class of B's is identical with the intersection of the class of M's with the class of N's, and the class of C's is identical with the intersection of the class of N's with the class of L's, the classes of A's, of B's and of C's being classes specified by the observable properties

* F. P. Ramsey, *The Foundations of Mathematics and other logical essays*, p. 231.

A, *B* and *C*. Nothing is said about what it is which determines the class which is the-class-of-*L*'s, or the class which is the-class-of-*M*'s, or the class which is the-class-of-*N*'s; the fact that these classes must satisfy the conditions imposed by the higher-level hypotheses which are the premisses in the deductive system imposes restrictions upon them; but, subject to these restrictions, they may be any classes whatever. To say that they exist is to say that each of them has some member or members (i.e. is not the null-class); and this is a logical consequence of the premisses combined with the existence of at least two out of the classes of *A*'s, of *B*'s, of *C*'s.

On this view, to say that theoretical concepts exist is to assert the truth of the theory in which they occur. "'There is such a quality as mass' is nonsense," wrote Ramsey, "unless it means merely to affirm the consequences of a mechanical theory."* But there is another way of answering the question open to those of us who, with knowledge of the work of symbolic logicians done since Ramsey's death in 1930, are able to make a sharper distinction than was made by Ramsey between a deductive system and a calculus representing it. This other way of answering the question, "Do electrons really exist? Do the classes of *L*'s, of *M*'s and of *N*'s of the three-factor theory really exist?" is to side-step the question, and to make no remark whatever about a theoretical concept itself as opposed to a theoretical term. Instead of answering the question directly, an answer is given which refers not to the concept of electrons or to the class of *L*'s, but to the word "electron" or to the symbol λ and to the use of these symbols in the initial formulae of the calculus which is interpreted, in the way described, from the bottom upwards. If the questioner complains that his question was about the concept and not about the symbol, and has not been answered, it will be necessary to go back and give Ramsey's answer.† But it is possible that he may be satisfied with an explanation of how the symbol is used, combined with a recognition of the fact that symbols can have meaning in the context of a calculus without having any meaning outside such a context.

Provided that the function of theoretical terms in a calculus which represents a scientific deductive system is understood, it is

---

* Op. cit. p. 261.

† The first answer must also be given to anyone who proposes to use concept in such a way that every separate symbol *ipso facto* designates a concept.

of little importance whether one chooses to answer the question "Do electrons really exist?" by saying "Yes, for the electron theory is true" or (without saying "No" to the question) by explaining the functions of the words and the other symbols used for electrons in the calculus which is interpreted as the deductive system of contemporary atomic physics. There is an advantage in formal logic in giving the second answer and of avoiding saying either that a theoretical concept exists or that it does not exist. For the first answer makes the set of the premisses of the deductive system one single existential proposition: There exist properties X, Y, Z, etc., which are such that, etc.; and consequently the notion of a scientific hypothesis as being a universal generalization would require to be modified. On the other hand, the first answer more than the second may be better for emphasizing the fact that the premisses of the deductive system, represented by the initial formulae of the calculus, must be taken together as a whole; and that the growth of the theory by the addition of a new premiss affects all the other premisses. In calculus terms this fact can be expounded by saying that the addition of a new initial formula to a calculus alters the possibility of interpreting the old formulae as well as providing a possibility of interpreting the new one.

It is instructive to notice that one of the most important theoretical terms used in contemporary physics—Schrödinger's wave-function —is frequently called the "ψ-function", being referred to merely by means of the symbol ψ which it is customary to use in the calculus of wave mechanics. I suspect that no physicist would wish directly to answer questions like: What is the concept denoted by the symbol ψ? Does Schrödinger's ψ-function really exist?* The physicist will almost certainly prefer to give the indirect answer of explaining how the symbol ψ is used in his calculus.† The physicist does not start with a scientific deductive system containing propositions which are 'about' concepts which he denotes by the symbol ψ, and represent these propositions in the deductive system by formulae in a calculus; he starts with a calculus whose derived formulae can be interpreted as the empirical generalizations which he is concerned to explain. For a physicist,

---

* No one supposes that ψ denotes a wave, in any ordinary sense of wave.

† Or, of course, some other symbol which is synonymous within the calculus to Schrödinger's ψ.

to think about ψ-functions is to use the symbol ψ in an appropriate way in his calculus. When he has explained this appropriate way, there is nothing further to say upon what the propositions expressed by the formulae containing ψ are about. Once the status within a calculus of a theoretical term has been expounded, there is no further question as to the ontological status of the theoretical concept.

THE INDIRECT MEANING OF GENERAL STATEMENTS

The use of theoretical terms in an advanced science like physics is the most striking example of the way in which meaning is given indirectly, but not directly. A close examination will show, however, that the same situation occurs in many places where at first glance it would not be suspected. Take the case, for example, of general statements. Up to this point we have assumed that, while there might well be difficulty in understanding how the formulae containing theoretical terms in a calculus could be empirically interpreted, there was no such problem in regard to the derived formulae of the calculus which were to be interpreted as empirical generalizations; indeed, it was on the basis of an interpretation of these derived formulae that the theoretical terms occurring higher up in the hierarchy of the science were to be understood. But, in fact, a similar problem arises with regard to the word "every" as arose with regard to the word "electron".

For how is it that meaning is given to a sentence expressing a universal generalization, e.g. "Every cat eats fish"? Not in the direct way in which meaning can be given to the sentence "Drake eats fish" used to express a proposition about a particular cat called "Drake", for the proposition that every cat eats fish is not an observable proposition in the way in which the proposition that Drake eats fish is an observable one. The expression "Every cat" is not used to designate a particular cat; it has a meaning in a more complicated way. To say that Drake eats fish is to attribute the property of being ichthyophagous to the particular thing called "Drake". To say that every cat eats fish is not to attribute this property to one particular thing called "Every cat", nor is it to attribute ichthyophagy collectively to the class of all cats, for classes cannot eat fish. What this statement does is to assert a proposition from which, in conjunction with a proposition asserting that a particular thing is a cat, there logically follows that the particular thing in question eats fish. The statement is given a meaning, not in the direct way in which meaning is given to "Every cat eats fish", but by virtue of its position in a calculus.

It would be out of place here to develop a calculus appropriate for representing the logical relationships between logically contingent generalizations and their instances with as much elaboration as was done for the calculi in the last chapter.* A very simple calculus suitable for this purpose may be constructed in the following way: Take as formulae both the series of marks which are formulae in the Huntington Calculus, i.e. series consisting of two elements of that calculus with a double-arrow sign ↔ between them, and also new series of marks consisting of the following four parts in order from left to right: left-hand square bracket, an element of the Huntington Calculus which is neither a variable nor contains a variable as a part of it, a roman letter in clarendon type, a right-hand square bracket. Examples of these new formulae are $[\alpha \mathbf{c}]$, $[(\beta\alpha)\,\mathbf{b}]$, $[(\alpha'\beta)'\,\mathbf{c}]$, but not $[\xi \mathbf{c}]$, $[(\alpha\eta)\,\mathbf{b}]$, since ξ and η are variables. The initial formulae and rules of play of the new calculus are to be those of the Huntington Calculus, with the addition of a new rule concerned with the new type of formula. This rule will permit the writing down of a formula of the form $[\beta \mathbf{c}]$ whenever two formulae of the forms of $\alpha \leftrightarrow (\beta\alpha)$ and of $[\alpha \mathbf{c}]$ are already written down.†

This calculus may be interpreted to represent a deductive system by taking elements to stand for classes, and the algebraic part of the calculus (which does not make any use of the new formulae and new rule of play, and which is in fact exactly the Huntington Calculus) to represent the logic of classes, and by taking $[\alpha \mathbf{c}]$ to express the proposition that the thing c designated by $\mathbf{c}$ is a member of the class designated by α, i.e. that c has the property A. Since

* The variables of the Second Calculus (pp. 36 f.) and of the Huntington Calculus (p. 60) provide a technique for deducing instances of logically necessary generalizations about classes from these generalizations.

† If the old calculus in question had been one which, unlike the Huntington Calculus, had contained a rule of detachment (p. 47), it could have been modified to represent instances of contingent generalizations without adding a new rule of play by adding a new initial formula and by distinguishing among the class of clarendon-type letters a subclass (e.g. **x**, **y**, **z**) of 'clarendon-type variables' to which the variable-substitution rule would apply, on the understanding that only elements would be substitutable for Greek-letter variables and only clarendon-type letters would be substitutable for clarendon-type variables.

with this interpretation $\alpha \leftrightarrow (\beta\alpha)$ expresses the proposition that every A is B, the new rule of play will express the logical truth that the proposition that c is B is a logical consequence of the pair of propositions «Every A is B» and «c is A». If c is Drake and A and B are the properties of being feline and of being ichthyophagous respectively, the new rule expresses the logical truth that «Drake eats fish» logically follows from «Every cat eats fish» and «Drake is a cat». On the assumption that direct meaning can be given to the sentences "Drake is a cat" and "Drake eats fish", which are taken to express observable propositions, indirect meaning can be given to the sentence "Every cat eats fish" by the way in which it functions in a calculus with the rule of play that the formula $[\beta\mathbf{c}]$ is derivable from the pair of formulae $\alpha \leftrightarrow (\beta\alpha)$ and $[\alpha\mathbf{c}]$ interpreted in the way described. Just as in the new symbolic calculus an interpretation of $[\alpha\mathbf{c}]$ and of $[\beta\mathbf{c}]$ determines the interpretation of $\alpha \leftrightarrow (\beta\alpha)$, so in the verbal calculus which is formally equivalent to it an interpretation of "Drake is a cat" and of "Drake eats fish" determines an interpretation of "Every cat eats fish" as representing the logically weakest proposition which is such that from the pair of propositions of which it is the one, and «Drake is a cat» is the other, there logically follows the proposition that Drake eats fish. The meaning of sentences containing the word "every" or its synonyms, i.e. the meaning of sentences expressing generalizations, is given indirectly by the characteristic way of functioning of these sentences in a calculus in which the lowest-level sentences are interpreted as expressing observable facts about particular observable things.

The older logicians recognized that a word like "every" had no meaning in itself, but only has a meaning in conjunction with other words. They expressed this by calling such a word *syncategorematical*. It is comparatively recently that it has been appreciated that not only has "every" no meaning apart from the sentences in which it occurs, but that these sentences have no meaning apart from their relationships to other sentences in a calculus interpreted as a deductive system. What is true of "every" is true of all logical words. Indeed, the very notion of proposition itself is not independent of its use in deductive systems. For it is part of the nature of a proposition to be subject to the first principles of propositional logic, and these first principles are especially concerned with the deducibility of propositions from one another. A proposition cannot be understood merely as being the meaning of an indicative sentence in English or in some other language (though this is the simplest way to introduce the notion of proposition): it must also obey the laws of propositional logic. Thus one sentence or formula cannot be interpreted as standing for a proposition in isolation from an interpretation of other sentences or formulae to stand for other logically related propositions.

Thus the feature in our use of the theoretical terms of a science which has caused so much puzzlement in recent years—the fact that the terms appear to have no meaning apart from their context in the representation of the scientific system—is paralleled by the behaviour of the logical terms involved in the expression of even the most *terre-à-terre* empirical generalization. A formula containing Schrödinger's ψ cannot be understood in isolation from other formulae, but neither can any general statement. All scientific thinking is general thinking concerned with connecting pieces of empirical knowledge with one another; theoretical concepts are only a particularly elaborate way of making these connexions, and theoretical terms only a particularly striking case of contextual meaning.

ARE SCIENTIFIC HYPOTHESES PROPOSITIONS AT ALL?

Even if a science is in an elementary stage and contains no theoretical concepts, its hypotheses, though expressed in everyday language (e.g. "Every European swan is white"), are all generalizations. As has been shown, the meaning of a sentence expressing a generalization is indirect as compared with that of a sentence expressing a possible observation. Thus the meanings of sentences expressing scientific hypotheses are always indirect. Some contemporary philosophers have wished to emphasize this fact by limiting the term proposition to cover only the direct meaning of a sentence, and thus by refusing to call a general hypothesis a proposition at all. Such a limitation is inconvenient, since hypotheses as well as propositions in the limited sense obey the laws of propositional logic, are capable of truth and falsity, are objects of belief or other cognitive attitudes, and are expressed by indicative sentences; they thus satisfy all the usual criteria for being a proposition. And the limitation is also arbitrary, since there would seem to be no good reason for excluding from the category of expressions of propositions sentences like "Every European swan is white", the indirectness of whose meaning in relation to the meanings of "This thing is a European swan" and "This thing is white" can be precisely explained, without also excluding these last two sentences, whose meaning may very well be indirect in a much more obscure way. If these philosophical puritans will only consent to apply the name "proposition" to absolutely direct and non-contextual meanings, it may well be that there is nothing to which they can apply the term (all meanings may be contextual); and then a useful word would be wasted.

Philosophers who decline to class general hypotheses as propositions would prefer to consider them as 'rules'. "Laws of nature," writes Schlick, " . . . are directions, rules of procedure enabling the investigator to find his way about reality, to discover true propositions, to expect with assurance particular events."[*] And Ramsey says similarly that «Every A is B» is not a judgment, but a rule for judging that if I meet an A, I shall regard it as a B.[†] This alternative classification would make it appear as if the role of generalizations in deductive systems was, not that of premisses from which observable conclusions could be deduced, but that of principles of inference according to which the conclusions follow. Now it is possible to view a general hypothesis in this way. The inference from «Drake is a cat» to «Drake eats fish» by means of «Every cat eats fish» can be regarded as drawing the conclusion from a single premiss by virtue of a special contingent principle of inference expressed by saying that every cat likes fish. But whereas the piece of logic involved in the inference to the proposition «c is B» from the pair of propositions «Every A is B» and «c is A» may properly be treated as a principle of inference (though it may also be treated as a new logically necessary proposition in the system[‡]), it would be most misleading to treat a logically contingent proposition like «Every cat eats fish» as a principle of inference. To do so would be to mix experience and the logical methods by which we think about experience in a very confusing way.

[*] Moritz Schlick, in *Die Naturwissenschaften*, vol. 19 (1931), p. 156, reprinted in *Gesammelte Aufsätze* (Vienna, 1938), p. 68 (my translation).
[†] *The Foundations of Mathematics and other logical essays*, p. 241.
[‡] See above, p. 83 footnote.

Ramsey and Schlick have done a service to philosophy by emphasizing the peculiar features of generalizations which arise from the fact that sentences expressing them can only be understood in connexion with the way in which they are used in deductive inference. It seems better to put the point which they wish to make by admitting that generalizations are propositions, and by pointing out that these propositions play a different role in deductive systems from that played by propositions which are directly testable by experience, and that consequently the sentences expressing them have meaning in a different way.

All the sentences and formulae occurring in a theoretical treatise on some branch of science (except those which are part of the pure mathematical section of the treatise, if it includes such a section) will, since they express general hypotheses, have indirect meaning in the sense in which this phrase has been used, for their meanings will be given by their contexts in relation to one another and to the sentences expressing directly testable propositions which the experimentalist can derive from them. We see now how it is that, in considering the logic of a science, we cannot avoid thinking about the sentences or other symbols used to express scientific hypotheses as well as about the hypotheses themselves. A practising scientist interested only either in the deductive organization of his system or in the empirical conclusions to be deduced within it can 'use' his symbols without 'mentioning' them. But immediately he becomes self-conscious about what he is doing, and tries to connect up the terms occurring in his deductive system with the observables in its conclusions, he will have to think about what he means by the words "electron" or "wave-function" or "mental energy", and thus will have to mention the symbols in order to explain how he proposes to interpret them.

CHAPTER IV

MODELS FOR SCIENTIFIC THEORIES; THEIR USE AND MISUSE

We have explained the meaning of the theoretical terms in a calculus representing a scientific theory by showing how the interpretation of the calculus is fitted to the observable facts from the bottom upwards. In the geometrical illustration (p. 63) which was given of the three-factor theory the terms occurring in the derived formulae of the calculus

$$(\alpha\beta)\leftrightarrow(\gamma(\alpha\beta)), \quad (\beta\gamma)\leftrightarrow(\alpha(\beta\gamma)), \quad (\gamma\alpha)\leftrightarrow(\beta(\gamma\alpha))$$

were first interpreted as standing for given classes of points and the formulae containing these terms as representing propositions about the classes of points common to two or to three of these given classes. The theoretical terms λ, μ, ν occurring in the initial formulae of the calculus

$$\alpha\leftrightarrow(\lambda\mu), \quad \beta\leftrightarrow(\mu\nu), \quad \gamma\leftrightarrow(\nu\lambda)$$

were subsequently interpreted as standing for classes of points which were subjected to certain restrictions by having to satisfy the conditions expressed by these formulae but which were otherwise indeterminate. This geometrical illustration of the three-factor theory was thus given as an interpretation of the theory's calculus in the same way as the original interpretation of the calculus was given to explain the associations of three observable properties A, B, C by the positing of three unobservable factors L, M, N.

The difference between the two interpretations is that in the geometrical interpretation, unlike the 'factor' interpretation, there are things for which the theoretical terms λ, μ, ν stand which are observable in the same way as the things for which α, β, γ stand, the peculiarity of the theoretical terms consisting in their indeterminateness. But the two deductive systems which are interpretations of the same calculus are alike in that, in each case in interpreting the calculus, it is derived formulae of the calculus and not initial formulae containing the theoretical terms that are first given a meaning. In both deductive systems the higher-level hypotheses containing the theoretical terms, though *logically prior* to the lower-level hypotheses in the sense that the former serve as premises for the deduction of the latter, are *epistemologically posterior* to them in the sense that the interpretation of the initial formulae of the calculus to represent the higher-level hypotheses follows on, and is dependent upon, the interpretation of derived formulae as representing lower-level hypotheses.

It is frequently possible, however, to give an interpretation of a particular calculus in such a way that the higher-level hypotheses are both logically and epistemologically prior to the lower-level hypotheses. An example of such an interpretation of the three-factor-theory's calculus was given on p. 58, where λ, μ, ν were taken as standing for the class of four-sided plane figures, the class of equilateral plane figures and the class of equiangular plane figures respectively. In this case the initial formulae $\alpha \leftrightarrow (\lambda\mu)$, $\beta \leftrightarrow (\mu\nu)$, $\gamma \leftrightarrow (\nu\lambda)$ serve to define α as standing for the class of equilateral four-sided plane figures, and to define β and γ similarly. If for the symbol α be substituted "the class of rhombuses", the first formula will serve to define "rhombus". If the word "rhombus" were not in my vocabulary, the formula would serve to introduce it into my vocabulary by stating that the word was interchangeable with a phrase "equilateral four-sided plane figure" to which I had already attached a meaning.

If the theoretical terms λ, μ, ν are first interpreted, and the initial formulae then used to define α, β, γ in terms of the meanings of λ, μ, ν, the initial formulae will be interpreted as expressing logically necessary propositions. If "rhombus" means simply equilateral four-sided plane figure, it is not a matter of empirical fact that the class of rhombuses is identical with the class of equilateral four-sided plane figures, for this depends upon the way language is used, in that the class of rhombuses has been defined in such a way that this class-identity must be satisfied.

If we have before us two deductive systems which are each interpretations of the same calculus, in the first of which the interpretation of the initial formulae containing the theoretical terms is epistemologically prior to that of the derived formulae not containing these theoretical terms, whereas in the second interpretation the reverse is the case, the derived formulae being the epistemologically prior, the first deductive system will be said to be related to the second deductive system as *model* is to *theory*. The first deductive system will be said to be a model for the theory which is the second deductive system, and the second deductive system a theory for which the first deductive system is a model. A theory and a model for it, or a model and a theory for which it is a model, have the same formal structure, since theory and model are each represented by the same calculus. There is a one-one correlation between the propositions of the theory and those of the model; propositions which are logical consequences of propositions of the theory have correlates in the model which are logical consequences of the correlates in the model of these latter propositions in the theory, and vice versa. But the theory and the model have different epistemological structures: in the model the logically prior premisses determine the meaning of the terms occurring in the representation in the calculus of the conclusions; in the theory the logically posterior consequences determine the meaning of the theoretical terms occurring in the representation in the calculus of the premisses. To use again the metaphor of a zip-fastener, the calculus is attached to the theory at the bottom, and the zip-fastener moves upwards; the calculus is attached to the model at the top, and the zip-fastener moves downwards.

I am using the word "model" because my description of the relation between the two deductive systems represented by the same calculus—between the *model* and the *theory*—is an attempt to make more precise the notion of a model for a scientific theory widely current in discussions of the philosophy of science. The term "model" has been used in different though related ways by different writers: the distinction between two deductive systems represented by the same calculus but differing epistemologically, which I am using the correlative words "model" and "theory" to express, is analogous to a distinction made by Heinrich Hertz, the most philosophically profound of the great nineteenth-century physicists who wrote on the philosophy of science.

Hertz says that our procedure in trying to make inferences about the future based upon knowledge of the past is always as follows:

"We make for ourselves internal pictures or symbols of external objects, and we make them of such a kind that the necessary consequences in thought of the pictures are always the pictures of the necessary consequences in nature of the objects pictured. . . . When on the basis of our accumulated previous experiences we have succeeded in constructing pictures with the desired properties, we can quickly derive by means of them, as by means of models, the consequences which in the external world would only occur in the course of a long period of time or as a result of our own intervention."* Hertz is at pains to emphasize that, although "the pictures are our conception of things" and must agree with the things in satisfying the condition that their necessary consequences in thought are the things' necessary consequences in nature, yet for his purpose—that of arranging the propositions of a science like mechanics in a unified deductive system—it is not necessary that they should agree with the things in any other respect whatever. Hertz asserts that whether or not there is any other respect in which our conceptions agree with the things is a matter which we neither know nor have any means of discovering by experience. It is not clear whether Hertz's 'pictures'† would correspond to what in the language of this book are the initial formulae of a calculus or to the interpretations of such formulae. On the former assumption their 'necessary consequences in thought' will be derived formulae in the calculus which represent the facts about the 'external world' which are consequences in the theory, the calculus being one which will also represent the workings of a model. On the latter assumption the 'necessary consequences in thought' of Hertz's 'pictures' will be propositions corresponding, in the way in which propositions about a model correspond, to propositions about the 'external world'. But in either case Hertz is maintaining, against those who would require that reality should resemble our pictures of it, that the only resemblance required is that of formal structure.

If we set the notion of model over against that of theory in the

* From the first paragraph of Hertz's Introduction to his posthumously published *Die Prinzipien der Mechanik* (Leipzig, 1894). In my quotation I have slightly changed the wording of its English translation *The Principles of Mechanics* (London, 1899), p. 1.

† *Bilder*, sometimes *Vorstellungen*. In the first sentence quoted *innere Scheinbilder oder Symbole*.

way proposed, we shall see that to think about a scientific theory by thinking about a model for it is an alternative to thinking about the theory by explicitly thinking about the calculus representing it. For to think about the model is to think about an interpretation of the calculus which works with respect to the order in which the interpretation is effected as well as with respect to the order of deduction in the same direction as the order of deriving formulae within the calculus. Thus to use the model is to use a quite straightforward interpretation of the calculus, and a comparison of this quite straightforward interpretation with the sophisticatedly inverted interpretation which is the scientific theory, will serve most of the purposes of a comparison of the calculus itself with the scientific theory.

To think in terms of the model is therefore frequently the most convenient way of thinking about the structure of the theory, for it avoids the self-consciousness required in order to have before the mind at the same time both the set of propositions arranged in a deductive system which is the theory, and the set of sentences or formulae arranged in order which is the calculus representing the theory. If, in expressing a comparison between a theory and its calculus, the theory is spoken of by using the symbols of its calculus, these symbols will have both to be 'used' (in the expression of the theory) and 'mentioned' (in the expression of the calculus); and this dual function of the symbols will be liable to cause confusion. Moreover, the derivability relationships between the formulae of the calculus, though they will correspond to, will not be identical with the logical-consequence relationships between the propositions of the theory. On the other hand, if comparisons are always made between the theory and the model instead of between the theory and its calculus, the sentences expressing both the theory and the model can be 'used' without having also to be 'mentioned'; and the relationships between the propositions of the model will be of the same kind as those holding between the propositions of the theory, namely relationships of logical consequence. Thus there are great advantages in thinking about a scientific theory through the medium of thinking about a model for it; to do this avoids the complications and difficulties involved in having to think explicitly about the language or other form of symbolism by which the theory is represented. The use of models allows of a philosophically unsophisticated approach to an understanding of the structure of a scientific deductive system.*

DISADVANTAGES OF MODELS

There are, however, serious dangers in the use of models, dangers sufficiently serious for me deliberately to have refrained from alluding to the possibility of considering scientific theories by means of considering models for them, until the exposition in the last chapter had been given, which was in terms of the relationship between a scientific deductive system and a calculus used to represent it.

The first danger is that the theory will be identified with a model for it, so that the objects with which the model is concerned—the model-interpretation of the theoretical terms λ, μ, ν of the theory's calculus—will be supposed actually to be the same as the theoretical concepts of the theory. To these theoretical concepts will then be attributed properties which belong to the objects of a model but which are irrelevant to the similarity in formal structure which is all that is required of the relationship of model to theory. Models for chemical theories with molecules as linked systems of atoms and models for physical theories with atoms as 'solar systems' of separate elementary particles (electrons, protons, etc.) have in the past led many people to suppose that atoms, or electrons, shared characteristics of the model other than those which made the model an appropriate one. This danger is less now that "the last trace of the old, hard, massy atom has disappeared"; but it lies latent whenever a model is used. Thinking of scientific theories by means of models is always *as-if* thinking; hydrogen atoms behave (in certain respects) as if they were solar systems each with an electronic planet revolving round a protonic sun. But hydrogen atoms are not solar systems; it is only useful to think of them as if they were such systems if one remembers all the time that they are not. The price of the employment of models is eternal vigilance.

* If there is a doubt as to the self-consistency of the highest-level hypotheses of a scientific theory, an interpretation of the calculus representing the theory by means of a model may serve to establish their consistency. P. A. M. Dirac, after saying that "the main object of physical science is not the provision of pictures" and that "whether a picture exists or not is a matter of only secondary importance", goes on to say: "One may, however, extend the meaning of the word 'picture' to include any *way of looking at the fundamental laws which makes their self-consistency obvious*" (*The Principles of Quantum Mechanics*, third edition (Oxford, 1947), p. 10; his italics).

But there is a second danger inherent in the use of models, a danger which is more subtle than that of projecting on to the concepts of the theory some of the empirical features of the objects of the model. This danger is that of transferring the logical necessity of some of the features of the chosen model on to the theory, and thus of supposing, wrongly, that the theory, or parts of the theory, have a logical necessity which is in fact fictitious. In its grossest form this temptation may be illustrated from models for our three-factor theory.

If a model for this theory is constructed by interpreting the theoretical terms λ, μ, ν as standing for three classes specified by three observable properties, the initial formulae of the calculus containing these terms may be interpreted as providing definitions of α, β, γ by means of λ, μ, ν to which meanings have already been attached. The propositions in the model which are represented by these initial formulae of the calculus will then be logically necessary propositions; and, since the other initial formulae of the calculus are being interpreted as expressing logical truths about the relationships of classes, all the initial formulae of the calculus will represent logically necessary propositions. But the logical consequences of premisses, all of which are logically necessary propositions, are themselves all logically necessary propositions; so all the propositions of the model will be logically necessary. Those of them which do not occur in the pure deductive system of classes which is the interpretation of the algebraic part of the calculus (the Huntington Calculus) will be logically necessary, not by virtue of being logical truths or instances of such truths, but as being logical consequences of propositions 'true by definition'. No part of the model will be contingent; although empirical things enter into its propositions, the truth of none of these depends upon what these particular empirical things are. There is no empirical or logical difference between the model for the three-factor theory given on p. 58 with λ, μ, ν standing for the classes of four-sided plane figures, of equilateral plane figures and of equiangular plane figures respectively, and a model in which λ, μ, ν stand for three classes specified by any three empirical properties L, M, N (of course of the same logical category) whatever.*

---

* If two of the classes, e.g. those designated by λ and by μ, are mutually exclusive, the class designated by α will be the null-class. That this class is the

Here the temptation is presented in such a gross form that one would have to be very simple-minded to fall into it, and to argue that, since the proposition of the model expressed by $(\alpha\beta) \leftrightarrow (\gamma(\alpha\beta))$ is logically necessary, so is the proposition in the theory expressed by the same formula. The temptation is gross in this case because all the variable-free initial formulae of the calculus are definitory formulae for α, β, γ; these formulae have been used to provide definitions of α, β, γ in terms of the empirical properties L, M, N. So it is clear that the reason why the initial propositions of the model are logically necessary is because they are true by definition; and logical necessity by virtue of definition is so obviously a matter of how the symbols are used that there is little risk of transferring it to the theory where the symbols are clearly being used in a different way.

Most calculi used in science are more complicated than those representing our three- and four-factor theories, where the formulae in which the theoretical terms occur are all definitory formulae. A calculus may contain initial formulae in which the theoretical terms occur in other ways. In particular, the calculus may contain initial formulae in which, apart from symbols which are being interpreted as standing for logical constants or relations and symbols which are being used as variables, the only terms which occur are theoretical terms. The occurrence of such formulae (which for convenience will be called λ-*formulae*) in a calculus greatly complicates the problem of separating the logically necessary from the logically contingent among the hypotheses of a scientific deductive system represented by such a calculus. For a λ-formula among the initial formulae imposes a restriction upon the interpretation of the theoretical term or terms which appear in it similar to that imposed by a definition, and a proposition represented by a λ-formula will therefore have claims to being considered true by definition. And in a model in which the premisses include propositions relating observable properties to one another, whether or not these propositions are logically necessary will depend upon what are the properties with which the propositions of the model are concerned.

null-class, however, will not be a conclusion in the deductive system. It would be a conclusion if the proposition represented by $(\lambda\mu) \leftrightarrow o$ were a premiss in the system. But the deductive system would then be represented by a different calculus and so would not be a model for the theory.

So there can be models with the same formal structure which differ one from another in that in some of them the premisses are logically necessary while in others they are logically contingent. So, in thinking of a theory by means of a model, it will make a great difference into which of these two categories the model falls.

Many philosophers, reluctant to admit that the necessity of a scientific law consists simply in its being a true generalization associating the occurrence of one empirical property with that of another, have sought for a necessity in scientific laws which is either logical necessity or is closely akin to it. Some of them have found support for their position in the features of logical necessity which are involved in the representation of the hypotheses of a scientific theory by means of formulae or other modes of expression which appear naturally to represent a deductive system containing logically necessary propositions. In order to reject (as I wish to do) the view that there is anything objective in causal necessity over and above constant conjunction, it is essential to disentangle the genuine elements of logical necessity involved in scientific deductive systems and in the models which it is natural to use in thinking about them. What is required is, in Hertz's words, "sharply to distinguish in our pictures between what arises from necessity in thought, what from experience and what from our arbitrary choice".* To do this it will be convenient to approach the matter by first examining a highly relevant fact about scientific thinking, namely, that alternative theories can frequently be constructed to explain the same set of empirical generalizations.

ALTERNATIVE BUT EQUIVALENT THEORIES

The three-factor theory developed in the last chapter explained the three empirical generalizations represented by the three formulae

$$(\alpha\beta) \leftrightarrow (\gamma(\alpha\beta)), \quad (\beta\gamma) \leftrightarrow (\alpha(\beta\gamma)), \quad (\gamma\alpha) \leftrightarrow (\beta(\gamma\alpha))$$

by three hypotheses represented by the formulae

$$\alpha \leftrightarrow (\lambda\mu), \quad \beta \leftrightarrow (\mu\nu), \quad \gamma \leftrightarrow (\nu\lambda).$$

This explanation was effected by exhibiting a calculus in which these three latter formulae were added to the initial formulae

---

* *Die Prinzipien der Mechanik*, p. 10 (see English translation, p. 8).

of an algebraic calculus and in which the three former formulae appeared as derived formulae in the calculus.

There are many different ways in which new formulae using the theoretical terms λ, μ, ν can be added as new initial formulae to the same algebraic calculus and in which the same formulae not involving these terms can be derived. For example, if to the Huntington Calculus (which is an algebraic calculus), instead of adding as new initial formulae

$$\alpha \leftrightarrow (\lambda\mu), \quad \beta \leftrightarrow (\mu\nu), \quad \gamma \leftrightarrow (\nu\lambda),$$

we add as new initial formulae

$$\alpha \leftrightarrow ((\lambda\mu) \cup (\nu(\lambda'\mu'))), \quad \beta \leftrightarrow ((\mu\nu) \cup (\lambda(\mu'\nu'))), \quad \gamma \leftrightarrow ((\nu\lambda) \cup (\mu(\nu'\lambda'))),$$

exactly the same formulae connecting only α, β, γ can be derived as are derivable in the original three-factor theory's calculus.* The two calculi may be said to be (α,β,γ)-*equivalent*; and the two deductive systems, if α, β, γ are taken to stand for classes given by empirical properties, may be said to be *empirically equivalent*. The theory which this new calculus represents is a theory in which the class of A's is identified with the class of things which are either both L and M or both N and both non-L and non-M (or both), and this is a much more complicated theory than the simple three-factor theory, so a positive reason would be required for preferring it.

Moreover, it could only be extended into a four-factor theory by altering the separate primitive formulae. And if the first formula were changed into either

$$\alpha \leftrightarrow ((\lambda\mu) \cup ((\nu\rho)\,(\lambda'\mu'))) \quad \text{or} \quad \alpha \leftrightarrow ((\lambda\mu) \cup ((\nu\cup\rho)\,(\lambda'\mu'))),$$

with corresponding changes in the other formulae, although the formulae we want would be derivable in either of these calculi, yet other formulae would be derivable which cannot be derived in the original four-factor theory's calculus. So a reason for selecting one particular theory to explain a set of generalizations is the possibility it offers for adding to it in a convenient way to explain further generalizations, as is done by the original three-factor theory, but not by the theory represented by this new calculus.

* The resultant of eliminating λ, μ, ν from these formulae is the same as the resultant of eliminating λ, μ, ν from the three formulae of the latter calculus

HYPOTHESES CONTAINING ONLY THEORETICAL CONCEPTS

Consider the calculus (which will be called C_1) obtained by adding to the Huntington Calculus, instead of the six variable-free initial formulae of the four-factor theory,

$$\alpha \leftrightarrow (\lambda\mu), \quad \beta \leftrightarrow (\mu\nu), \quad \gamma \leftrightarrow (\nu\lambda),$$
$$\delta \leftrightarrow (\lambda\rho), \quad \epsilon \leftrightarrow (\mu\rho), \quad \kappa \leftrightarrow (\nu\rho),$$

the six variable-free initial formulae

$$\alpha \leftrightarrow ((\lambda\mu)\,(\nu' \cup \rho')), \quad \beta \leftrightarrow ((\mu\nu)\,(\lambda' \cup \rho')), \quad \gamma \leftrightarrow ((\nu\lambda)\,(\mu' \cup \rho')),$$
$$\delta \leftrightarrow ((\lambda\rho)\,(\mu' \cup \nu')), \quad \epsilon \leftrightarrow ((\mu\rho)\,(\nu' \cup \lambda')), \quad \kappa \leftrightarrow ((\nu\rho)\,(\lambda' \cup \mu')).$$

In this calculus C_1 there can be derived all the formulae not containing the theoretical terms λ, μ, ν, ρ which are derivable in the original four-factor theory's calculus (call this C_0) together with the formula $(\alpha\kappa) \leftrightarrow o$ and the formulae derivable from the addition of this formula to C_0, which include $(\beta\delta) \leftrightarrow o$, $(\gamma\epsilon) \leftrightarrow o$.

We can, however, construct another calculus, C_2, which differs from C_0 in respect to its derivable formulae not containing theoretical terms in exactly the same way as C_1 differs from C_0, by adding to the initial formulae of C_0 instead of by changing them. If we add to the six variable-free initial formulae of C_0 the new initial formula $(\lambda(\mu(\nu\rho))) \leftrightarrow o$,

we obtain a calculus C_2 in which occur $(\alpha\kappa) \leftrightarrow o$ and all the other formulae of C_1 which contain no theoretical terms. The calculus C_2 is in fact stronger than the calculus C_1 in the sense that it includes all and more than all the formulae of C_1.* But these extra formulae of C_2 which are not included in C_1 contain some or all of the theoretical terms; so far as formulae containing none of these terms are concerned all those derivable in C_1 are derivable in C_2, and vice versa.† The calculi C_1 and C_2 are $(\alpha, \beta, \gamma, \delta, \epsilon, \kappa)$-equivalent.

There is a difference, important in many contexts, between the calculi C_1 and C_2. Each of the variable-free initial formulae of C_1, like those of all the calculi of the three-factor theory, the four-factor

---

* From $(\lambda(\mu(\nu\rho))) \leftrightarrow o$ we can derive, by the rules of the Huntington Calculus, $(\lambda\mu) \leftrightarrow ((\lambda\mu)((\lambda' \cup \mu') \cup (\nu' \cup \rho')))$, whence $(\lambda\mu) \leftrightarrow ((\lambda\mu)(\nu' \cup \rho'))$ may be derived.

† The resultant of eliminating λ, μ, ν, ρ from the six variable-free initial formulae of C_1 is the same as the resultant of eliminating λ, μ, ν, ρ from the seven variable-free initial formulae of C_2.

theory and the calculus of p. 97 contains both symbols α, β, etc., interpretable as standing for observable entities, and theoretical terms λ, μ, etc., not susceptible of such direct empirical interpretation. In each of them one of the symbols α, β, etc., stands alone as the left element in the formula, and only theoretical terms λ, μ, etc., occur in the right element. Each of them is thus a definitory formula for one of the symbols α, β, etc. The primitive formulae of the calculus C_2, however, include, besides six definitory formulae for α, β, γ, δ, ϵ, κ respectively, a formula $(\lambda(\mu(\nu\rho))) \leftrightarrow o$ which is of a type which we have not previously encountered. For in it, besides the element o which occurs in the algebraic part of the calculus (the Huntington Calculus), there occur only theoretical terms as primitive elements.

Because this new type of formulae occurs in the calculus C_2, C_2 may be considered as having been constructed at a higher level of sophistication than the calculi in which the theoretical terms only occur in definitory formulae for elements standing for observable entities. These two types of formula represent two types of proposition occurring in deductive systems represented by calculi containing formulae of both types. These two types of propositions are related to those which have been distinguished by N. R. Campbell and, following him, by F. P. Ramsey, as the 'hypotheses' and the 'dictionary' of a scientific theory. Campbell's 'hypotheses' consist of "statements about some collection of ideas which are characteristic of the theory"; he contrasts them with the 'dictionary', consisting of "statements of the relation between these ideas and some other ideas of a different nature", these latter ideas being such that "it must be possible to determine, apart from all knowledge of the theory, whether certain propositions involving these ideas are true or false".* Campbell's 'hypotheses' would be represented by formulae containing only theoretical terms (the λ-formulae); his 'dictionary' by definitory formulae for elements standing for observable entities or by any other formulae that can be used to derive formulae representing propositions about observables from formulae containing theoretical terms.

* *Physics, The Elements* (Cambridge, 1920), p. 122. F. P. Ramsey, *The Foundations of Mathematics and other logical essays*, p. 215, follows Campbell in a dichotomy into 'axioms and theorems' and 'dictionary'; unlike Campbell's dictionary, Ramsey's one would be represented only by definitory formulae.

The distinction between Campbell's 'hypotheses' (expressed by λ-formulae) and his 'dictionary' (expressed, in our examples, by definitory formulae for elements standing for observables) is not as absolute as might appear from his, or from Ramsey's, discussion. As we have seen, exactly the same set of empirical generalizations can sometimes be explained either by a set of dictionary-propositions without any separate theoretical-term hypotheses or by a set of dictionary-propositions along with a separate theoretical-term hypothesis. Indeed, when the algebraic part common to the two calculi is of a suitable character, e.g. the Huntington Calculus or any other calculus containing the same formulae (i.e. any Boolean algebra), a calculus with only a dictionary can be constructed which is $(\alpha, \beta,$ etc.)-equivalent to any calculus containing λ-formulae as well as a dictionary.*

Nevertheless, hypotheses which are only about theoretical concepts (which for convenience will be called *Campbellian hypotheses*) and which are expressed by λ-formulae in the calculus representing the applied deductive system play a very large part in most scientific theories which use theoretical concepts, whether or not sufficient mathematical ingenuity could enable us to dispense with them.

ARE CAMPBELLIAN HYPOTHESES LOGICALLY NECESSARY?

The theoretical terms of a scientific system represented by a calculus may, as we have seen, be regarded as being implicitly defined by the set of initial formulae which contain them, the implicit definition being in terms of the observable properties or other observable things which enter into the lowest-level hypotheses of the system. But if some of the initial formulae are λ-formulae (to be interpreted as expressing Campbellian hypotheses), it seems paradoxical to say that they serve to give implicit definitions of λ, μ, etc., by means of the observables, since symbols α, β, etc., standing for the observables do not occur in these λ-formulae. The inclination then arises to separate them from the other initial formulae containing theoretical terms, and to say

* See A. N. Whitehead, *Universal Algebra*, p. 60. In Whitehead's language the initial formulae of calculus C_1 are *unlimiting* with respect to λ, μ, ν, ρ simultaneously, whereas the initial formulae of calculus C_2 are not unlimiting with respect to λ, μ, ν, ρ simultaneously, since these formulae impose a restriction upon the sets of values of λ, μ, ν, ρ that can simultaneously satisfy the formulae.

that, whereas these latter formulae serve implicitly to define the theoretical terms by means of the observables, the λ-formulae serve to express relationships of the theoretical concepts to one another. And it is tempting to go on to say that, since the λ-formulae contain no symbols standing for observables, the Campbellian hypotheses asserting the relationships of the theoretical concepts to one another are subject to no empirical test, and are therefore logically necessary propositions.

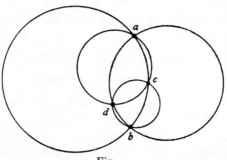

Fig. 4.

The plausibility of this argument is increased by the fact that a calculus containing λ-formulae among its initial formulae can frequently be interpreted as a pure deductive system in which the λ-formulae represent propositions which, without question, are logically necessary. So there can be a *pure* model for a scientific theory containing Campbellian hypotheses. Consider, for example, the calculus C_2 whose initial formulae include the λ-formula

$$(\lambda(\mu(\nu\rho))) \leftrightarrow o.$$

Interpret the calculus in the following way. Let a, b, c, d be four points on a plane which are such that no three of them lie on the same straight line and that the four of them do not lie on the same circle.* Then the four points will determine four distinct circles the circle $\{abc\}$ passing through the points a, b, c; the circle $\{abd\}$ passing through the points a, b, d; the circle $\{acd\}$ passing through the points a, c, d; and the circle $\{bcd\}$ passing through the points b, c, d (see fig. 4).

* The non-collinearity and non-concyclicity are to be regarded as logically necessary propositions about a, b, c, d; i.e. a, b, c, d are to be regarded as being determined not by empirical criteria but, for example, by their Cartesian coordinates being (0, 0), (0, 1), (1, 2), (1, 0) respectively.

Take λ as standing for any particular class of objects, all of which are situated at points on the circle $\{abc\}$, and similarly μ, ν, ρ as standing for a particular class of objects, all of which are situated at points on the circle $\{abd\}$, on the circle $\{acd\}$, on the circle $\{bcd\}$ respectively. Since two distinct circles have at most two points in common, the class of objects represented by $(\lambda\mu)$ will consist of objects situated either at a or at b, the class of objects represented by $(\mu\nu)$ will consist of objects situated either at a or at d, and so on.

The six variable-free initial formulae of the calculus, other than the λ-formula, namely,

$$\alpha \leftrightarrow (\lambda\mu), \quad \beta \leftrightarrow (\mu\nu), \quad \gamma \leftrightarrow (\nu\lambda),$$
$$\delta \leftrightarrow (\lambda\rho), \quad \epsilon \leftrightarrow (\mu\rho), \quad \kappa \leftrightarrow (\nu\rho),$$

will then serve to define each of α, β, γ, δ, ϵ, κ in terms of the objects situated at two of the four points a, b, c, d where the four circles intersect. The derived formulae will represent logically necessary propositions imposing restrictions upon the positions of the objects with which the model is concerned. Thus the formula $(\alpha\beta) \leftrightarrow (\gamma(\alpha\beta))$ will represent the proposition that the class of the particular objects which are situated both either at a or at b and either at a or at d is identical with the class of the particular objects which are situated both either at a or at c and both either at a or at b and either at a or at d. And the formula $(\alpha\kappa) \leftrightarrow o$ will represent the proposition that there are no objects which are situated both either at a or at b and either at c or at d. These propositions are all logically necessary if the relation of situation is taken to be one which is such that it is logically impossible for an object to be situated at more than one distinct point, and if (as was supposed) a, b, c, d are all distinct points.

The logical necessity of these propositions springs from the fact that the initial λ-formula of the calculus C_2, namely $(\lambda(\mu(\nu\rho))) \leftrightarrow o$, represents, according to the interpretation just given, a proposition which is a logical consequence of a theorem of pure geometry. For $(\lambda(\mu(\nu\rho))) \leftrightarrow o$ represents the proposition that there is no object which is both situated on the circle $\{abc\}$ and both situated on the circle $\{abd\}$ and both situated on the circle $\{acd\}$ and situated on the circle $\{bcd\}$, which is equivalent to saying that there is no object situated on the common intersection of all these four circles. But if it is impossible for these four circles to intersect in the same point, *a fortiori* no object can be situated on this point of intersection. And the proposition that four distinct circles, each of which passes through a different triad of points selected from a set of four points, have no common point of intersection is a theorem of pure geometry which in most geometrical deductive systems would be deduced from the fundamental theorem that a circle is determined by three points lying on it, i.e. that two distinct circles cannot have three points in common.*

If to the calculus C_2 there are added sufficient additional initial formulae and rules of play for it to include a calculus capable of representing a deductive system of pure geometry, it will be possible to discard the λ-formula $(\lambda(\mu(\nu\rho))) \leftrightarrow o$ as an initial formula, since this will be derivable from one of the formulae in the part of the calculus interpretable as pure geometry. These formulae will all express general propositions about all points on a plane, all circles, etc., satisfying certain conditions; it is possible for these formulae to be put into forms in which all the symbols which are not interpretable as logical constants are variables,† so that this part of the calculus will be an algebraic calculus in a natural extension of the meaning of this term given on p. 38.‡ If the formulae are taken in these forms, those initial formulae of this new calculus, C_3, which are not included in its algebraic part will be only the six definitory formulae for α, β, γ, δ, ϵ, κ in terms of λ, μ, ν, ρ, and these will be the same as those initial formulae of the calculus C_0 which are not included in its algebraic part. Thus the two calculi, though agreeing in their variable-free initial formulae, will contain different variable-free derived formulae; the formulae $(\lambda(\mu(\nu\rho))) \leftrightarrow o$ and $(\alpha\kappa) \leftrightarrow o$, for example, will be derivable in calculus C_3, but not in calculus C_0.

* For suppose that the four circles had a common point of intersection e. If e were to be distinct from a and from b, the two circles $\{abc\}$, $\{abd\}$ would have the three points a, b, e in common and would be identical. If e were to be a or b, the circle $\{bcd\}$ would pass through a, or the circle $\{acd\}$ would pass through b; and in either case the four circles would be identical.

† Various sorts of variables will be required, with variously different variable-substitution rules.

‡ It may seem strange to call the representation of a pure geometrical deductive system an *algebraic* calculus. But theorems of pure plane geometry can be proved without using any specifically geometrical initial propositions by treating them as being theorems about pairs of numbers and thus incorporating the geometry into ordinary algebra.

There is, however, nothing surprising in this. The calculus-game C_3 is provided with a more elastic method of play than is C_0, so that moves can be made in the former which are not permissible in the latter. The λ-formulae of C_3 derive from the greater formal strength of the calculus C_3. And their interpretation as logically necessary propositions does not arise out of nothing, but out of the fact that the theoretical terms of the calculus are given such an interpretation that the algebraic part of the calculus applies. In this interpretation the logical necessity of the Campbellian hypothesis expressed by $(\lambda(\mu(\nu\rho))) \leftrightarrow o$ comes, not from the particular classes of things designated by λ, μ, ν, ρ but from the fact that these particular classes are taken to be of a certain logical category. The model is a pure model, but the purity of the model is due to the material out of which it has been constructed and is not a characteristic of the theory for which it is a model.

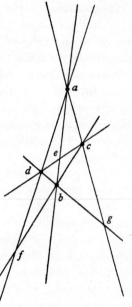

Fig. 5.

That this is the case may be seen by giving a similar but different interpretation to calculus C_2 in which the propositions represented by the λ-formulae are logically contingent. Interpret the calculus starting the same way as before (p. 101) by supposing a, b, c, d to be four points on a plane no three of which are collinear. Instead of taking the four points to determine four circles, consider instead the four points as determining four sets of three lines each (called *line-triads*): the line-triad $\langle abc \rangle$ consisting of the three straight lines passing through a and b, through b and c, through c and a respectively; the line-triad $\langle abd \rangle$ consisting of the lines ab, bd and da; the line-triad $\langle acd \rangle$ consisting of the lines ac, cd and da; and the line-triad $\langle bcd \rangle$ consisting of the lines bc, cd and db (see fig. 5). The line-triads are triangles having their sides indefinitely extended. Let e, f, g be the intersections of the lines ab and cd, of ad and bc, and of ac and bd, respectively (the 'diagonal points' of the quadrilateral $adbc$). As in the earlier interpretation, take λ as standing for any particular class of objects all of which are situated at points on the line-triad $\langle abc \rangle$, and similarly μ, ν, ρ, as standing for a particular class of objects all of which are situated at points on the line-triad $\langle abd \rangle$, on the line-triad $\langle acd \rangle$, or on the line-triad $\langle bcd \rangle$ respectively. Then the class of objects represented by $(\lambda\mu)$ will consist of objects situated either on the line ab which is common to the two line-triads or at the point f or at the point g, the class of objects represented by $(\mu\nu)$ will consist of objects situated either on the line ad or at e or at g, the class of objects represented by $(\nu\lambda)$ of objects situated either on the line ac or at e or at f, and so on. The six variable-free initial formulae of the calculus, other than the λ-formula, will then serve to define each of α, β, γ, δ, ϵ, κ in terms of the objects situated either as lying on one of the six lines or at one of the points e, f, g. Those derived formulae whose derivation does not make use of the initial λ-formula of the calculus will represent logically necessary propositions which impose restrictions upon the positions of the objects. Thus the formula $(\alpha\beta) \leftrightarrow (\gamma(\alpha\beta))$ will represent the proposition that the class of those particular objects which are situated both either on the line ab or at f or at g and either on the line ad or at e or at g is identical with the class of those particular objects which are situated both either on the line ac or at e or at f and both either on the line ab or at f or at g and either on the line ad or at e or at g. But the proposition represented by the initial λ-formula $(\lambda(\mu(\nu\rho))) \leftrightarrow o$ is not a logically necessary proposition. For there are points which are common to all four line-triads—namely, the points e, f, g; so the proposition represented by $(\lambda(\mu(\nu\rho))) \leftrightarrow o$ is or is not true according as there are or there are not objects of the prescribed class of objects situated at one or more of these three points. This is an empirical question if (as we are supposing) it is an empirical question whether or not an object of the prescribed sort is at one of the positions which it is possible for it to occupy. Similarly, the derived formulae whose derivation makes use of the initial λ-formula—$(\alpha\kappa) \leftrightarrow o$, for example—will represent propositions which are logically contingent.

A scientific theory in which the empirical generalizations about six observable properties are explained by displaying them as logical consequences of hypotheses relating the six observable properties to four theoretical factors together with a Campbellian hypothesis asserting an interconnexion between these four factors may thus have corresponding to it either a model in which the Campbellian hypothesis of the model is a logically necessary truth or a model in which this Campbellian hypothesis is a logically contingent proposition. The intersections of our four circles is an example of the first model, the intersections of our four coplanar line-triads an example of the second model.* The difference in the two otherwise very similar models is due to a geometrical difference between a system of four circles and a system of four line-triads where every circle, or line-triad, passes through three out of four coplanar points. A system of circles satisfying this condition can have no common point, but a system of line-triads satisfying the condition has three common points—the diagonal points of the quadrilateral. This geometrical difference is a consequence of the fact that, unlike a pair of line-triads, a pair of circles having two points in common cannot have a third point in common while preserving their distinctness.

If we use the intersecting-circle model for thinking about the theory, we shall think of the Campbellian hypothesis of the theory by thinking of a logically necessary consequence of a geometrical proposition, whereas if we use the intersecting-line-triad model our corresponding thinking will be of a contingent proposition. According as we prefer one or the other of these alternatives shall we choose one or other of the models (if, that is, we wish to think about the theory by means of a model at all). A reason for choosing the former alternative would be if we hoped to extend the theory by adding to it new initial propositions of the same form as the initial propositions of the theory but which would be concerned with a greater number of theoretical concepts. For in this case it might be possible for the new Campbellian hypotheses in the model to be consequences of the geometrical theorem of which the old Campbellian hypothesis in the model was a consequence. In which case it would not be necessary in the new model to add these new Campbellian hypotheses as new initial propositions; the geometry of the model would provide them just as it provided the old

---

* If the four points determining the four line-triads are not coplanar (i.e. if in the figure on p. 104 the point a be imagined raised above the plane of the paper containing b, c, d), there will be no intersections e, f, g of the lines ab and cd, ad and bc, ac and bd respectively; and the model will be similar to the intersecting-circle model in having a logically necessary Campbellian hypothesis.

Campbellian hypothesis.* But if we hope to extend the theory by adding to it as initial propositions Campbellian hypotheses of a new logical form, it will probably be more convenient to choose the latter alternative and to use a model in which the old Campbellian hypothesis, as well as the new ones, appear as contingent propositions. For otherwise some of the Campbellian hypotheses in the model would be logically necessary and some contingent (unless by chance the new Campbellian hypotheses, though differing in logical form from the old one, were still deducible from the logic of the model). This would be highly confusing, since it would tempt us to think that some of the Campbellian hypotheses in the theory were of a different status from the others.†

These considerations show that the fact that some or all of the Campbellian hypotheses of a theory correspond to logically necessary propositions in a model in no way distinguishes these hypotheses from the other Campbellian hypotheses of the theory. The fact that there may be convenient models to hand in which the Campbellian hypotheses are logically necessary features of the model no more makes the corresponding Campbellian hypotheses of the theory logically necessary than does the fact that the definitory initial formulae of the calculus common to theory and to model represent, in the model, propositions which are true by definition make the interpretations of these definitory formulae in the theory logically necessary. Even what is represented by a formula of the form $\theta \leftrightarrow \phi$ may not be logically necessary, for it may be used, not to give an alternative name to the same theoretical concept, but

---

* This feature is not well illustrated by the intersecting-circle model, which was designed to be diagrammatically intelligible. A model suitable for showing this feature may be constructed by taking λ, μ, ν, ρ to stand for any particular subclass of the class of circles passing through a, through b, through c, through d respectively. This model can be extended to include a fifth factor by taking σ to stand for any particular subclass of the class of circles passing through a fifth point e which is not concyclic with any three of the four points a, b, c, d. The new λ-formulae

$$(\lambda(\mu(\nu\sigma))) \leftrightarrow o, \quad (\lambda(\mu(\rho\sigma))) \leftrightarrow o, \quad (\lambda(\nu(\rho\sigma))) \leftrightarrow o, \quad (\mu(\nu(\rho\sigma))) \leftrightarrow o,$$

will, like the old λ-formula $(\lambda(\mu(\nu\rho))) \leftrightarrow o$, represent logically necessary Campbellian hypotheses which are consequences of arithmetical identities about the coordinates of the points a, b, c, d, e.

† This temptation lies latent in Minkowski's four-dimensional geometrical model for Relativity Theory, where the velocity of light appears as the ratio between the units of measurement along three of the axes and that along the fourth axis (multiplied by $\sqrt{-1}$), and thus as a logically necessary feature of the model.

to express the fact that the theoretical concept used in one scientific deductive system (e.g. optics) is the same as the theoretical concept used in another scientific deductive system (e.g. electromagnetic theory) whose lowest-level hypotheses have a different subject-matter from those of the first system. Maxwell's identification of waves of light with electromagnetic waves of a certain range of wave-length had the effect of including the deductive system of optics as a subsystem of the deductive system of electromagnetic theory; this unification is no less an empirical hypothesis because, once it is made, the theoretical term "light-wave" becomes redundant.*

If it is possible to construct a pure mathematical model for a scientific theory there are many advantages in doing so. For it may well be easier to deduce new lowest-level propositions in the model, to which will correspond new empirical generalizations in the theory which will then be available for observational test, if the higher-level propositions are mathematical theorems which can be conflated with other known mathematical theorems to form a very powerful pure deductive system. And it may be possible to deduce these higher-level propositions from theorems concerning more fundamental mathematical concepts, which will suggest an extension of the theory upwards by taking hypotheses about theoretical concepts corresponding to these more fundamental mathematical concepts as the initial propositions of the theory.†

Frequently there may be advantages in constructing a model in which some, but not all, of the initial propositions are logically necessary. This will happen when we wish to consider together various scientific theories which have some highest-level hypotheses

---

* Sir Arthur Eddington said that "the identity of light with electromagnetic waves cannot be counted as an internal hypothesis of physics, since it is altogether outside the province of physics to consider how the stimulation of the optic nerve by electromagnetic waves awakens in consciousness the sensation called light" (*The Philosophy of Physical Science* (Cambridge, 1939), p. 44). But the light-waves which Maxwell's theory identifies with electromagnetic waves are theoretical concepts posited in optics to explain empirical generalizations concerned (among other things) with 'the sensation called light'. It is part of the province of physics to explain such empirical generalizations: Maxwell's theory, by subsuming optical laws under the laws of electromagnetism, extended the empirical reference of electromagnetic theory to cover the observable facts which had previously been the special province of optics.

† In the science of mechanics the theoretical concept of *action* was introduced in this way.

in common but differ as to the rest. It may then be convenient to think of them by means of a model in which the initial propositions corresponding to the common highest-level hypotheses are logically necessary propositions, whereas those corresponding to the hypotheses which differ from theory to theory are contingent. This consideration is strengthened if a highest-level hypothesis common to a number of theories is of a type which Sir Edmund Whittaker has called "postulates of impotence".* These are propositions asserting the impossibility, physical not logical, of doing something, e.g. of building a perpetual-motion machine or of measuring simultaneously the exact position and the exact momentum of a physical particle. Such propositions play a very large part at the present time in the fundamental theories of physics, and many subsidiary physical theories are obtained by combining one of these impotence hypotheses with less general hypotheses. Moreover, these impotence hypotheses are of a form very convenient for mathematical representation. The statement of physical impossibility in the theory would then correspond to a statement of logical impossibility in the model, and it is frequently easy to construct a model with this correspondence.†

Nevertheless, however convenient and mathematically elegant it might be to use a model with logically necessary propositions, unless such a model corresponds to the theory with its empirically testable consequences, it will not serve as a model for the theory. Sir Arthur Eddington, in maintaining in an extreme form the thesis that the fundamental laws of physics are 'epistemological', 'wholly subjective' and capable of being known *a priori*, said that "all that we require from observation is evidence of identification— that the entities denoted by certain symbols in the mathematics are those which the experimental physicist recognises under the names

---

* *From Euclid to Eddington* (Cambridge, 1949), pp. 58 ff. Whittaker, while distinguishing postulates of impotence from logically necessary propositions ("we can quite readily imagine a universe in which any physical postulate of impotence would be untrue"), speaks of them as '*a priori* principles'. By this he appears only to mean hypotheses of a high degree of generality at a distant remove from 'experimental fact'.

† E.g. the intersecting-circle model to represent the factual impossibility of conjoining the four properties, L, M, N, R (p. 101) and the Minkowskian four-dimensional-cone model for Relativity Theory, where logical impossibility corresponds to the physical impossibility of sending a message faster than with the velocity of light.

'proton' and 'electron'".* But this requirement is more stringent than was suspected by Eddington. For it requires that it is the experimentalist's criterion for the application of the theoretical terms (i.e. the experimental testing of the empirical generalizations deduced from the hypotheses containing the theoretical terms) that settles which theory is to be adopted, and hence which mathematical model is appropriate.

The upshot of this somewhat involved discussion is that no consideration as to the logically necessary or contingent status of the highest-level propositions of a model for a theory is any reason for supposing that any highest-level hypothesis in the scientific theory itself is logically necessary. Campbellian hypotheses in a scientific theory, though not relating observable entities directly, serve to relate them indirectly in that they impose restrictions upon the theoretical concepts and thereby enable empirical generalizations about the observables to be deduced which would not be deducible from the hypotheses relating the observables to the theoretical concepts alone. As we have seen, it is frequently possible to construct a theory with no Campbellian initial hypotheses which is empirically equivalent to a theory containing Campbellian hypotheses in the sense that in each theory there can be deduced exactly the same empirical generalizations about observable properties. The initial propositions of the theory with the Campbellian hypotheses will, taken together, be logically stronger than those of the empirically equivalent theory without such Campbellian hypotheses.† But this additional strength does not arise from any part of the initial propositions of the former theory being logically necessary; it arises from our 'arbitrary choice' to use the stronger theory rather than the weaker theory which is empirically equivalent to it.

In constructing a theory to explain certain accepted empirical generalizations, we have to propound a set of hypotheses, some of

---

* *Relativity Theory of Protons and Electrons* (Cambridge, 1936), p. 3. See also *The Philosophy of Physical Science*, p. 134. It would be out of place here to discuss in detail the 'selective subjectivism' which was Eddington's philosophy of physics; my comments on some of his arguments may be found in *Mind*, n.s., vol. 38 (1929), pp. 409ff. and vol. 49 (1940), pp. 455ff.

† In the calculus C_1 (p. 98) there can be derived the formula

$$((\lambda(\mu(\nu\rho)))\,(\alpha\cup(\beta\cup(\gamma\cup(\delta\cup(\epsilon\cup\kappa))))))\leftrightarrow o,$$

which is weaker than the formula $(\lambda(\mu(\nu\rho)))\leftrightarrow o$ of the calculus C_2.

which may be Campbellian hypotheses, sufficiently strong logically for the known empirical generalizations to be logical consequences of the set of hypotheses taken together. If we make the set stronger than it need be, either by way of allowing the deduction of more empirical generalizations, or by way of allowing the deduction of more propositions relating the theoretical concepts with the observables, that is our choice; we may do it because the mathematics for the stronger theory is simpler or because there is a more intuitively satisfying model for the stronger theory or because we prefer to have a stronger theory for possible extension in the future. It is true that if the supererogatorily strong theory is empirically equivalent to the other theory, its supererogatoriness does not arise from 'experience'. But this does not involve that it should arise from 'necessity in thought'; Hertz's third alternative* is the true one—it arises from a choice which, though not arbitrary in the sense that no good reasons can be given for it, is an 'arbitrary choice' in the sense that another choice would have been equally consistent with the demands of pure logic and of experience.

FUNDAMENTAL HYPOTHESES AND STERILE FORMULAE

We have seen that there is no good reason for picking out some of the hypotheses which are the initial propositions of a scientific theory and taking them to be logically necessary. The whole set of them taken together is clearly contingent, since the theory containing that set of initial hypotheses would have to be abandoned if any of the empirical generalizations which are the consequences of this set were found by observation to be false. The new theory that would then be constructed to take the place of the rejected theory might well be obtained from this latter theory by substituting one new hypothesis for an old one in the set of initial propositions and leaving the rest unchanged. In the case of an advanced science like physics almost every theory will contain a large number of hypotheses, several of which will be common to theories covering different branches of physics. If one of these theories has to be rejected as a result of experience while the others stand, it will probably be convenient not to reject the hypotheses which are common to the theories. A hypothesis (e.g. the law of the conservation of energy) may occur in so many theories, and

* See above, p. 96.

play such an apparently irreplaceable part in many of them, that we may decide to make almost any other change in a theory containing it rather than to abandon this hypothesis. Within the context of the theories in which it occurs this hypothesis will exhibit the characteristic way of functioning of a logically necessary proposition—that of being subject to no empirical test.* But though within this particular context it behaves as if it were logically necessary, it is not genuinely logically necessary. Under sufficiently different circumstances we shall discard the hypothesis, however 'fundamental' we may have thought it, as not being profitable in explaining observed facts (as has, indeed, been happening recently with the 'law' of the conservation of energy). So the fact that we may properly have a great resistance to throwing a fundamental hypothesis out of what we regard as our body of scientific knowledge does not imply that we should hold to it through thick and thin. The corrigibility of all scientific hypotheses is less likely to be forgotten now than it was fifty years ago before the successive revolutions effected in physics by the advent of Relativity Theory and of quantum mechanics.

The matter is complicated by the fact that the acceptance of a scientific hypothesis as part of our body of scientific knowledge frequently has the effect of causing us to redefine the terms occurring in the expression of the hypothesis, so that the hypothesis becomes a proposition true by definition. For example, I use the sentence "Every hydrogen atom consists of one proton and one electron" to express a contingent proposition, since I learnt the use of the phrase "hydrogen atom" as a theoretical term used within a deductive system of chemistry, grounded upon Dalton's Atomic Theory, with different chemical atoms as the units out of which molecules were compounded. "Hydrogen atom" for me therefore means nothing about its internal constitution; the meaning is determined by the place of the phrase as a theoretical term in a theory which explains empirical generalizations about specimens of a gas obtained by putting bits of zinc into hydrochloric acid, a gas which will burn with an almost invisible flame and will produce water as the product of combustion. A man who had learnt a lot of physics before he had learnt any chemistry might

* For this reason Arthur Pap calls such hypotheses "functionally" or "contextually" *a priori* (*The A Priori in Physical Theory* (New York, 1946)).

well have learnt first the use of "proton" and "electron" as theoretical terms of atomic physics, and subsequently have learnt the use of the phrase "hydrogen atom" as meaning the atomic system which consists of one proton and one electron. For this man the sentence "Every hydrogen atom consists of one proton and one electron" would express a proposition logically necessary because true by definition.

What will happen when this man learns chemistry? There are two possibilities. Either he will learn his chemistry independently of his knowledge of atomic physics. In this case he will treat the phrase "hydrogen atom" used in a chemical context as having a different meaning from the same phrase used in a context of atomic physics; he will learn it as a theoretical term of chemistry as a separate science, and will subsequently accept the contingent proposition that the chemical hydrogen atom is identical with the hydrogen atom of atomic physics in a similar way to that in which Maxwell accepted the contingent proposition that light-waves are electromagnetic waves. Or the man will learn only such chemistry as can be deduced from accepted theories of atomic physics. In this case the theoretical terms of that branch of atomic physics which he calls chemistry will be the theoretical terms of atomic physics itself—"proton", "electron", etc.; there will be no specifically chemical theoretical terms, and no concepts which are chemical as contrasted with physical. The phrase "hydrogen atom" will continue to be used by him as synonymous with "system consisting of one proton and one electron"; he will use it as an abbreviation for a phrase expressing a compound physical concept (a logical construction out of physical concepts) and not as a separate theoretical term at all.

I do not wish to deny that some sentences or formulae may properly occur in the representation of a scientific deductive system whose function is that of introducing symbols which enable other sentences or formulae to be written more shortly than would otherwise be possible. But such sentences or formulae are *sterile*, and the symbols introduced by them are formally redundant; a calculus could be constructed which would represent the scientific deductive system equally well and which would not contain these formulae or symbols. In a calculus containing no sterile formulae with redundant symbols none of the formulae expressing hypotheses of the scientific deductive system express propositions which are logically necessary.

Confusion arises because exactly the same calculus can frequently be interpreted either as representing a scientific deductive system by a representation in which one set of symbols is formally redundant or as representing another system by a representation in which another set of symbols is formally redundant. An inspection of the calculus by itself will not, of course, indicate which way it is being interpreted. As a science develops, and a calculus appropriate for representing the scientific system at a certain stage of knowledge is incorporated in a more extensive calculus appropriate for representing a more developed system, some of its symbols will almost certainly change to being redundant, and some of its formulae to being sterile formulae. For example, "hydrogen atom" may cease to function as a theoretical term in a chemical deductive system, and function instead redundantly, not as a theoretical term, but as an abbreviation for a logical construction out of other theoretical terms in a physico-chemical deductive system with a logically necessary proposition among the hypotheses which is true by definition. Usually when such a change occurs and a symbol which previously has been used as a theoretical term becomes redundant, there is also a change in the opposite direction, and a symbol which previously has been redundant comes to be used as a genuine theoretical term. The 'conventionalists' who have emphasized the fact that a sentence which at one time expresses a contingent scientific hypothesis may come to express a logically necessary proposition true as a matter of 'convention' or of 'disguised definition' have not always attached sufficient importance to the fact that reverse changes also take place; if any sentence may be so used as to express a proposition true by convention, equally well it may not be so used. But whether or not there are such compensating changes, it is impossible to change the interpretation of the calculus representing a scientific system so that all the symbols become redundant. At any moment a calculus interpreted as one containing sterile formulae may be translated into one in which every formula, not part of the pure part (if any) of the calculus, fulfils its proper function of expressing a logically contingent proposition.

CHAPTER V

STATISTICAL HYPOTHESES, PROBABILITY STATEMENTS AND CLASS-RATIO ARITHMETIC

The scientific deductive systems with which we have so far been concerned are those in which all the hypotheses, at whatever level in the hierarchy they may occur, are generalizations either of the form «Everything which is A is also B» or of the form «Nothing which is A is also B».* In the lowest-level hypotheses of these forms the properties A and B are observable properties, and the hypotheses are refuted by the observations of a thing which is A without being B, or the observation of a thing which is both A and B, respectively.

Many of the generalizations which occur in science are not of the simple form of a *universal hypothesis* asserting that 100 %, or that 0 %, of the things which are A are also B, but are of the form of a *statistical hypothesis* asserting that a certain proportion between 100 and 0 % of the things which are A are also B. Many of the generalizations occurring in the social sciences are of this form, and many of those occurring in the biological sciences (e.g. 51 % of children born are boys). In physics the Kinetic Theory of Gases explained the lowest-level generalizations about the properties of gases by treating them as statistical hypotheses about a proportion of all gases which is very nearly 100 %, and by constructing a deductive system in which these statistical hypotheses were explained as being deducible from statistical higher-level hypotheses about impacts of the molecules of which the gas is composed. Max Born's statistical interpretation of quantum mechanics (1926) interpreted the initial formulae containing Schrödinger's wave-functions as theoretical terms expressing hypotheses from which there followed only statistical hypotheses about observables. The Kinetic Theory of Gases giving a statistical explanation of the behaviour of a gas is

---

* This form can be reduced to the first form by considering the negative property non-B. "Nothing which is A is also B" is then synonymous with "Everything which is A is also non-B".

compatible with a theory which would give a non-statistical explanation of such behaviour. But the theory of quantum mechanics is incompatible with any theory which would give a non-statistical explanation of the field covered by quantum mechanics, which is the whole of atomic physics. This incompatibility does not imply, of course, that it is logically impossible that there should be laws of atomic physics which are non-statistical (i.e. what are frequently called "deterministic"); for the whole theory of quantum mechanics may be found to be inadequate, and may be superseded by another theory allowing of deterministic laws. But while quantum mechanics stands, it cannot be subsumed under a deterministic theory, and a deterministic theory cannot be tacked on to it. Since it seems more plausible to suppose that the line of development in physics will be based upon quantum mechanics, and will share with it its essential statistical character, than that physics will start somewhere quite fresh, it looks as if statistical explanations which are irreducible to non-statistical ones have come to stay in physics, and will thus be the fundamental hypotheses for the other physical sciences.

While statistical generalizations were current only in the social and biological sciences and in physical theories like the Kinetic Theory of Gases where non-statistical explanations were not excluded, it was reasonable for a philosopher of science to regard the statistical hypotheses as being acceptable *faute de mieux*, and only until they could be displaced by the non-statistical hypotheses which might well be discovered when the social and biological sciences had reached a more advanced stage and when physics knew more in detail about the individual molecules and atoms. But now that the most advanced of the sciences in the most sophisticated and far-reaching of its theories postulates an irreducibly statistical form of explanation, it will be unreasonable for a philosopher of science to ignore the special problems presented by statistical hypotheses. On the contrary, it will be safer for the philosopher to take statistical hypotheses as being the normalcy, and to regard universal (i.e. non-statistical) hypotheses as being extreme cases of statistical hypotheses when the proportions in question are 100 or 0 %. So we must now consider what modifications will be required in our account of scientific systems as deductive hierarchies in which empirical generalizations are deducible from higher-level hypotheses in order to allow for the empirical generalizations being statistical in character.

It should be remarked here that the question as to whether or not theories which yield only statistical generalizations will in the future either be rejected as not fitting the observable facts or will be subsumed under theories which yield universal generalizations is not one which it is necessary for a philosopher of science to answer. He will be wise to give an account which takes statistical hypotheses very seriously indeed, for the present opinion of physicists is that the statistical theories of quantum mechanics will not be superseded by non-statistical ones. And even if in fact all statistical theories were destined to be superseded sooner or later by deterministic ones, they will represent, before they are superseded, the best scientific thought of the time, and this cannot be ignored. Besides these considerations, there is the general philosophical point that, if a sentence is significant, it is significant whether or not what it expresses is true, so that the question of how a sentence has meaning does not depend on whether or not in fact it expresses a true proposition. So the way in which sentences like "51 % of children born are boys" or "50 % of the atoms constituting a piece of radium will disintegrate within 1700 years" have meaning requires investigation; and this investigation, though it will of course depend upon what are the empirical conditions under which the propositions expressed by these sentences are true or false, will not depend upon whether in fact the propositions are true or are false.

The doctrine of the way of functioning in our thought of scientific deductive systems will be expanded in this and the next chapter into a more general doctrine which will cover systems in which the lowest-level hypotheses are statistical generalizations. This more general doctrine will include as extreme cases systems in which the statistical generalizations are universal generalizations. The account to be given will not make use of any logical notions of an essentially novel character; the epistemological notion of *rejecting* a hypothesis will have to be extended to allow for the possibility of cancelling a rejection and of reinstating a hypothesis which had been provisionally rejected; but there will be no peculiar logical relations like that of being a partial logical consequence (as in the probability theory of Harold Jeffreys or of J. M. Keynes) and no apparently empirical hypotheses that are in fact not empirically testable (as in the Limiting-Frequency view of Richard von Mises). The justification for the extended notion of rejecting hypotheses will require treating what seem to be lowest-level hypotheses (e.g. that 51 % of children born are boys) as highest-level hypotheses in an unendingly descending deductive hierarchy, the deductive principles used in this deduction being those of a special branch of mathematics (to be called *class-ratio arithmetic*). It is the difficulty in disentangling the various relevant considerations, some mathematical, some epistemological, as well as the importance now assumed by statistical laws in science, that is my excuse for the length of the three chapters in this book devoted to statistical hypotheses.

PROBABILITY: ITS DIFFERENT SENSES

One of the natural ways for expressing a statistical hypothesis is by means of the word "probability" and its cognates. Instead of saying that the proportion of male births among births is 51 % we may say that the probability, or the chance, of a birth being a male birth is $\frac{51}{100}$. The advantage of using the special language of probability rather than that of proportions is that it is usually not possible to take the proportion language literally. The literal use of the term "proportion" is that in which to say that the proportion among the members of a class β* of those which are also members of a class α is a certain number h is to assert the proposition that the number of members of the intersection of α with β (i.e. the number of things which are both members of α and members of β) divided by the number of members of β is h, or—to express this in the ordinary arithmetical way of writing fractions—$N'\theta$ denoting the number of members of the class θ,

$$\frac{N'(\alpha\beta)}{N'\beta} = h.$$

But when we say that the proportion (in a non-literal sense) of male births among births is 51 %, we are not saying of any particular class of births that 51 % of them are births of males, for the actual

---

* The small Greek letters will be used henceforth in this book to denote classes, and their adjunction to represent the intersection of the classes, the calculi in which they occur being always interpreted in this way. They will not be used, as in previous chapters, as marks in an uninterpreted calculus. Nor will they be used in a calculus interpreted in a different way.

proportion might differ very widely from 51 % in a particular class of births, or in a number of particular classes of births, without our wishing to reject the proposition that the proportion (in the non-literal sense) is 51 %. Since the account of the nature of statistical hypotheses to be given here will make great use of, and indeed will be in terms of, proportions in the simple arithmetical literal sense, it would be inconvenient to use the same word also in a non-literal sense.

There are, indeed, objections to using the word "probability" in expounding the character of statistical hypotheses. For the word and its cognates are used in different senses in different sorts of contexts; and, what is worse, used in two distinct senses in connexion with scientific hypotheses. For besides speaking of the hypothesis that the probability of a radium atom disintegrating within a period of 1700 years is $\frac{1}{2}$, we also speak of Einstein's theory of gravitation being probable, or of its being more probable than Newton's theory, where it is a whole scientific theory that is asserted to be probable, or to be more probable than another, and not an event to which the theory itself assigns a probability. The importance of this distinction between probability of kinds of events and probability of hypotheses has been emphasized by many recent logicians.* It is the first sense of probability which we use in translating statistical laws; it is probability as used within the system of a science, probability which enters into the hypotheses themselves. That the probability of a radium atom disintegrating within 1700 years is $\frac{1}{2}$ is a hypothesis within the body of physics; since it is reasonable to believe this hypothesis at the present time, the hypothesis itself may be said to be probable in the second sense of probability. Confusion is sometimes caused by a conflation of the two senses of probability in a sentence in which both senses are involved but only one word is used. For example, the sentence "It is highly probable that a radium atom will not disintegrate within one year" conflates both senses, for it is equivalent to saying that the hypo-

---

* Karl Popper, *Logik der Forschung*, §§ 80 ff.; Rudolf Carnap, in *Philosophy and Phenomenological Research*, vol. 5 (1945), pp. 513 ff.; *Logical Foundations of Probability*, Chapter II; J. O. Urmson, in *Analysis*, vol. 8 (1947), pp. 9 ff.; Bertrand Russell, *Human Knowledge, its scope and limits* (London, 1948), Part v; William Kneale, *Probability and Induction* (Oxford, 1949), p. 22. The distinction was foreshadowed by F. P. Ramsey (*The Foundations of Mathematics and other logical essays*, p. 157).

thesis that there is a high probability (in the first sense) of a radium atom not disintegrating in one year is itself probable (in the second sense). But when the two senses have been disentangled, it seems clear that they are different senses; the probability occurring within a scientific hypothesis is a probability which has a value measurable by a definite number, but it is by no means certain that probability as the reasonableness of a hypothesis has characteristics which permit a number to be attached to it.* Probability within a science is an objective concept, which is independent of the position of the hypothesis asserting this probability inside the body of scientific knowledge. The hypothesis about the probability of disintegration of a radium atom contains no reference to my or to anyone else's belief; the probability, in the second sense, of this hypothesis itself refers to the position of the hypothesis in relation to a body of reasonable belief. To say that the hypothesis is probable implies that if the hypothesis is inside this body of belief it should stay there, and if it is not, it should be incorporated in it. The second sense of probability is thus epistemological rather than purely logical.

Many of the philosophers who have recently emphasized the distinction between the two concepts of probability (and these philosophers have held quite different views as to the analyses of the two concepts) have proposed to use different names for the two concepts. Probability in the first sense—probability within a science —has been renamed by Rudolf Carnap *relative frequency*, by Bertrand Russell *mathematical probability* and by William Kneale *chance*. It has also been called *statistical probability* and *empirical probability*. There seem to me to be objections to all these alternative names;† so I shall use *probability* without any qualification for probability within a scientific hypothesis. The other sort of probability, which Carnap calls *confirmation*, Russell *credibility* and Kneale *acceptability*, I shall straightforwardly call *reasonableness*, and shall neither call it probability nor suppose that it is a quantity measurable by a number.

* See below, pp. 354 ff.

† Probability within a science is not identical with relative frequency, though explicable in terms of it. To call it mathematical *par excellence* is to prejudge the question as to whether the second sort of probability is not also mathematical. And mathematical probability is sometimes used to mean a pure mathematical notion satisfying the theorems of the pure mathematics of probability. *Chance* is the best alternative, but it has other valuable uses, e.g. a chance event.

There are other senses which the word "probability" and its cognates may have in contexts in which scientific hypotheses are not concerned. Many of these senses are closely related to one or other of my *probability* or my *reasonableness*. The proposition that the probability that this particular die will fall five uppermost is $\frac{1}{6}$, if it is understood not as a statement about the next or any particular fall of the die but as a general proposition about any fall of this die, is a statistical hypothesis, its only difference from the scientific case being that the hypothesis is not sufficiently general to be counted as a scientific hypothesis.* Indeed, since the probability in this case is logically similar to probability within a scientific hypothesis, it is frequently convenient to take propositions about throws of a particular die or draws from particular bags of balls, or cuts of particular packs of cards, as examples of statistical hypotheses, since specific values of probabilities can then plausibly be taken.

The statement that the probability is $\frac{1}{6}$ that a particular die will fall five uppermost on the next occasion it is thrown may have several meanings. It may be intended as an assertion of the general proposition about any fall of the die by reference to the particular case of the next fall, or it may be intended as an assertion that it is reasonable to apply this general proposition to this particular case, i.e. that it is reasonable to suppose the next fall of the die to be a typical one. Or it may mean both these propositions together. It may also be used to express a rate at which the asserter, or some ideal reasonable man, would be prepared to bet on the die falling five uppermost next time; this sense is related to the other meanings, but may not be identical with any of them. Similarly, the statement that it will probably freeze to-night may be taken in various different ways.

It would be irrelevant to my purpose in discussing the philosophy of science to examine in detail the uses of probability concepts in non-scientific contexts. What a philosopher of science is concerned with is, first, probability as it occurs within the hypotheses of a science, i.e. with the concept of probability as the characteristic of

---

* If "a die" is substituted for "this particular die" the statement may be taken to express a proposition not about any actual die but about an ideal die. In this case it will be a logically necessary proposition in the pure deductive system of probability, the ideal die being used in a model for this system. Remarks at greater length about probabilities in the context of games of chance will be found in the next chapter.

statistical hypotheses; and, secondly, probable in the sense of reasonable as applied to scientific hypotheses themselves or to belief in them. For the rest of this book the word will be used only for the empirical, numerically measurable concept characteristic of statistical generalizations; the reasonableness of belief in scientific hypotheses will never, in the rest of this book, be spoken of in terms of probability. It will be considered in Chapter VIII.

PROBABILITIES IN TERMS OF CLASS-RATIOS

The statistical hypothesis that $100p\%$ of the members of the class β are also members of the class α will thus be expressed in terms of probability in the form of "The probability of a member of β being a member of α is p", or of "The probability of a β-specimen being an α-specimen is p". The problem of the meaning of a sentence expressing a statistical hypothesis will be treated as the problem of the meaning of a probability statement of this form.

On the assumption that the class β is neither the null-class nor is an infinite class (i.e. that the number of members of β is a finite number which is not zero), the probability of a β-specimen being an α-specimen can be identified with the proportion among the members of β of those which are members of α, i.e. with

$$\frac{N'(\alpha\beta)}{N'\beta} = h.$$

This fraction will be called a *class-ratio*: it is the ratio which the number of members of a subclass bears to the number of members in the class which includes the subclass, i.e. the proportion among the members of the class of those which are members of the subclass.

When probabilities can be identified with class-ratios whose denominators are finite numbers which are not zero, no further problem arises about them. All these probabilities are rational numbers lying between 0 and 1 inclusive, and all the logically necessary laws connecting related probabilities appear as arithmetical propositions connecting related fractions. For example, the characteristic law of the pure deductive system of probability, that the probability of a β-specimen being both an α-specimen and a γ-specimen is equal to the probability of a β-specimen being an α-specimen multiplied by the probability of a thing which is both an α-specimen and a β-specimen being a γ-specimen, appears as the proposition

$$\frac{N'((\alpha\gamma)\,\beta)}{N'\beta} = \frac{N'(\alpha\beta)}{N'\beta} \frac{N'(\gamma(\alpha\beta))}{N'(\alpha\beta)},$$

which is an arithmetical truism, the class $((\alpha\gamma)\,\beta)$ being identical with the class $(\gamma(\alpha\beta))$. The probability of a child born in Cambridge during 1950 being both male and blue-eyed is equal to the probability of a child born in Cambridge during 1950 being male multiplied by the probability of a male child born in Cambridge during 1950 being blue-eyed.

But this identification of a probability with an actual class-ratio can only be made if it be assumed that the class whose number provides the denominator in the fractional expression of the class-ratio is a finite class which is not an empty class. This assumption is required in order that the fraction expressing the class-ratio should be significant; if the assumption is false the fraction has no meaning.* In the case of probability statements which are scientific hypotheses, however, this assumption can never be made. For if it were assumed that the class of reference had only a finite number of members, the generalization would be restricted so as to apply only to a limited number of its instances, and would thus not have the generality which we require of a generalization for it to be ranked as a scientific hypothesis. It may be the case that there have been, are, and will be only a limited number of β-specimens; whether or not this is so must be irrelevant to the meaning of a scientific probability statement. Just because of this irrelevance, it is impossible to accept any account of such a probability statement which would make the meaning of the statement depend upon whether or not the class of reference was finite.

The point can be put in another way. If the probability statement were interpreted according to a mode of interpretation which depended upon the class of reference being a finite class, a finite set of observations—namely, observations of the members of this finite class—would suffice conclusively to establish the truth of the probability statement. But this is just what cannot happen with scientific hypotheses; however many observations we have made a further observation may serve to refute the hypothesis. Thus the class of

---

* In ordinary algebra a/b is only defined when a and b are finite numbers, and b is not equal to zero.

reference in a scientific hypothesis must be a class which is not limited in advance by the way in which the expression of the hypothesis is interpreted.*

We therefore cannot treat scientific probability-statements as making assertions about actual class-ratios. But since the class-ratio in a set of observations is the direct test for a statistical hypothesis just as the class-ratio in one observation is the direct test for a universal hypothesis (if this class ratio is o, the hypothesis must be rejected), it is natural to suppose that meaning is given to these probability statements somehow by reference to observable class-ratios. That meaning is given to probability statements *somehow* in this way is implied by the use made of these statements by nearly all scientists and statisticians. The difficulty for the philosopher lies in elucidating this *somehow*.

There are two ways of dealing with the matter which have been explicitly propounded.

PROBABILITIES AS LIMITING VALUES OF CLASS-RATIOS

One way is to make the probability statement refer not to any one actual class-ratio but to the limit of the values of the class-ratio in an infinite series of finite classes of reference, each of which includes all its predecessors in the series. To put this view more precisely, an infinite series of finite classes β_1, β_2, ..., β_j, ... is chosen such that each of these classes is included in β, and β_j is included in β_{j+1}, for every j. Then if the infinite sequence of class-ratios

$$\frac{N'(\alpha\beta_1)}{N'\beta_1}, \quad \frac{N'(\alpha\beta_2)}{N'\beta_2}, \quad ..., \quad \frac{N'(\alpha\beta_j)}{N'\beta_j}, \quad ...$$

tends to a limiting value, this limiting value is to be taken to be the probability that a β-specimen should be an α-specimen.

This view of probability—the Limiting-Frequency view—has been elaborated with slight variations and defended by many recent logicians, among whom are Richard von Mises, Karl Popper and Hans Reichenbach.† Apart from the details, which differ from one

---

* The possibility of the class of reference having no members will be considered, under the title of 'subjunctive conditionals', in Chapter IX.

† Richard von Mises, *Probability Statistics and Truth* (London, 1939) (original German edition, Vienna, 1928); Karl Popper, *Logik der Forschung*, Chapter VI; Hans Reichenbach, *Theory of Probability* (Berkeley, 1949)(original German edition, Leyden, 1935). Von Mises's form of the theory is criticized by William Kneale; *Probability and Induction*, and Reichenbach's form by Bertrand Russell, *Human Knowledge, its scope and limits*.

exponent to another, all forms of the theory have two defects which are, to my mind, so serious as to vitiate it. The first defect is that, since the notion of limit is an ordinal notion, the theory makes the value of a probability depend upon the order in which the observations are grouped into the infinite series of classes of reference $\beta_1, \beta_2, \dots$; a different grouping would yield a different sequence of class-ratios with quite possibly a different limiting value. The theory does not, it is true, require that the observations should be ordered in time, but it does require that there should be some principle according to which the classes of reference can be assigned successive numbers so that they can be ordered in a series. But any notion of order seems quite irrelevant to the scientific notion of probability in general, which is concerned with the significance of statements like "51 % of children born are boys" in which there is no reference to any order whatever.*

The second objection is that, since a sequence of numbers can start with any finite set of terms and still tend to any limit whatsoever, not only is the limiting value of the class-ratio not determined by the class-ratio in any finite set of classes of reference, but—and this is more serious—a limiting value does not determine what will be the class-ratio in any finite class of reference. The former lack of connexion implies that the limiting value cannot be regarded as a logical construction out of class-ratios in observable classes; the latter lack of connexion implies that it cannot be regarded as a theoretical concept either, since no conclusion as to class-ratios in finite classes (which are all that are observable) can be drawn from it.

The Limiting-Frequency view of probability, propounded by Venn as an analysis of probability statements in terms of series of actual events, has a tendency to become transposed in the hands of all its exponents into a theory in which 'ideal' series of events are substituted for actual series, so that a probability statement becomes a Campbellian hypothesis about a theoretical concept, the limit of class-ratios in an ideal series, with which the probability is to be identified.† There is no objection to such a transposition; indeed, to treat probability statements as higher-level hypotheses in a

---

* The special case of 'probabilities in chains', where the order is relevant, though of great importance in many parts of physics, is nevertheless a special case.

† John Venn, *The Logic of Chance*, third edition (London, 1888), p. 95.

deductive system is involved in my exposition of them. But the effect of such a change is to make the probabilities of the Limiting-Frequency view less directly rooted in observable facts than the exponents of the view wish to maintain.

If probability statements are treated as higher-level hypotheses in a scientific theory, the Limiting-Frequency view is equivalent to describing the theory of probability by exhibiting a pure model in which the propositions corresponding to probability statements are logically necessary propositions about the existence of limits of sequences given by a mathematical rule, in the way in which the sequence of numbers $1\frac{1}{2}$, $\frac{3}{4}$, $1\frac{1}{8}$, $\frac{15}{16}$, ..., whose nth term is given by the formula $1-(-\frac{1}{2})^n$, tends to a limit (namely, 1) as n increases without limit. The objection to using such a model is not that the propositions in it are logically necessary whereas those of the statistical system are contingent, but is simply that the model will not do what it is required to do, namely, to enable propositions to be deduced in the model about finite slices of the sequence of numbers, to which correspond in the theory propositions about actually observable class-ratios. For from the mathematical fact that an infinite sequence of numbers tends to a limit nothing can be deduced as to the behaviour of any finite set of numbers selected from the infinite sequence.* This mathematical model is therefore inappropriate, since it does not enable the proper deductions to be drawn in it.

PROBABILITIES AS CLASS-RATIOS IN 'HYPOTHETICAL INFINITE POPULATIONS'

We have seen that a probability cannot be taken to be in a literal sense a class-ratio in an infinite class. But is it not possible to regard a probability as a number associated with an infinite class (what mathematicians would call a *parameter* of the class) which resembles a class-ratio in that, if sets of observations are regarded as samples

---

* Mathematicians sometimes use probability language to express theorems about limits of mathematically determined sequences. For example, they use the sentence "The probability that a positive integer is square-free (i.e. has no repeated prime factor) is $6/\pi^2$" to mean that the function of n defined as the proportion among the integers not exceeding n of those which are square-free tends to the limit $6/\pi^2$ as n increases without limit. There cannot be deduced from this proposition anything about the actual proportions of square-free integers in the integers from 1 to 100, or from 100 to 1000, or from $1,000,000$ to $1,001,000$, etc.

from the infinite class, conclusions about the class-ratios to be found in these samples can be drawn from the probability statement about the value of this parameter?

This is Sir R. A. Fisher's way of looking at the matter, and he has been followed by the majority of contemporary statisticians. According to Fisher, "the actual data are regarded as constituting a random sample" of "a hypothetical infinite population". "The *probability* of a certain object fulfilling a certain condition...is a parameter which specifies a simple dichotomy [of fulfilliṇg or of not fulfilling the condition] in an infinite hypothetical population, and it represents neither more nor less than the frequency ratio which we imagine such a population to exhibit. For example, when we say that the probability of throwing a five with a die is one-sixth, we must not be taken to mean that any six throws with that die one and one only will necessarily be a five; or that of any six million throws, exactly one million will be fives; but that of a hypothetical population of an infinite number of throws, with the die in its original condition, exactly one-sixth will be fives."* Here Fisher speaks of a frequency ratio (i.e. a class-ratio) in an infinite population; since such a statement has no meaning if taken *au pied de la lettre* Fisher has sometimes, mistakenly, been supposed to be propounding a Limiting-Frequency view.† But though Fisher and his disciples have not expressed themselves as precisely as could be desired upon the logical relationship between the 'hypothetical infinite population' and the 'random samples' of it, what they do with their infinite hypothetical populations leaves us in no doubt that they are treating these as being theoretical concepts, and the ascription of a probability parameter to a hypothetical infinite population as being a Campbellian hypothesis (containing only theoretical concepts) which functions as a higher-level hypothesis in a deductive system in which empirically testable propositions appear as logical consequences.

---

* *Philosophical Transactions of the Royal Society*, series A, vol. 222 (1922), pp. 311 f. See also his *Statistical Methods for Research Workers* (Edinburgh, 1925), p. 7, where, however, the probability parameters are not specifically stated to represent frequency ratios.

† "My own definition is not based on the limit of frequencies, if by this [is meant] experimental frequencies, for I believe we have no knowledge of the existence of such limits" (R. A. Fisher, in *Journal of the Royal Statistical Society*, vol. 98 (1935), p. 81).

On this interpretation the philosophy of probability of Fisher and the mathematical statisticians who have followed him differs only in pictorialness from that of statistical mathematicians who have approached the subject from the direction of pure mathematics.* What these mathematicians have done is, in my language, to construct a calculus containing an algebraic part which is interpreted as a piece of pure mathematics (theory of measure of sets of points); to give no direct interpretation to the numerical parameters, taken as standing for probabilities, appearing in the non-algebraic initial formulae of their calculus; but to give an indirect interpretation to these symbols for probabilities by taking some of the derived formulae in the calculus to represent propositions which can be tested against experience. Fisher's hypothetical infinite population may be regarded as a model for such a theory of probability, the model being visualized as that of a bag containing an infinite number of balls from which random draws are to be made. To the probability p of the theory there is to correspond a proportion p of the balls which are black.

A model in which there are an infinite number of balls in the bag has the following merits: (1) Every proportion of black balls can be represented in such a model, whereas if the bag contains only m balls the only representable proportions are multiples of $1/m$. (2) The drawing of a ball, whether it be white or black, does not alter the proportion of black balls in the bag; hence separate draws from the bag will be independent (in the sense that the probability of each draw being black will be the same) without the balls drawn being replaced after each draw, and thus a set of n independent draws can be obtained by drawing a handful of n balls all at once. (3) The number n of this sample can be as large as is desired; hence the same model will serve for any size of sample and for any finite set of samples. (4) Such samples will be finite subclasses of the infinite class of balls in the bag; the infinite class will therefore be of the same logical type as that of the samples, and to imagine making draws from an infinite bag presents no psychological difficulty.

The defects of the infinite-bag model are, however, serious.

---

* A. Kolmogoroff, *Foundations of Probability* (New York, 1950) (original German edition, Berlin, 1933), Chapter I, §2; Harald Cramér, *Mathematical Methods of Statistics* (Princeton, 1946), p. 151; A. C. Aitken, *Statistical Mathematics* (Edinburgh, 1939), Chapter I.

Since the notion of the proportion of black balls among the infinity of balls in the bag cannot be taken literally, to make the probability p of the theory correspond with this proportion in the model is not very illuminating. What happens in the minds of those using this bag model is, I believe, that when they start by making the probability correspond to the proportion of balls in the bag, they are thinking of the bag as finite so that the proportion has a literal meaning. The size of the bag then has to become infinite in order that the proportion should not be altered by the removal of one ball, and in order that as many balls as are desired can be drawn from the bag without the bag becoming exhausted. This confusion of thought as to the size of the bag is imposed by the attempt to use drawings *without replacement* of balls from a bag as a model for the theory of probability; to serve this purpose the bag must be both finite and infinite. The model suggested by the hypothetical-infinite-population view is therefore inappropriate, through being self-contradictory.*

A 'BRIAREUS' MODEL FOR PROBABILITY THEORY

We see that what is wrong with the infinite-bag model is that the drawings are made without replacement. If each ball which is drawn is replaced after its colour has been noted, the constitution of the bag will be the same as it was before the draw, and it will not be necessary for the bag to be infinite in order that the draws should be independent. And if the essence of a draw is regarded as the noting of the colour of a ball rather than that of the securing of the ball itself, a finite bag will be inexhaustible by any number of draws. So the model of a finite bag to which each ball which is drawn is returned before another draw takes place escapes from the self-contradiction of the infinite-bag model. Since the object of replacing the drawn ball in the bag is to reproduce the constitution of the original bag, an equivalent model for the case of n draws will be obtained by considering n similar bags each containing the same number, m, of balls and the same proportion, p, of black balls, and of supposing that the n draws are made by drawing one ball from

---

* By taking the bag as containing a very large number, but not an infinite number, of balls, the proportion of black balls in the bag will have a good sense. But then the draws will only be approximately independent, the approximation improving as the number of balls in the bag is increased. So a large finite bag model, though not self-contradictory, will make use of complicated mathematics

each of the n bags. Whether the balls are subsequently replaced in the bag or not is then irrelevant, since once the ball has been drawn the particular bag from which it has been drawn has played its part and disappears from the model. This equivalent model of draws from a set of similar bags has two advantages over that of draws with replacement from one bag: the minor advantage that it is easier to expound the working of a model if each draw is made from a separate bag; the major advantage that with the second model the irrelevance to the model of the order in which the draws are made can be represented by supposing an n-handed Briareus making the n draws simultaneously. With the drawing-with-replacement model the order will have to be deliberately ignored; and there will always be the temptation, which will have to be resisted, of taking the order of draws into account, and thus of treating the set of draws as an ordered series and of falling into one of the errors of the Limiting-Frequency view.

The model I therefore propose for probability theory consists of the following correspondences. To the probability statement that the probability of a child born being a boy is $\frac{51}{100}$ and to the observed class-ratio statement that among 1000 children born 503 are boys, there is to correspond, respectively, a set-up of the model with a set of 1000 bags each containing 100 balls 51 of which are black, and a draw, one from each of the 1000 bags, of 1000 balls 503 of which are black. To the proposition about the observed class-ratio there corresponds not a draw of balls forming a *subclass* of a class whose proportion of black balls is $\frac{51}{100}$, as in the hypothetical-infinite-population model, but a draw of balls forming a *selection* from a class of classes, the proportion of black balls in each of which is $\frac{51}{100}$. The new model allows for the possibility of a draw of an increased size by increasing the number of bags while keeping the number of balls in each bag constant.

This 'Briareus' model will only serve for a probability theory in which every probability is either a proper fraction or 0 or 1. But the higher mathematical methods of probability mathematics, requiring continuous probabilities, can perfectly well be regarded as convenient techniques for arriving at, or approximating to, probabilities which are rational numbers.* So this limitation is

---

* 'Infinite probability fields appear only as idealized schemata of real random processes" (A. Kolmogoroff, op. cit. Chapter II, §1).

not a blemish in the model; on the contrary, the fact that the model requires a restriction of probabilities to rational numbers makes probabilities more analogous to class-ratios (which are subject to this restriction) than they would otherwise be.

There is, however, one feature in the model to which nothing corresponds in the theory. As will be seen, the number, m, of balls in each bag cancels out in the calculations within the model which correspond to deductions in the theory. So m may be taken as any number which is such that pm is an integer (i.e. such that it is possible for there to be exactly a proportion p of the m balls which are black); it is convenient to take it as the smallest number with this property, as has been done in the example just given.*

In the statistical theory which has as premiss that the probability of a child born being a boy is $\frac{51}{100}$, neither the proposition that in a particular observed set of 1000 children born 503 are boys, nor any other proposition asserting an actual class-ratio in any set of observations is deducible. Correspondingly, in the model it does not follow from each of the 1000 bags each containing 100 balls 51 of which are black that, if 1000 balls are drawn, one from each bag, 503 (or any other number) will be black. The distinguishing characteristic of a statistical hypothesis, properly reflected in the model, is just exactly that propositions about observable class-ratios are tests for the falsity of the statistical hypothesis, and hence determine the meaning of the probability statement itself, without being logical consequences of the hypothesis. It is this essential characteristic of statistical hypotheses which must now be investigated. For the significance of probability statements consists in the fact that they occur in deductive systems which are fitted to experience in a looser way than are deductive systems in which probability statements do not occur.

This looser way of fitting will be elaborated in the next chapter. It depends upon the fact that, though propositions about class-ratios in particular selections cannot be deduced from probability statements, yet propositions about the proportions among the possible selections of those selections which have certain class-ratios can be

* As will be seen when the pure theory is developed, for the model to correspond to the theory it is not necessary that the number of balls in each of the bags should be equal, provided that the proportion of black balls in each bag is the same. The numbers $m_1, m_2, ..., m_n$ of balls in the bags cancel out in all the calculations required.

so deduced. To take a very simple example of the bag model: three bags each containing two balls, one black and one white. There are eight possible ways of drawing a selection of one ball from each bag, namely,

BBB   BBW   BWB   BWW   WBB   WBW   WWB   WWW

[one of these eight selections—*WBB*—is shown by the dotted lines in Fig. 6]. Of these eight ways of drawing a selection

1 selection yields a black-ball ratio of 1
3 selections yield a black-ball ratio of $\frac{2}{3}$
3 selections yield a black-ball ratio of $\frac{1}{3}$
1 selection yields a black-ball ratio of 0

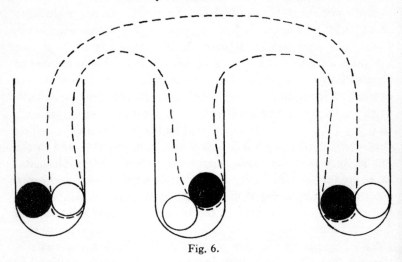

Fig. 6.

Though it is impossible to deduce from the constitution of the bags that a selection having a particular black-ball ratio (e.g. $\frac{2}{3}$) will be drawn, what can be deduced is that the proportion among the selections which might be drawn of those selections whose black-ball ratio is $\frac{2}{3}$ is $\frac{3}{8}$. Similarly, the proportion among the selections of those whose black-ball ratio is $\frac{1}{3}$ is $\frac{3}{8}$; and therefore the proportion among the selections of those whose black-ball ratio deviates from $\frac{1}{2}$ by not more than $\frac{1}{6}$ is $\frac{3}{8} + \frac{3}{8} = \frac{3}{4}$, and the proportion among the selections of those whose black-ball ratio deviates from $\frac{1}{2}$ by more than $\frac{1}{6}$ is $1 - \frac{3}{4} = \frac{1}{4}$. It is small numbers like the $\frac{1}{4}$ calculated in this way that provide the rationale for the interpretation of probability statements in terms of observable class-ratios; and sufficient of the pure mathematics required (which will be called *class-ratio arithmetic*) must be expounded to display this rationale.

THE NOTION OF A RANDOM SAMPLE

Before doing so, there is one point that is liable to cause confusion. What corresponds in the set-of-bags model to the observed set of *n* instances in the theory is a selection of balls drawn simultaneously one from each bag by an *n*-handed Briareus. Writers using as model draws from a bag or set of bags usually require that the draws should be 'random' draws. It is usually not thought sufficient that Briareus should put one hand into each bag and should seize hold of any ball; he must be careful to secure that he is making a 'random' selection from each bag. Now this notion of randomness is a notion which has a perfectly good meaning if probability statements are already understood; a random selection will then be a selection which it was as probable should be made as any alternative selection. But since the model of draws from a bag is being used to elucidate the notion of probability, this explanation of randomness would be circular if the draws had to be specified as being random draws. The truth is that the theory of probability to which the balls-in-bags model has to correspond does not require the selections from the bags to be limited in any way. For what corresponds to the conclusion in the deductive system of the theory are propositions (such as those given in the last paragraph) about the proportions among the possible selections of those selections which have certain properties. And that in calculating these proportions each of the possible selections must be counted once and once only is a proposition which is true by virtue of the definition of "proportion". So it is otiose, and, indeed, misleading, to speak of the selections or samples or draws as being 'random', or as being made from a class of 'equally probable' selections or samples or draws, since the equality of treatment which these epithets demand has already been provided by the set-up of the model. Of course if this set-up will not enable conclusions to be deduced corresponding to conclusions in the theory which can pass the proper test of experience, the set-up will have to be modified. But it will be modified not by making the selections more random but by adjusting the number of black and white balls in the bags. On the view of probability within a science maintained in this book, the adjective "random" applied to a draw in reference to a model, or applied to an experiment to test a statistical theory, has merely the grammatical function of making it clear that the peculiar features of statistical hypotheses are involved in the statements in which the adjective occurs.

CLASS-RATIO ARITHMETIC

We must now treat the pure part of a probability deductive system formally and generally without reference to the balls-in-bags model. What will be given will be the smallest portion of class-ratio arithmetic* which is required for the interpretation of probability statements within a science. One result of elementary algebra will be cited without proof (the number of combinations of n things taken s at a time); but everything else will be proved, though the proofs will not be given in the fully formalized manner of Chapter II.

The fundamental notions are those of what will be called a *probability hyperclass* and of a *selection* from such a hyperclass. These notions are defined as follows:

Two finite non-empty classes β_1, β_2 will be said to be *α-equiproportional* if the proportion among the members of β_1 of those members which are members of a class α (which may be either finite or infinite and may be empty) is the same as the proportion among the members of β_2 of those which are members of α; i.e. if

$$\frac{N'(\alpha\beta_1)}{N'\beta_1} = \frac{N'(\alpha\beta_2)}{N'\beta_2}.$$

Every finite non-empty class is α-equiproportional to itself, whatever the class α may be.

Let n be any positive integer. Consider n finite non-empty classes β_1, β_2, ..., β_n which are (1) all included in a class β (which may be either finite or infinite), (2) mutually exclusive by pairs, (3) α-equiproportional by pairs. From (3) it follows that

$$\frac{N'(\alpha\beta_1)}{N'\beta_1} = \frac{N'(\alpha\beta_2)}{N'\beta_2} = \ldots = \frac{N'(\alpha\beta_n)}{N'\beta_n} = p, \quad \text{say.}$$

These classes β_1, β_2, ..., β_n may be considered as the n members of a class of the second order (which will be called a *hyperclass* and represented by a Greek capital letter).

A hyperclass whose members are a finite number of finite non-

* Sometimes called *combinatorial analysis* or *combinatorics*.

empty classes will be called an $\{\alpha, \beta\}$-*probability hyperclass* if its member-classes are mutually exclusive and α-equiproportional subclasses of β. The proportion p of members of α in each of its member-classes will be called the α-*parameter*, or sometimes the *probability parameter*, of the probability hyperclass. It follows from the definition of p that p is a proper fraction unless either every member-class is included in α, in which case it is 1, or unless every member-class is exclusive to α, in which case it is 0. In all cases p is a rational number and $0 \leqslant p \leqslant 1$.

It should be noticed that, whereas α and β may be infinite classes, every member of a probability hyperclass is by definition a finite class, and the probability hyperclass itself is also by definition a finite class. Since the class-ratio arithmetic will deal only with probability hyperclasses and with their members, finite numbers only will be concerned, and no question of infinite sequences of numbers or of limits of such sequences will arise.

Every subhyperclass of an $\{\alpha, \beta\}$-probability hyperclass will be an $\{\alpha, \beta\}$-probability hyperclass, and every such subhyperclass (except the null-hyperclass, i.e. the hyperclass with no member-classes) will have the same α-parameter, p.

Now consider a class having n members which comprises one and only one member out of each of the n member-classes of a probability hyperclass B. Such a class will be called a *selection* from the probability hyperclass B. Since a selection may be obtained by taking any one of the m_1 members of β_1, any one of the m_2 members of β_2, ..., any one of the m_n members of β_n, the number of selections from the probability hyperclass is $m_1 m_2 ... m_n$. The hyperclass of all the selections from B will be called the *omniselectional hyperclass* of B.* An omniselectional hyperclass is a finite class of the second order like the probability hyperclass which determines it; the latter has n member-classes, the former $m_1 m_2 ... m_n$ member-classes.

---

* "Selection" is the term used by Bertrand Russell in *Introduction to Mathematical Philosophy* (London, 1919), p. 119, where an omniselectional hyperclass is called a "multiplicative class", since it was used by A. N. Whitehead to define multiplication (*American Journal of Mathematics*, vol. 24 (1902), pp. 383ff.). In Whitehead and Russell's *Principia Mathematica* (Cambridge, 1910–13) selections from hyperclasses of classes are derived from the more general notion of selections from relations (see vol. 1, Part 11, Section D). It would be possible to develop class-ratio arithmetic in terms of this more general notion, but there would not be sufficient advantage to compensate for the increased complexity of such a development. See below, pp. 194f.

Let us consider now, instead of the hyperclass of all selections, the hyperclass of those selections which contain exactly s members of α, where $0 \leqslant s \leqslant n$.

Dichotomize the probability hyperclass B into two exclusive subhyperclasses B_1, B_2, which together contain all the members of B. Suppose that B_1 has the s members $\beta_1, \beta_2, ..., \beta_s$; then B_2 will have the $n-s$ members $\beta_{s+1}, \beta_{s+2}, ..., \beta_n$. Consider those selections from B_1 all of whose members are members of α. There are pm_1 members of β_1 which are members of α, pm_2 members of β_2 which are members of α, ..., pm_s members of β_s which are members of α. Since a selection from B_1 all of whose members are members of α is obtained by taking any one of these pm_1 members of β_1, any one of these pm_2 members of β_2, ..., any one of these pm_s members of β_s, the number of such selections is $(pm_1)(pm_2) ... (pm_s)$. Now consider those selections from B_2 none of whose members are members of α. Writing $1 - p$ as q,* there are qm_{s+1} members of β_{s+1} which are not members of α, qm_{s+2} members of β_{s+2} which are not members of α, ..., qm_n members of β_n which are not members of α. So the number of selections from B_2 none of whose members are members of α is $(qm_{s+1})(qm_{s+2}) ... (qm_n)$. Each of these selections from B_2 can be combined with each of the $p^s m_1 m_2 ... m_s$ selections from B_1 to give a selection from B which contains s members of α, one from each of the s member-classes of B_1, and $n-s$ members of α'—the complement of α, i.e. the class of things which are not members of α—one from each of the $n-s$ member-classes of B_2. So the number of such selections from B is

$$(pm_1)(pm_2) ... (pm_s)(qm_{s+1})(qm_{s+2}) ... (qm_n),$$

i.e. $p^s q^{n-s} m_1 m_2 ... m_n$.

However, unless s is 0 or is n, there is more than one way of dichotomizing B into two subhyperclasses one of which contains s members and the other $n-s$ members. The number of such dichotomies is the number of combinations of n things taken s at a time (indeed to be a way of dichotomizing is the definition of a combination) ^nC_s, and this can be proved to be $\dfrac{n!}{s!(n-s)!}$.†

---

* q, like p, is either a proper fraction or is 0 or is 1.

† $n! = 1.2.3 ..... n$. In old-fashioned books $n!$ is written $\underline{|n}$. In modern books ^nC_s is frequently written $\dbinom{n}{s}$. $0!$ is defined to be equal to 1; so ^nC_n, $^nC_0$, $^0C_0$ are all equal to 1.

Consequently in order to obtain the number of selections from B containing s members of α taken from members of any s-fold subhyperclass of B together with $n-s$ members of α' taken from the other member-classes of B, we must multiply $p^s q^{n-s} m_1 m_2 \dots m_n$ by ^nC_s. But these selections are all the selections from B which contain exactly s members of α; so the number of such selections is therefore

$$^nC_s p^s q^{n-s} m_1 m_2 \dots m_n.$$

The proportion within the hyperclass of all selections from B—the omniselectional hyperclass of B—of those selections which contain exactly s members of α is therefore obtained by dividing this number by the number of members of the omniselectional hyperclass, which is $m_1 m_2 \dots m_n$. So the proportion out of all the selections from B of those selections which comprise exactly s members of α is

$$\frac{^nC_s p^s q^{n-s} m_1 m_2 \dots m_n}{m_1 m_2 \dots m_n} = {}^nC_s p^s q^{n-s}.$$

The remarkable thing about this theorem is not so much what numbers it is about as what numbers it is not about. The numbers $m_1, m_2, \dots, m_n$—the sizes of the member-classes of the probability hyperclass—have cancelled out; and we have the result that the proportion of selections from a probability hyperclass which comprise exactly s members of α depends upon two and only two numbers associated with the hyperclass—its α-parameter p and its number of member-classes n. The proportion depends in no way upon the sizes of these member-classes, only upon how many of them the hyperclass contains. Since the further theorems to be proved are deduced from this theorem and also make no reference to $m_1, m_2, \dots, m_n$, in applying this pure deductive system it will not be necessary to attach definite empirically given values to $m_1, m_2, \dots, m_n$. They can be used as uninterpreted symbols, like the phases of the wave-equation in quantum mechanics.

To proceed with the class-ratio arithmetic we require: since a selection which comprises exactly s members of α does not comprise exactly t members of α, where $s \neq t$, the proportion within the omniselectional hyperclass of those selections which contain a number of members of α which either is s_1 or is s_2 or is some number lying between s_1 and s_2, where $0 \leqslant s_1 < s_2 \leqslant n$, is calculated by adding together the values of the proportions for each value of s lying in the interval $s_1 \leqslant s \leqslant s_2$. This sum is

$$^nC_{s_1}p^{s_1}q^{n-s_1}+{}^nC_{s_1+1}p^{s_1+1}q^{n-(s_1+1)}+\ \ldots\ +{}^nC_{s_2}p^{s_2}q^{n-s_2},$$

which will be abbreviated into $\sum\limits_{s=s_1}^{s_2}{}^nC_sp^sq^{n-s}$.* Since every selection contains either 0 or 1 or 2 or ... n members of α, the proportion within the omniselectional hyperclass of those selections which contain either 0 or 1 or 2 or ... n members of α is 1, and $\sum\limits_{s=0}^{n}{}^nC_sp^sq^{n-s}=1$. This is a special case of the binomial theorem of ordinary algebra.

The average number of members of α per selection is calculated by multiplying each number from 0 to n inclusive by the proportion of selections which comprise exactly that number of members of α, and adding together these weighted proportions. This average number is thus $\sum\limits_{s=0}^{n}s\,.\,{}^nC_sp^sq^{n-s}$, which is proved in the Appendix to this chapter (Theorem II) to be np. Since each selection has n members, the average proportion of members of α per selection is $np/n=p$.

What we wish to calculate for our logical purposes is the proportion within the omniselectional hyperclass of those selections whose number s of members of α lies outside a certain interval surrounding this average number np. That is, we wish to calculate

$$\sum_{s=0}^{n_1}{}^nC_sp^sq^{n-s}+\sum_{s=n_2}^{n}{}^nC_sp^sq^{n-s},$$

where n_1 is less than the average number np by one specified amount and n_2 is greater than np by another specified amount, the interval outside which the number of members of the required selection fall being the interval from n_1 to n_2. A general method for calculating approximately this (and any similar) proportion was propounded by De Moivre (1733) and developed by Laplace (1812). This method requires use of the infinitesimal calculus and will not be expounded here. Instead a simple method, due originally to Bienaymé (1853) but independently elaborated by Tchebichef (1867), will be used to

---

* $\sum\limits_{s=s_1}^{s_2}f(s)$, where $f(s)$ is any function of s, has only been defined when $s_1<s_2$. It is convenient to define it when $s_1=s_2$ as equal to $f(s_1)$ and when $s_1>s_2$ as equal to 0.

provide, not an approximation to the value itself, but an upper bound which the value cannot exceed. And this will suffice for our purpose.

The form in which it will be most convenient to assert the theorem, which is proved in the Appendix to this chapter (Theorem IV), is the following one:

Take any positive number k, however small. Consider the interval* stretching from $np - \sqrt{(npq/k)}$ (its left-hand point) to $np + \sqrt{(npq/k)}$ (its right-hand point); i.e. the interval whose centre point is np and whose total length is $2\sqrt{(npq/k)}$. This interval will be expressed as the interval $[np - \sqrt{(npq/k)}, np + \sqrt{(npq/k)}]$. Let n_1 be the greatest integer less than the left-hand point of the interval, n_2 be the least integer greater than its right-hand point. Then the theorem states that

$$\sum_{s=0}^{n_1} {}^nC_s p^s q^{n-s} + \sum_{s=n_2}^{n} {}^nC_s p^s q^{n-s} < k,$$

i.e. the proportion among all the selections of those selections whose number of members of α lies outside the interval

$$[np - \sqrt{(npq/k)}, \ np + \sqrt{(npq/k)}]$$

is less than k, however small a positive number k may be.

The theorem may be restated in a form which is even more convenient for the use which we shall make of it. Instead of considering s, the number of members of α in an n-fold selection, let us consider $r = s/n$, which is the ratio of the number of members of α to the total number of members of the n-fold selection. r will be called the α-ratio of the selection. Since $0 \leqslant s \leqslant n$, $0 \leqslant r \leqslant 1$. Then to say that s lies between $np - \sqrt{(npq/k)}$ and $np + \sqrt{(npq/k)}$ is equivalent to saying that r lies between $p - \sqrt{(pq/nk)}$ and $p + \sqrt{(pq/nk)}$. So the theorem may be restated in the form that *the proportion among all the selections of those selections whose α-ratio lies outside the interval $[p - \sqrt{(pq/nk)}, p + \sqrt{(pq/nk)}]$ is less than k, however small a positive number k may be.* If the amount by which the α-ratio of a selection differs from p be called its *deviation* from p, the proportion among the selections of those selections whose α-ratio deviates from p by more than $\sqrt{(pq/nk)}$ is less than k; and the complementary

* An interval will always be taken to include its end-points; all our intervals will thus be what mathematicians call "closed intervals". 'Within' an interval or 'within' an ellipse or lying 'between' two points will always be taken to include the boundary points or lines.

proportion among the selections of those selections whose α-ratio deviates from p by not more than $\sqrt{(pq/nk)}$ is greater than $1-k$, however small a positive number k may be.

This theorem in the pure deductive system of class-ratio arithmetic is the keystone which holds together the account of the meaning of probability statements to be propounded in the next chapter. For this reason a rigorous proof of it, which uses only elementary algebra, is given in the Appendix to this chapter. Because of its fundamental importance it is desirable that what the theorem states and what are its immediate consequences should be fully appreciated. One way of doing this is to represent what is stated in the theorem in a graphical diagram. This can be done in an elegant manner by a representation devised by Jerzy Neyman for a slightly different purpose, and this representation will be of great assistance to us in the exposition of probability theory in the next chapter.

THE SYSTEM OF PROBABILITY ELLIPSES

Take rectangular cartesian coordinates, and plot the values of r— the α-ratio s/n in an n-fold selection—along the horizontal axis stretching from the points o to 1 inclusive, and plot the values of p— the α-parameter of the probability hyperclass—along the vertical axis from the points o to 1 inclusive. Complete the square by drawing a horizontal line from the point with coordinates $r=0$, $p=1$ to the point $r=1$, $p=1$, and a vertical line from the point $r=1$, $p=0$ to the point $r=1$, $p=1$ (see Fig. 7). Any point lying within the square with a rational p-coordinate and an r-coordinate which is a multiple of $1/n$ will represent a possible combination of a value of p with a value of r. Now draw the system of ellipses given by the equation

$$(r-p)^2 = \frac{1}{b}\left[p(1-p)\right],$$

where b is any positive number.* [To avoid confusion in the diagram only two of these ellipses have been drawn, the outer ellipse being that for which $b=4$, the inner ellipse being that for which $b=25$.] There are an unlimited number of these ellipses; they all touch one another and the straight line which is given by

* A summary of the geometrical properties of these ellipses is given at the end of the Appendix to this chapter.

the equation $p=1$ at the top right-hand corner of the square—the
corner $(1, 1)$—and they also touch one another and the straight line
$p=0$ at the bottom left-hand corner of the square—the corner $(0, 0)$.
For any given value of b, say b_1, all the ellipses with values of b
greater than b_1 lie inside the b_1-ellipse; thus the ellipses form a

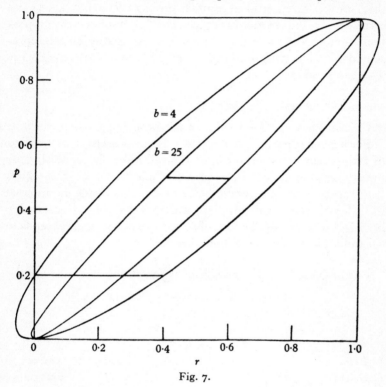

Fig. 7.

'Chinese-box' system getting flatter and flatter as b increases. As
b increases without limit, the ellipses flatten themselves on to the
diagonal straight line whose equation is $r=p$.

For any given values of b and of p, say b_1 and p_1, the interval
$[p_1 - \sqrt{\{p_1(1 -p_1)/b_1\}}, p_1 + \sqrt{\{p_1(1 -p_1)/b_1\}}]$ will be represented by the
part of the straight line $p=p_1$ lying within the b_1-ellipse.* [In the

* The values for r for given values b_1 of b and p_1 of p are obtained by solving
(for r) the quadratic equation
$$(r-p_1)^2=p_1(1 -p_1)/b_1,$$
whose solutions are $p_1 \pm \sqrt{\{p_1(1 -p_1)/b_1\}}$.

diagram two straight lines have been drawn to represent these intervals in the cases $b = 25$, $p = \frac{1}{2}$ and $b = 4$, $p = \frac{1}{5}$.] So for a given value b_1 of b, the region within the b_1 ellipse will represent the whole set of these intervals for all values of p lying between 0 and 1.

The Bienaymé-Tchebichef theorem may now be put in the form that, for any given rational number p lying between 0 and 1, any given positive integer n and any given positive number k, the proportion among all the selections of those selections which are represented by points lying within the square but outside the b-ellipse is less than k, where $b = nk$.

THE LAW OF GREAT NUMBERS

Consider any one fixed value of the probability parameter p.

For a given value k_1 of k, the intervals outside which the α-ratios of fewer than a proportion k_1 of the selections lie diminish continually as n increases; they form a Chinese-box set of intervals with the interval for a greater value of n always inside the interval for a lesser value of n. And since the size of the interval varies inversely with the square root of n, the size of the interval tends to the limit 0 as n increases without limit.

TABLE IV. *Values of pq/nd^2 for different values of n and of d*

(Spaces left blank when $pq/nd^2 > 1$.)

Case in which $p = \frac{1}{2}$, $q = \frac{1}{2}$.

$n =$	100	1000	10,000	100,000	1,000,000
$d = 0 \cdot 1$	0·25	0·025	0·0025	0·00025	0·000025
$d = 0 \cdot 01$	—	—	0·25	0·025	0·0025
$d = 0 \cdot 001$	—	—	—	—	0·25

Case in which $p = \frac{1}{5}$, $q = \frac{4}{5}$.

$n =$	100	1000	10,000	100,000	1,000,000
$d = 0 \cdot 1$	0·16	0·016	0·0016	0·00016	0·000016
$d = 0 \cdot 01$	—	—	0·16	0·016	0·0016
$d = 0 \cdot 001$	—	—	—	—	0·16

If now another value k_2 of k be taken which is less than k_1, the intervals form a similar Chinese-box set with the interval which for the k_1 case corresponded to n now corresponding to $(k_1/k_2)n$. The Chinese-box set is the same, but its n-numbering is k_1/k_2 times as great.

So if, instead of fixing k, we fix the size of the interval with p as centre, we can find an upper bound for the proportion of selections whose α-ratio falls outside this fixed interval. This upper bound is given by the theorem, proved in the Appendix to this chapter (Theorem VI), that the proportion among all the selections of those selections whose α-ratio lies outside the interval $[p-d, p+d]$, d being any positive number however small, is not greater than pq/nd^2.

Two numerical examples of this theorem are given in Table IV.

From this theorem there follows immediately what in books on the theory of probability is usually called the Law of Great Numbers. For the upper bound of the proportion pq/nd^2 tends to the limit o as n increases without limit, and consequently the proportion of selections whose α-ratio deviates from p by more than d (which proportion cannot be greater than pq/nd^2) also tends to the limit o as n increases without limit. The complementary proportion among the selections of those whose α-ratio deviates from p by not more than d (which proportion cannot be less than $1-pq/nd^2$) tends to the limit 1 as n increases without limit. This statement is the expression in terms of the notion of mathematical limit of the fact that, however small an interval with centre at p is taken, a number n can be found which is such that the proportion in n-fold selections of those selections whose α-ratio falls within this interval approaches 1 as closely as may be desired. If "almost all" be taken as meaning all but an arbitrarily small proportion k, and "nearly equal to p" be taken as meaning deviating by not more than an arbitrarily small amount d from p, the Law of Great Numbers can be expressed by saying that, if n is taken to be large enough,* *almost all n-fold selections have α-ratios nearly equal to p.*

Gamblers who apply this mathematical result to games of chance frequently simplify it in two erroneous ways. (1) The "almost" before the "all n-fold selections" is omitted, so that the theorem is supposed to state that, if n is taken to be large enough, *all* the n-fold selections have α-ratios nearly equal to p. This is obviously false except in the two extreme cases $p = 1$ and $p = 0$, when it is obviously true, since in these two cases all n-fold selections have

---

* Taken, in fact, to be larger than pq/kd^2.

α-ratios exactly equal to 1 or equal to 0, whatever positive number n may be. (2) The "nearly" before the "equal to p" is omitted, so that the theorem is supposed to assert that, if n is taken to be large enough, almost all the n-fold selections have α-ratios *exactly equal to p*. This, like the first simplification, is obviously true in the two extreme cases. In other cases it is obviously false if np is not an integer; if np is an integer, the proportion of such selections is $^nC_{np}p^{np}q^{nq}$, and this can be shown to tend to the limit 0 as n increases without limit.[*] So, in fact, a contrary theorem holds that (unless p is 0 or 1) if n is large enough, almost none of the n-fold selections have α-ratios exactly equal to p. One or other or both of these two mathematical errors lie at the root of most 'infallible gambling systems'.

APPENDIX

In all the following proofs l is to be restricted to being a non-negative integer, n to being a positive integer, p and q to being non-negative rational numbers subject to the further restriction $p+q=1$.

Theorem I has been proved in the preceding chapter for the case in which $l>0$; its extension to the case in which $l=0$ is required for the proofs of Theorems II and III, which are steps in the proof of Theorem IV.

THEOREM I. [The binomial expansion of $(q+p)^l$ for a non-negative integral exponent for the case in which p is rational, $0 \leqslant p \leqslant 1$, $q=1-p$.]

$$\sum_{s=0}^{l} {}^lC_s p^s q^{l-s} = 1.$$

This has been proved (p. 138) when l is a positive integer; when $l=0$,

$$\sum_{s=0}^{l} {}^lC_s p^s q^{l-s} = {}^0C_0 p^0 q^0 = 1 ;[†]$$

and the theorem is proved whenever l is a non-negative integer.

THEOREM II. [The first moment about zero (the mean) in a binomial distribution.]

$$\sum_{s=0}^{l} s \cdot {}^lC_s p^s q^{l-s} = lp.$$

---

[*] Since this result is not used in the sequel, its proof, which is not simple, is not given.

[†] $a^0 = 1$ for every a (including 0).

Proof. Consider first the case in which l is a positive integer:

$$\sum_{s=0}^{l} s \cdot {}^{l}C_{s} p^{s} q^{l-s} = 0 \cdot q^{l} + 1 \cdot lpq^{l-1} + 2 \cdot \frac{l(l-1)}{1 \cdot 2} p^{2} q^{l-2}$$

$$+ 3 \cdot \frac{l(l-1)(l-2)}{1 \cdot 2 \cdot 3} p^{3} q^{l-3} + \ldots + l \cdot p^{l}$$

$$= lp \left[q^{l-1} + (l-1) pq^{l-2} + \frac{(l-1)(l-2)}{1 \cdot 2} p^{2} q^{l-3} + \ldots + p^{l-1} \right]$$

$$= lp \cdot \sum_{s=0}^{l-1} {}^{l-1}C_{s} p^{s} q^{(l-1)-s}$$

$$= lp,$$

by Theorem I, since $l-1$ is a non-negative integer.

Now consider the case in which $l=0$:

$$\sum_{s=0}^{l} s \cdot {}^{l}C_{s} p^{s} q^{l-s} = 0 \cdot {}^{0}C_{0} p^{0} q^{0} = 0 = lp;$$

and the theorem is proved whenever l is a non-negative integer.

THEOREM III. [The second moment about the mean (the variance, the square of the standard deviation) in a binomial distribution.]

$$\sum_{s=0}^{n} (np-s)^{2} \cdot {}^{n}C_{s} p^{s} q^{n-s} = npq.$$

Proof.

$$\sum_{s=0}^{n} (np-s)^{2} \cdot {}^{n}C_{s} p^{s} q^{n-s} = n^{2} p^{2} \sum_{s=0}^{n} {}^{n}C_{s} p^{s} q^{n-s} - 2np \sum_{s=0}^{n} s \cdot {}^{n}C_{s} p^{s} q^{n-s}$$

$$+ \sum_{s=0}^{n} s^{2} \cdot {}^{n}C_{s} p^{s} q^{n-s}$$

$$= n^{2} p^{2} \cdot 1 - 2np \cdot np + \sum_{s=0}^{n} s^{2} \cdot {}^{n}C_{s} p^{s} q^{n-s},$$

by Theorems I and II.

$$\sum_{s=0}^{n} s^{2} \cdot {}^{n}C_{s} p^{s} q^{n-s} = 0^{2} \cdot q^{n} + 1^{2} \cdot npq^{n-1} + 2^{2} \cdot \frac{n(n-1)}{1 \cdot 2} p^{2} q^{n-2}$$

$$+ 3^{2} \cdot \frac{n(n-1)(n-2)}{1 \cdot 2 \cdot 3} p^{3} q^{n-3} + \ldots + n^{2} \cdot p^{n}$$

$$= np \left[q^{n-1} + 2 \cdot (n-1) pq^{n-2} + 3 \cdot \frac{(n-1)(n-2)}{1 \cdot 2} p^{2} q^{n-3} \right.$$

$$\left. + \ldots + n \cdot p^{n-1} \right]$$

$$= np \left[\left\{ q^{n-1} + (n-1)pq^{n-2} + \frac{(n-1)(n-2)}{1 \cdot 2} p^2 q^{n-3} \right. \right.$$

$$+ \dots + p^{n-1} \Big\} + \Big\{ 0 \cdot q^{n-1} + 1 \cdot (n-1)pq^{n-2}$$

$$\left. \left. + 2 \cdot \frac{(n-1)(n-2)}{1 \cdot 2} p^2 q^{n-3} + \dots + (n-1) \cdot p^{n-1} \right\} \right]$$

$$= np \left[\sum_{s=0}^{n-1} {}^{n-1}C_s p^s q^{(n-1)-s} + \sum_{s=0}^{n-1} s \cdot {}^{n-1}C_s p^s q^{(n-1)-s} \right]$$

$$= np[1 + (n-1)p],$$

by Theorems I and II, since $n-1$ is a non-negative integer.

Hence
$$\sum_{s=0}^{n} (np-s)^2 \cdot {}^nC_s p^s q^{n-s} = n^2 p^2 - 2n^2 p^2 + np + n^2 p^2 - np^2$$

$$= np(1-p)$$

$$= npq.$$

[Theorem III also holds when $n = 0$, since in that case

$$\sum_{s=0}^{n} (np-s)^2 \cdot {}^nC_s p^s q^{n-s} = 0^2 \cdot {}^0C_0 p^0 q^0 = 0 = npq.$$

But this extension is not required in the sequel.]

THEOREM IV. [A form of the Bienaymé-Tchebichef inequality in the case of a binomial distribution.]

If m_1 is the greatest integer less than $np - \sqrt{(npq/k)}$ and m_2 is the least integer greater than $np + \sqrt{(npq/k)}$, where k is any positive number, then
$$\sum_{s=0}^{m_1} {}^nC_s p^s q^{n-s} + \sum_{s=m_2}^{n} {}^nC_s p^s q^{n-s} < k.$$

Proof. Consider first the case in which $p > 0$, $q > 0$. Each of the sums
$$\sum_{s=0}^{m_1} {}^nC_s p^s q^{n-s}, \quad \sum_{s=m_2}^{n} {}^nC_s p^s q^{n-s}$$

is either zero or is positive.

If they are both zero, the theorem is proved.

Suppose that one or other (or both) of these are positive.

$$\sum_{s=0}^{m_1} {}^nC_s p^s q^{n-s} + \sum_{s=m_2}^{n} {}^nC_s p^s q^{n-s}$$

$$< \frac{k}{npq} \left[(np-m_1)^2 \cdot \sum_{s=0}^{m_1} {}^nC_s p^s q^{n-s} + (np-m_2)^2 \cdot \sum_{s=m_2}^{n} {}^nC_s p^s q^{n-s} \right],$$

since $(np-m_1)^2 > \frac{npq}{k}$ and $(np-m_2)^2 > \frac{npq}{k}$.

Hence
$$\sum_{s=0}^{m_1} {}^nC_s p^s q^{n-s} + \sum_{s=m_2}^{n} {}^nC_s p^s q^{n-s}$$

$$< \frac{k}{npq} \left[\sum_{s=0}^{m_1} (np-s)^2 . {}^nC_s p^s q^{n-s} + \sum_{s=m_2}^{n} (np-s)^2 . {}^nC_s p^s q^{n-s} \right],$$

since $(np-s)^2 \geqslant (np-m_1)^2$ for $s = 0, 1, 2, \ldots, m_1,$

and $(np-s)^2 \geqslant (np-m_2)^2$ for $s = m_2, m_2+1, m_2+2, \ldots, n.$

Hence
$$\sum_{s=0}^{m_1} {}^nC_s p^s q^{n-s} + \sum_{s=m_2}^{n} {}^nC_s p^s q^{n-s}$$

$$< \frac{k}{npq} \left[\sum_{s=0}^{n} (np-s)^2 . {}^nC_s p^s q^{n-s} \right],$$

since $\sum_{s=m_1+1}^{m_2-1} (np-s)^2 . {}^nC_s p^s q^{n-s} \geqslant 0.$

Hence
$$\sum_{s=0}^{m_1} {}^nC_s p^s q^{n-s} + \sum_{s=m_2}^{n} {}^nC_s p^s q^{n-s} < \frac{k}{npq} . npq, \quad \text{by Theorem III;}$$

and
$$\sum_{s=0}^{m_1} {}^nC_s p^s q^{n-s} + \sum_{s=m_2}^{n} {}^nC_s p^s q^{n-s} < k.$$

Now consider the case in which $p = 0, q = 1.$ Since in this case $np - \sqrt{(npq/k)} = np + \sqrt{(npq/k)} = 0,$ $m_1 = -1$ and $m_2 = 1$:

$$\sum_{s=0}^{m_1} {}^nC_s p^s q^{n-s} = 0, \quad \text{since } m_1 < 0;*$$

and
$$\sum_{s=m_2}^{n} {}^nC_s p^s q^{n-s} = 0, \quad \text{since } 0^s = 0 \text{ for } s = 1, 2, \ldots, n.$$

Hence $\sum_{s=0}^{m_1} {}^nC_s p^s q^{n-s} + \sum_{s=m_2}^{n} {}^nC_s p^s q^{n-s} = 0 < k.$

Finally, consider the case in which $p = 1, q = 0.$ Since in this case $np - \sqrt{(npq/k)} = np + \sqrt{(npq/k)} = n,$ $m_1 = n-1$ and $m_2 = n+1$:

$$\sum_{s=0}^{m_1} {}^nC_s p^s q^{n-s} = 0, \quad \text{since } 0^{n-s} = 0 \text{ for } s = 0, 1, 2, \ldots, n-1;$$

and
$$\sum_{s=m_2}^{n} {}^nC_s p^s q^{n-s} = 0, \quad \text{since } m_2 > n.$$

Hence $\sum_{s=0}^{m_1} {}^nC_s p^s q^{n-s} + \sum_{s=m_2}^{n} {}^nC_s p^s q^{n-s} = 0 < k;$

and the theorem is proved in all cases.

* $\sum_{s=s_1}^{s_2} f(s)$ has been defined (p. 138) as equal to 0 when $s_1 > s_2.$

THEOREM V. [A corollary to Theorem IV which will be used in Chapter VI.]

If $p \leqslant \frac{1}{2}$ and m_2 is the least integer greater than $np + \sqrt{(npq/k)}$, where k is any positive number, then, for every number y which is such that $0 \leqslant y \leqslant p$,

$$\sum_{s=m_2}^{n} {}^nC_s y^s (1-y)^{n-s} < k.$$

Proof. Let m_3 be the greatest integer less than $ny - \sqrt{\{ny(1-y)/k\}}$ and m_4 be the least integer greater than $ny + \sqrt{\{ny(1-y)/k\}}$.

$$\begin{aligned} pq - y(1-y) &= p(1-p) - y(1-y) \\ &= (p-y) - (p^2 - y^2) \\ &= (p-y)[1 - (p+y)] \\ &\geqslant 0, \quad \text{since } y \leqslant p \leqslant \tfrac{1}{2}. \end{aligned}$$

Hence $$pq \geqslant y(1-y)$$

and $$np + \sqrt{\frac{npq}{k}} \geqslant ny + \sqrt{\frac{ny(1-y)}{k}};$$

whence $m_2 \geqslant m_4$.

$$\begin{aligned} \sum_{s=m_2}^{n} {}^nC_s y^s (1-y)^{n-s} &\leqslant \sum_{s=m_4}^{n} {}^nC_s y^s (1-y)^{n-s} \\ &\leqslant \sum_{s=0}^{m_3} {}^nC_s y^s (1-y)^{n-s} + \sum_{s=m_4}^{n} {}^nC_s y^s (1-y)^{n-s} \\ &< k, \quad \text{by Theorem IV.} \end{aligned}$$

[The restriction $p \leqslant \frac{1}{2}$ is unnecessary; but the proof is more complicated without it, and depends upon the fact that $ny + \sqrt{\{ny(1-y)/k\}}$ is an increasing function of y for $0 \leqslant ny + \sqrt{\{ny(1-y)/k\}} \leqslant n$.]

THEOREM VI. [Another form of the Bienaymé-Tchebichef inequality in the case of a binomial distribution.]

If n_1 is the greatest integer less than $n(p-d)$ and n_2 is the least integer greater than $n(p+d)$, where d is any positive number, then

$$\sum_{s=0}^{n_1} {}^nC_s p^s q^{n-s} + \sum_{s=n_2}^{n} {}^nC_s p^s q^{n-s} \leqslant \frac{pq}{nd^2}.$$

Proof. Consider first the case in which $p > 0$, $q > 0$. Take any positive number d. Let k be pq/nd^2, which is a positive number whenever $p > 0$, $q \geqslant 0$. d is then the positive square root of pq/nk, i.e. $\sqrt{(pq/nk)}$; and $n(p-d) = np - \sqrt{(npq/k)}$, $n(p+d) = np + \sqrt{(npq/k)}$.

The conditions of Theorem IV are thus satisfied for $k = pq/nd^2$; hence

$$\sum_{s=0}^{n_1} {}^nC_s p^s q^{n-s} + \sum_{s=n_2}^{n} {}^nC_s p^s q^{n-s} < k$$

$$< \frac{pq}{nd^2}.$$

Now consider the case in which $p = 0$, $q = 1$. In this case, since $n_1 < -nd$, $n_1 < 0$ and, since $n_2 > nd$, $n_2 > 0$:

$$\sum_{s=0}^{n_1} {}^nC_s p^s q^{n-s} = 0, \quad \text{since } n_1 < 0;$$

and

$$\sum_{s=n_2}^{n} {}^nC_s p^s q^{n-s} = 0, \quad \text{since } 0^s = 0 \text{ for } s \geqslant n_2 \geqslant 0.$$

Hence

$$\sum_{s=0}^{n_1} {}^nC_s p^s q^{n-s} + \sum_{s=n_2}^{n} {}^nC_s p^s q^{n-s} = 0 \leqslant \frac{pq}{nd^2}.$$

Finally, consider the case in which $p = 1$, $q = 0$. In this case, since $n_1 < n - nd$, $n_1 < n$ and, since $n_2 > n + nd$, $n_2 > n$:

$$\sum_{s=0}^{n_1} {}^nC_s p^s q^{n-s} = 0, \quad \text{since } 0^{n-s} = 0 \text{ for } s \leqslant n_1 < n;$$

and

$$\sum_{s=n_2}^{n} {}^nC_s p^s q^{n-s} = 0, \quad \text{since } n_2 > n.$$

Hence

$$\sum_{s=0}^{n_1} {}^nC_s p^s q^{n-s} + \sum_{s=n_2}^{n} {}^nC_s p^s q^{n-s} = 0 \leqslant \frac{pq}{nd^2};$$

and the theorem is proved in all cases.

[The proof shows that

$$\sum_{s=0}^{n_1} {}^nC_s p^s q^{n-s} + \sum_{s=n_2}^{n} {}^nC_s p^s q^{n-s} < \frac{pq}{nd^2}$$

unless either $p = 0$ or $p = 1$.]

Note. None of the proofs of these theorems depend upon the restriction of p and q to being rational numbers. Theorems without this restriction follow by exactly the same proofs from the Binomial Theorem for positive integral exponent of ordinary algebra, in which this restriction does not occur.

THE PROBABILITY ELLIPSES

Properties of the system of conics given by the equation

$$(r - p)^2 = \frac{1}{b} [p(1 - p)],$$

i.e.

$$br^2 - 2brp + (b + 1)p^2 - p = 0.$$

149

If $b > 0$, these conics are all ellipses, since the discriminant

$$b(b+1) - b^2 = b > 0.$$

The solutions for r in terms of p are

$$r = p \pm \sqrt{\frac{p(1-p)}{b}}.$$

The solutions for p in terms of r are

$$p = \frac{2br + 1 \pm \sqrt{\{4br(1-r)+1\}}}{2(b+1)}.$$

When $p = 0$, $r = 0$ (twice), and when $p = 1$, $r = 1$ (twice); so all the ellipses touch the line $p = 0$ (and therefore touch each other) at the point $(0,0)$, and all touch the line $p = 1$ (and therefore touch each other) at the point $(1, 1)$.

When $r = 0$, $p = 0$ or $\dfrac{1}{b+1}$.

When $r = 1$, $p = 1$ or $\dfrac{b}{b+1}$.

CHAPTER VI

THE MEANING OF PROBABILITY STATEMENTS WITHIN A SCIENTIFIC SYSTEM

The deductive system of class-ratios developed in the previous chapter is a pure deductive system in which all the theorems are of the hypothetical form: If a probability hyperclass has a certain probability parameter, then various consequences follow about proportions in a related hyperclass. In order to incorporate this pure deductive system into an applied system containing logically contingent propositions which may be confronted with experience, we must add to the pure system a proposition assigning a definite value to the probability parameter of the probability hyperclass. We shall then be able to deduce within this applied system propositions about classes related in various ways to the probability hyperclass.

In this chapter probability statements will be interpreted as being propositions assigning values to probability parameters of probability hyperclasses, and these propositions (statistical hypotheses) will function as highest-level hypotheses in an applied deductive system. The peculiarity of these statistical hypotheses is that they are not conclusively refutable by any experience; as we shall see, empirical criteria can be given for rejecting the hypotheses, but the rejections will always be provisional and liable to cancellation. The epistemological function of the deductive system determining the meaning of a statistical hypothesis which serves as highest-level hypothesis is different from that of a scientific deductive system in which non-statistical hypotheses about theoretical concepts are higher-level hypotheses. In the latter it is the lowest-level hypotheses that determine the meaning of the statements containing the theoretical term. In the former the meaning of the probability statements expressing the statistical hypotheses is given directly by an extended rejection rule; but the reason for using this rejection rule depends upon the working of the statistical hypothesis as the highest-level hypothesis in a probability deductive system. The examination of the meaning of probability statements is thus distinct from that of statements containing theoretical terms, and requires a separate chapter for its exposition.

UNIVERSAL HYPOTHESES AND STATISTICAL HYPOTHESES

A necessary preliminary to such an examination is to realize that all general statements are in fact probability statements, since to say that all A's are B's is the same thing as to say that 100% of A's are B's, which is (on my thesis) the same thing as to say that the probability of an A being a B is 1. Similarly, to say that no A's are B's is to say that 0% of A's are B's and to say that the probability of an A being a B is 0. Universal generalizations, whether affirmative or negative, are special cases of probability statements. The fact that, unlike other probability statements, they can be logically refuted by experience, arises not from their not being probability statements but from the probabilities which they ascribe taking the extreme values 1 and 0.*

So correct doctrine as to the logical functioning of probability statements in probability deductive systems must apply also to universal generalizations as limiting cases. And this may guide us in the direction in which to seek a solution.

A sentence expressing an empirical universal generalization on the lowest hypothesis level is, as we have seen, given its meaning by its position in a calculus interpreted as expressing an applied deductive system. For example, the sentence "All births are male births" means exactly that proposition (of minimum logical strength) which is such that, from the conjunction of it and the proposition that any particular event is a birth, there logically follows the proposition that that particular event is the birth of a boy. But to say that a proposition c is a logical consequence of

* The identification of probability statements ascribing probabilities of 1 or of 0 with universal generalizations will be disputed by those who wish to use an infinite-bag model for their probability theory. If, it is said, there is at least one black ball, but not an infinite number of black balls, in the bag containing an infinite number of balls, this fact about the model will correspond to a zero probability of an A being a B, but it will not correspond to no A being a B. My reply is that this is an additional reason for the inappropriateness of the infinite-bag model, since, in the way in which probability is used within a science, ascriptions of zero probability are taken to be indistinguishable from negative universal generalizations. The pure mathematicians who have subsumed class-ratio arithmetic under the more general theory of measure of infinite sets of points deal with the difficulty by such devices as omitting from consideration sets of points whose 'measure' is zero.

the conjunction of two propositions a and b is equivalent to saying that a is logically incompatible with the conjunction of b with not-c. Hence the meaning of the sentence "All births are male births" may equally well be taken to be determined as being that proposition (of minimum logical strength) which would be falsified by any birth being found which is not the birth of a boy.* The empirical meaning of the sentence is then given by the observations on the basis of which the sentence is taken to express a falsehood; the proposition expressed by the sentence is specified by the observations on the basis of which it is to be *rejected*. The *rule of rejection* which gives the empirical meaning of "All A's are B's" is: On the basis of an observation of a particular A reject the proposition expressed by the sentence "All A's are B's" if and only if the particular A observed is not a B. It is logically possible for any observed A either to be B or not to be B; the hypothesis that every A is B reduces these two possibilities to one; so if that unique possibility is not actualized, the hypothesis is false. If I follow this rule of rejection with respect to a universal generalization, I shall never reject a universal generalization which is true, and so shall never make a *mistaken rejection*.

A RULE OF REJECTION FOR PROBABILITY STATEMENTS

Consider now the sentence "The probability of a birth being the birth of a male is $\frac{51}{100}$" or, as it is synonymously expressed, "51% of births are male births"—a case of the general form of proposition expressed by "The probability of a member of β being a member of α is p", where p denotes a proper fraction, i.e. a rational number between 0 and 1. This form of sentence is given a meaning by the following rule of rejection (to be called a *k-rule-of-rejection*): Choose any small positive number k (e.g. $\frac{1}{20}$). On the basis of a set of n observations of members of β, reject the hypothesis that the probability is p that a member of β is a member of α if the α-ratio in this set of n observations (i.e. the number of members of α in

---

* Why this second, alternative way of giving a meaning to a universal statement was not mentioned when the first way was discussed (pp. 83 f.) was because I did not wish to complicate the calculus there used for discussing the matter by adding to it a symbol, and rules governing it, which would express the logic of negation. This second way of giving meaning to general statements is involved in Popper's falsification criterion for the 'empiricalness' of a system of hypotheses (see above, p. 15 footnote).

this set divided by n) is either less than $p - \sqrt{(pq/nk)}$ or is greater than $p + \sqrt{(pq/nk)}$.

This rule of rejection is equivalent to saying in the language of the last chapter: Regard the set of n observations as being an n-fold selective class from an n-fold $\{\alpha, \beta\}$-probability hyperclass, and take the probability sentence as expressing the hypothesis that the α-parameter of this probability hyperclass has the value p. Look at the probability ellipse whose value of b is nk. If r is the α-ratio in the set of n observations, reject the hypothesis if the point (r, p) falls outside the ellipse.

The argument for using this k-rule-of-rejection is as follows. Either the hypothesis which we are considering is true or it is false. If it is false no rejection of it is mistaken. If it is true, every rejection of it is mistaken.* But if it is true the proportion among the selective classes of those whose α-ratio falls outside the interval $[p - \sqrt{(pq/nk)}, p + \sqrt{(pq/nk)}]$ is less than k, and so the rule will only reject the hypothesis in less than a proportion k of the possible occasions of rejection, and a mistaken rejection will be made in less than a proportion k of such occasions. There is no logical impossibility in the following of the k-rule-of-rejection leading to a rejection of the hypothesis when it is true and thus leading to a mistaken rejection, but if the hypothesis is true such mistaken rejections will occur in a small proportion of the occasions—only in 5% of the possible occasions if k is taken as $\frac{1}{20}$. By choosing k to be small enough, these mistaken rejections can be made to occur in such a small proportion of cases that (to use the expression favoured by many writers on mathematical statistics) it is 'practically certain' that a particular rejection will not be one of this very rare class. Thus it will be 'practically certain' that to reject a hypothesis in accordance with the rule on a particular occasion will not be to commit the mistake of rejecting a proposition which in fact is true.

There is, however, a serious hiatus in this argument which will have to be bridged. The logical consequence of p being the α-parameter of the $\{\alpha, \beta\}$-probability hyperclass is that fewer than a proportion k of the selections from this hyperclass have an α-ratio lying outside a certain interval around p. But it is only one of these

---

* My mistaken rejections are equivalent to the 'type I errors' introduced into statistical literature by J. Neyman and E. S. Pearson (*Proceedings of the Cambridge Philosophical Society*, vol. 29 (1933), p. 493).

selections which is observed, and only one rejection on the basis of this one selection which is made or not made. To say that fewer than a proportion k of such selections have a certain property, or that fewer than a proportion k of rejections made on the basis of such observed selections are mistaken, is to make a remark not about the observed selection itself but about the hyperclass of possible selections. It is, in fact, a probability statement about the probability of a selection having a certain property (of its α-ratio lying outside a certain interval), and it states that this probability is less than k. So, as it may well be said, it is circular to explain the original probability statement in terms of the circumstances under which it would be rejected, when rejection itself has to be explained in terms of probability.

This is the familiar and valid objection made to most attempts to define probability in terms of frequency. From a premiss about probabilities there follow, by the class-ratio arithmetic, conclusions about other related probabilities, but never conclusions which are not probability statements. From the premiss about the α-parameter of a probability hyperclass there follows a conclusion about the proportion within the omniselectional hyperclass of selections which have certain α-ratios, but never a conclusion about any particular selection. To use any particular selection as the empirical criterion for rejecting the premiss is to suppose that in some way this particular selection typifies the omniselectional hyperclass, and the fact that very few of the other possible selections would resemble it in having an α-ratio with the property in question if the premiss were true would seem scarcely relevant, since we have not got the other possible selections to observe. It seems hardly rational to reject a hypothesis because a conclusion follows from it whose truth or falsity we have no means of testing.

This defect in the argument is serious enough for those who already think they know what probability hypotheses are, and who employ the argument in order to reject values of the probabilities propounded in these hypotheses. But the defect is even more serious for a person like myself who would propose to use the empirical criteria for rejecting a probability hypothesis for the purpose of attaching its meaning to the statement of the hypothesis. There would seem to be an unbreakable circle in such a method of definition.

Nevertheless, let us persist in thinking along our lines and see where we arrive. Suppose that we have rejected the original probability hypothesis by following the k-rule-of-rejection, having observed a set of n β-specimens whose α-ratio falls outside the proper interval. The justification for such a rejection is that, were the hypothesis to be true and were we to have applied the rule to all the selections from the probability hyperclass, we should have mistakenly rejected the hypothesis in less than a proportion k of the cases. We cannot, *ex hypothesi*, examine more than one of the classes which are selections from this probability hyperclass; but we can examine one of the hyperclasses which are selections from a probability hyperhyperclass, and use the result of such examination either to confirm the first rejection of the original probability hypothesis or to cancel this rejection and to restore the hypothesis to its *status quo ante*.

How we do this is, instead of considering one n-fold $\{\alpha, \beta\}$-probability hyperclass from which the set n of observations is a selection, to consider n_1 such probability hyperclasses and to treat a set of n_1 sets each of n observations as a selection (which is a hyperclass) from the probability hyperhyperclass of the n_1 omni-selectional hyperclasses corresponding to these n_1 probability hyperclasses. To put the matter more formally:

Consider n_1 n-fold $\{\alpha, \beta\}$-probability hyperclasses each of which has p for its α-parameter. Suppose that every member-class of any one of these hyperclasses is exclusive to every member-class of any other of these hyperclasses. Now consider the n_1 omni-selectional hyperclasses corresponding to these n_1 n-fold probability hyperclasses. Then, if B is the hyperclass including all hyperclasses defined in this way, and A is the hyperclass of all n-fold classes whose α-ratio lies outside the interval $[p - \sqrt{(pq/nk)}, p + \sqrt{(pq/nk)}]$, the n_1 omniselectional hyperclasses will be (1) all included in B, (2) mutually exclusive by pairs, (3) A-equiproportional in pairs. The hyperhyperclass of these n_1 hyperclasses is thus an n_1-fold $\{A, B\}$-probability hyperhyperclass with an A-parameter, p_1, which has been proved by the class-ratio arithmetic of the last chapter to be less than k.

Now consider the n_1-fold selections (which are hyperclasses) from this $\{A, B\}$-probability hyperhyperclass. By using the same class-ratio arithmetic we can deduce that the proportion amongst these selections of those whose A-ratio lies outside the interval $[p_1 - \sqrt{\{p_1(1-p_1)/n_1 k\}}, p_1 + \sqrt{\{p_1(1-p_1)/n_1 k\}}]$ is less than k. Since $p_1 < k$, we can further deduce (by Theorem V of the Appendix to the last chapter, when $k \leqslant \frac{1}{2}$) that the proportion among these selections of those whose A-ratio lies outside the interval $[0, k + \sqrt{\{k(1-k)/n_1 k\}}]$ is less than k; and, *a fortiori*, that the proportion is less than k of those selections whose A-ratio is greater than $k + \sqrt{(1/n_1)}$.

This argument can be continued by considering n_2 n_1-fold $\{A, B\}$-probability hyperhyperclasses, forming the n_2 omniselectional hyperhyperclasses corresponding to them and noting that, under suitable exclusiveness conditions, these are the members of an n_2-fold probability hyperhyperhyperclass whose A-parameter p_2 is less than k, where A is the hyperhyperclass of n_1-fold hyperclasses whose A-ratio is greater than $k + \sqrt{(1/n_1)}$. By the same class-ratio arithmetic it follows that the proportion among the n_2-fold selections (which are hyperhyperclasses) of those whose A-ratio is greater than $k + \sqrt{(1/n_2)}$ is less than k

In general: Let H_0 be the hypothesis that the α-parameter of the original probability hyperclass is p, H_1 the hypothesis that the A-parameter of the derived probability hyperhyperclass is less than k, H_2 the hypothesis that the A-parameter of the derived probability hyperhyperhyperclass is less than k, and so on. Then each of these hypotheses is a logical consequence of the preceding one, and so all are logical consequences of the original hypothesis H_0. They form a descending hierarchy of hypotheses in the probability deductive system with H_0 as highest-level hypothesis; since the falsity of any of these hypotheses is a consequence of the falsity of any succeeding one, to reject any of the hypotheses in the system involves rejecting all the hypotheses of higher level. Thus the rejection of H_0 can be effected by rejecting any of the hypotheses which are its logical consequences.

The k-rule-of-rejection which has been proposed for rejecting on the basis of a set of n observations the probability hypothesis H_0 can now be extended to reject H_0 on the basis of a hyperset of n_1 sets each of n observations, or on the basis of a hyperhyperset of n_2 hypersets each of n_1 sets each of n observations, and so on. These k-rules take the form:

K_1. On the basis of a set of n observations of members of β, reject H_0 if the α-ratio in this set lies outside the interval $[p - \sqrt{(pq/nk)}, \ p + \sqrt{(pq/nk)}]$.

K_2. On the basis of n_1n observations of members of β, regarded as a hyperset of n_1 sets each of n observations, reject H_1, and therefore also reject H_0, if the A-ratio in this hyperset is greater than $k + \sqrt{(1/n_1)}$.

K_3. On the basis of n_2n_1n observations of members of β, regarded as a hyperhyperset of n_2 hypersets each of n_1 sets each of n observations, reject H_2, and therefore also reject both H_1 and H_0 if the A-ratio in this hyperhyperset is greater than $k + \sqrt{(1/n_2)}$.

And so on.

The justification given for the first rejection test, K_1, is that H_1 follows from H_0, and that if H_1 were true the observed set would be a very rare specimen. Similarly, the justification given for the second rejection test, K_2, is that H_2 follows from H_1, which follows from H_0, and that if H_2 were true the observed hyperset would be a very rare specimen. And so on.

Now suppose that we have made n observations, applied the first rejection test, K_1, and in accordance with it have rejected the hypothesis. We can go on to examine a further n_1n observations, and by dividing them into n_1 sets each of n observations we can apply a second rejection test, K_2.

If this also rejects the hypothesis, the first rejection is confirmed.

But if this second test, K_2, does not reject the hypothesis the observations upon which it is based exhibit a hyperset of sets of observations in which sets of observations similar, in the relevant respect, to the first set of observations upon which the first test, K_1, was based occur as infrequently as (or only slightly less infrequently than)* was predicted by the hypothesis. Consequently the first set of observations must be regarded as atypical; sets of observations like this set would be expected to occur infrequently, and the second batch of observations to exhibit sets of observations like this set occurring with the appropriate infrequency. So the non-rejecting result of the second test, K_2, shows that the rejection of the hypothesis in accordance with the first test, K_1, must be

---

* This qualification is necessary because the theorem of class-ratio arithmetic used does not make a statement about the proportion of classes having an exact value of a class-ratio, but makes a statement about the proportion of classes whose class-ratio lies within (or without) a certain interval surrounding an exact value—which, in the case of the K_2-test, is the interval $[0, \ k + \sqrt{(1/n_1)}]$.

revised in the light of the evidence provided by the second batch of observations. Since this evidence does not reject the hypothesis, the rejection of the hypothesis on the result of the first test must be cancelled, and the hypothesis allowed to stand as unrejected by the available evidence.

The situation can be explained in another way by turning it round and by supposing that the second test, K_2, were made before the first test, K_1. The second test does not directly reject the original hypothesis H_0; it directly rejects the derived hypothesis H_1, and the rejection of H_1 involves rejection of H_0. Now H_1 states that the A-parameter of the probability hyperhyperclass a selection from which we have observed is less than k, and this has not been rejected by the test K_2 on the basis of the A-ratio in the observed selection not being greater than $k + \sqrt{(1/n_1)}$. Suppose we now apply the first test K_1 to each of the n_1 sets of n observations which are the members of this observed selection. Since a K_1-test rejects H_0 whenever the set of observations upon which it is based has an α-ratio lying outside the interval $[p - \sqrt{(pq/nk)}, p + \sqrt{(pq/nk)}]$, i.e. whenever this set of observations is a member of A, applying the K_1-test to each of the sets of observations which are members of the observed selection will result in not more than $n_1\{k + \sqrt{(1/n_1)}\}$ rejections. But the hypothesis H_0 predicts that the K_1-test will (mistakenly) reject the hypothesis in approximately a proportion which is less than k of the occasions on which it is applied; so the fact that the proportion of rejections which have been made on the basis of the n_1 K_1-tests is not greater than $k + \sqrt{(1/n_1)}$ cannot be held to militate against a hypothesis which predicted, approximately, a proportion less than k. It will be ridiculous to persist in the rejection of a hypothesis which predicted that on almost every occasion upon which a test would reject the hypothesis the rejection would be mistaken.

The result of all this argument is that, if the hypothesis which has been rejected is subjected to the second test and fails to be rejected by this test, the first rejection must be cancelled. For the purpose of rejecting the hypothesis the second test will have superseded the first test.

Similarly, if the second test, K_2, rejects the hypothesis (whether or not the first test K_1 has rejected it) but the third test, K_3, is applied and fails to reject the hypothesis, the rejection must be cancelled. In order that a rejection by a test at any stage should never be cancelled, the hypothesis must be rejected by tests at all later stages. The condition for a hypothesis to be *definitively rejected* is that there should be some stage at and after which all the tests reject the hypothesis.

Since we can never know at any one stage whether or not further observations, which we have not made, would reject the hypothesis at a later stage, we can never know that a hypothesis is to be *definitively* rejected. The series of k-rules-of-rejection are rules for making provisional rejections which are always subject to revision. On the basis of a set of observations the k-rule either rejects or fails to reject the hypothesis; the rejection is provisional, and is to be cancelled if on the basis of a hyperset of sets of observations, or of a hyperhyperset of hypersets of sets of observations, or etc., the rule fails to reject the hypothesis. A failure to reject is also, of course, provisional; if a hypothesis H_0 is not rejected by the rule on the basis of a set of observations it may well be rejected by the rule on the basis of a hyperset of sets of observations, or on the basis of a hyperhyperset of hypersets of sets of observations, or etc.; since, as H_1, H_2, etc., are all logical consequences of H_0, the rejection of H_0 must follow from the rejection of any of the hypotheses which are logical consequences of it. Thus, according to the rule, no batch of observations, however large, either definitively rejects or definitively fails to reject the hypothesis H_0.

Does this provisional character of the k-rule-of-rejection imply that it will not serve to give an empirical meaning to statements of statistical hypotheses, i.e. to probability statements? I think not. It would have this implication if a rigorist criterion for empirical meaning were accepted, as it was by the earlier proponents of the 'Verification Principle', according to whom an empirical statement would only have meaning if empirical conditions could be specified under which it would be definitively accepted or definitively rejected. (In the extreme rigorist form of the Principle, only if there were empirical conditions for its definitive acceptance; but this form is clearly untenable, since it would make nonsense of all universal hypotheses, which can be falsified but never completely verified by experience.) But it seems to me that we perfectly well can take as the criterion for empirical meaningfulness that there should be empirical conditions under which the statement in question would be accepted or empirical conditions under which it would be rejected—irrespective of whether such acceptance or rejection is definitive or is provisional and capable of revision on the basis of further experience. With this extended 'verification' criterion, the k-rule-of-rejection gives empirical meaning to probability statements. The statement that the probability that a member of the class β is a member of the class α is p expresses that proposition which is to be rejected if the α-ratio in a set of n observed members of β lies outside the interval $[p - \sqrt{(pq/nk)},\ p + \sqrt{(pq/nk)}]$— irrespective of the fact that this rejection may have to be cancelled if observation of a hyperset of sets of n members of β causes us not to reject the derived probability statement that the probability that a class of n members of β has an α-ratio lying outside this interval is less than k. And this derived probability statement expresses the proposition which would be rejected in a similar way—irrespective of the fact that this rejection might have to be cancelled if further observations required it.*

The extended verification criterion for the meaningfulness of an empirical statement has been expressed in the form that there are empirical conditions under which the statement would be rejected, even though not definitively rejected (or accepted, etc.—but this is not relevant to the case of probability statements, which are general propositions). The criterion must, however, be qualified in one way. There must be no logical impossibility in a rejection at any one stage never being cancelled at any later stage. For if this were to be logically impossible, it would be logically necessary for every rejection to be cancelled at some later stage, so that it would be logically impossible for the statement to continue to be rejected

---

* It is important to bear in mind throughout in reading this chapter that the series of rejection tests is proposed for use in an explanation of the *meaning* of a probability statement, and is not proposed as the best method for deciding whether or not a suggested value for a probability parameter should be rejected, still less as the best method for discriminating between two or more suggested values. Probability mathematics shows that, for all finite values of n, n_1, n_2, etc., there will be a range of values of the probability parameter around the 'true' value which is such that there will be a zero probability of *any* hypothesis which asserts that the parameter has a particular value falling within this range remaining rejected after some stage in the series of tests. This range will contract to a point as n, n_1, n_2, etc., increase without limit. But the fact that, for finite values of n, n_1, n_2, etc., there will be a range of incompatible statistical hypotheses which the series of tests will 'almost certainly' fail to reject does not vitiate my use of the series of tests in explaining the nature of a statistical hypothesis.

by all further observations after it had once been rejected. We should then know in advance before subjecting the hypothesis to the test of observation that, over a sufficiently long series of stages of tests, we should be no better off than we were to start with. In the case of our rule for giving meaning to probability statements, there is no logical impossibility of this kind. The sets of observations upon which the successive rejection tests are based are all separate sets; the rule's criterion for rejection is at each stage based upon empirical properties of these separate sets, and the possession by one set of the appropriate property is logically independent of the possession by another set of its appropriate property. So it is perfectly possible for successive tests in the series always to continue rejecting the probability statement. Though we can never be certain that rejection by one test will not be followed by a cancellation of this rejection by the failure to reject of some subsequent test, we are never certain that it will be so cancelled. Although every rejection may be cancelled, yet it also may not be; and the possibility of an unending series of empirical tests each of which rejects the statement serves to give, I maintain, an empirical meaning to the statement.

This point is connected with one involved in the form of the 'Verification Principle' which is applicable to empirical propositions which are not probability statements. Rash persons have occasionally stated the principle in the form that the meaning of a statement consists in a method by which it can be empirically verified with the means of observation now at our disposal. But this is a ridiculous form of the principle; we have at present no means for deciding whether or not there is a crater on the other side of the moon larger than the largest one on the side of the moon visible to us from the earth, but this statement is a perfectly significant one. The verification criterion must be put in the form, not of a practical possibility but of a logical possibility of verification; there is no logical impossibility in voyaging round the moon in a rocket, or in 'observing' objects at the back of the moon by a development of radar, or indeed in the moon turning round so that the other side faces the earth. Similarly, in considering the verifiability of probability statements, it is a logical possibility and not a practical possibility that we require; and there is no logical impossibility in our series of empirical tests, which are logically independent of one another, all rejecting the probability hypothesis.

A metaphor may perhaps be helpful. Imagine the hypotheses that we have to consider as papers in a basket labelled 'For consideration'. Imagine another basket labelled 'Rejected', and a procedure for considering the papers and for moving a paper from the first basket to the second (corresponding to rejection of the statistical hypothesis) and of moving a paper back again to the first basket for reconsideration (corresponding to cancellation of a previous rejection of a statistical hypothesis). The fact that every paper that has been moved into the 'Rejected' basket may have to be moved back again into the 'For consideration' one does not make the consideration procedure futile. What would make it futile would be if it were logically necessary that every paper which got into the 'Rejected' basket were always to find its way back to where it was at the beginning. The consideration procedure is useful if it is possible for papers which get into the 'Rejected' basket to stay there. The function of a waste-paper basket is not that of a lobster trap; the trap is inefficient if any lobster manages to escape, but a waste-paper basket serves a useful purpose if any papers manage to stay there.

IS THE RULE OF REJECTION AN ARBITRARY ONE?

The thesis maintained in this chapter is that the adoption of the k-rule-of-rejection gives an empirical meaning to probability statements in terms of an observable characteristic (the α-ratio) of an observable set of observations knowledge of which will serve to reject the statement. That the rejection may be cancelled by knowledge of another observable characteristic of further observations does not deprive the rejection test of value, since it is logically possible for a rejection to stand however many rejection tests in the hierarchy of tests are applied. The criterion is an empirical one, since the rejection test cannot be used *in vacuo* but only on the basis of observed knowledge. It is true that a rejection, or any series of rejections, never 'yields certainty' in the sense that all these rejections may be cancelled by a further test; but what John Dewey called "the quest for certainty" is, in the case of empirical knowledge, a snare and a delusion.

So far so good. But we are not out of our troubles. For a critic will reasonably enough say that the complicated criterion we have given—that a probability statement acquires its meaning from the rule that it is to be rejected if a certain ratio in a set of n observations falls outside a certain interval—is highly arbitrary. What is the reason for using this criterion rather than any other? And in particular the critic may make the radical criticism that the criterion in no way connects the rejection of the hypothesis with its falsity rather than with its truth, and that it is the distinction between the truth and the falsity of a hypothesis with which an empirical criterion of meaningfulness must be concerned, and not the circumstances under which I, or indeed the 'culture circle' of mathematical statisticians, agree to reject it. Unless some connexion can be shown between rejection and falsity, the criterion is worse than useless, since to use it would delude us into thinking that there was some such connexion.

This radical objection cannot be met by saying, as most statistical writers say, that, if k is taken to be small enough, it is 'practically certain' that we shall not mistakenly reject a true hypothesis, and that consequently it is 'practically certain' that any hypothesis which we reject will be false. For it is the meaning of the phrase "practically certain" that is itself in question. Similarly, it is no good to say, in a more sophisticated manner, that it is 'practically certain' that we shall mistakenly reject a true hypothesis in less than a proportion k of the times we apply the rejection test, since this is also a probability statement whose meaning has been given in terms of rejection. Fortunately, a quite satisfactory answer can be given by considering the extreme cases when the probability hypothesis states either that the probability of every member of β being a member of α is 0 or that this probability is 1. For if this probability is 0, the hypothesis will be refuted if any observed member of β is found to be a member of α, since such an observation is logically incompatible with no member of β being a member of α, which is what is asserted by the probability statement. Consequently finding one member of the class $(\alpha\beta)$ will prove the probability hypothesis to be false. But applying the k-rule-of-rejection will reject the hypothesis if the α-ratio in a set of n observations falls outside the interval

$$\left[0 - \sqrt{\frac{0(1-0)}{nk}}, \quad 0 + \sqrt{\frac{0(1-0)}{nk}}\right],$$

i.e. if the α-ratio differs from o. Since this degenerate interval, which consists only of the point o, does not depend upon the values of n and of k (provided that neither is o), a k-rule-of-rejection will reject the hypothesis if, in any number n of observations, at least one member of α is found, whatever positive number k may be. Thus, for any positive n and for any positive k, the rule of rejection will reject those hypotheses asserting a probability of o which are false. Similarly, the hypothesis stating that the probability of a member of β being a member of α is 1, which will be refuted if any member of β is found not to be a member of α, will be rejected by a k-rule-of-rejection (for any positive value of k) if the α-ratio in any set of n observations differs from 1, i.e. if at least one of this set is not a member of α, whatever be the number of observations. Thus, those hypotheses asserting a probability of 1 which are false will be rejected by the rule. So in the cases in which the probability is either o or 1, to use the k-rule-of-rejection comes to the same thing as using the rule of rejecting statements known to be false. This establishes a connexion between rejection and falsity in the two extreme cases, and thus seems to me to establish the required connexion in the other cases also. Except in the two extreme cases there is no logical necessity for a probability statement rejected by the k-rule on the basis of a set of observations to be false. But the fact that for the extreme values of p there is this logical necessity shows that rejection 'corresponds' to falsity in a way in which it does not 'correspond' to truth; and this is all, I think, that we want. What we require is that the rule should not do what we want of it if the word "accept" were substituted in it for the word "reject"; were this the case, the rule would in the extreme cases tell us to accept statements known to be false, which would be absurd.

We have now got to the position that meaning is attached to a probability statement by virtue of its rejectability on the basis of observations in accordance with a k-rule-of-rejection, this rejectability corresponding to falsity rather than to truth. An empirical meaning having thus been given to probability statements, an excellent reason can be given for making use of the k-rule-of-rejection rather than some other rule; namely, the 'justification' given when the k-rule was first expounded (p. 154). This reason, to repeat it, is that, if the probability statement is true, the probability of rejecting it when using the k-rule is less than k. Or, expressed with the symbols we have used, the reason for using the k-rule-of-rejection is that the proposition that the probability of rejecting H_0 by using the k-rule is less than k is logically equivalent to H_1, which is a logical consequence of H_0. Now, as was pointed out originally, it would be circular to use the fact that H_1 logically follows from H_0 to define the 'probability' occurring in the statement of H_0 by means of the terms occurring in H_1, since one of these terms is a probability (the probability that an α-ratio lies outside a certain interval). But when we have already given a meaning to a probability statement by a general rejection procedure based upon the result of observation, there is no circularity in propounding another probability statement (whose meaning is similarly given) as the reason for using this general rejection procedure. So there is no objection to using the fact that H_1 is a logical consequence of H_0 to justify defining the statement of H_0 by the k-rule-of-rejection, provided that this rule does not itself use H_1, which it does not.

I do not wish to deny that there is a type of circularity involved in giving as the justification for the mode of definition of a probability statement another probability statement. But it is not a vicious circularity; it does not vitiate the definition. What has happened is that we first give a rule according to which meaning is given to probability statements by virtue of their being such statements as are rejectable according to the rule on the basis of empirical knowledge; this provides the definition of probability statements as a particular sort of empirical statements. There is no circularity of any kind in using this method of definition. If then we ask why this method of definition should be used rather than any other, we are then asking for the reasons why probabilities defined in this way do what we want them to do; and the reason for this is another probability statement. Herein lies the circularity. But it vitiates nothing that has been said.

There would of course be a vicious circularity—a circularity in inference—if the reason for believing (or for disbelieving) a probability statement were to be belief (or disbelief) in another probability statement. But this is not involved in our method of definition. The reason for disbelieving a probability statement is the empirical knowledge on the basis of which it has to be rejected in accordance with the k-rule-of-rejection which determines the meaning of the statement; that it has to be rejected on the basis of such knowledge is not the reason for disbelieving it, but is a statement of how probability statements are used. That the reason why probability statements are used in this way can only be given by another probability statement is irrelevant to the reasons for disbelieving the original statement.

This question of circularity cannot be fully treated without writing a whole treatise on definition and the related notion of analysis. But it is necessary to mention that the term "definition" is used by some philosophers in a narrower sense, by others in a wider sense.* Those who use the narrower sense would confine definition of a word or phrase to the production of a synonym, and the definition of a sentence of a certain sort to the production of a sentence having the same meaning in which this feature is lacking. In this narrower sense the specific probability words as they occur in probability statements are indefinable. For it is impossible to translate a probability statement into a statement in which there are no probability words; and in this sense it can truly be said that the notion of probability is unanalysable. But, in this narrow sense of definition and analysis, there are no important questions in the philosophy of science—nor in any branch of philosophy—that can be solved by giving definitions or analyses. For the important questions are always those concerned with the use we make of statements of a particular sort, and the way in which we think of concepts of a particular sort; and the questions would not be puzzling if it were possible to settle the questions by making straightforward translations of the statements into statements about which we were not puzzled, or to analyse the concepts into a complex of other concepts related together in simple ways. In Chapter III it has been shown that the theoretical concepts of an advanced science like physics would not fulfil their function in thought if they were analysable (in this narrower sense) in terms of direct observation. And probabilities would not serve the purpose we require of them if statements about them could be translated into statements about directly observable frequencies. Something subtler, and more interesting, than mere translation is required in all the philosophically important cases.

* This wider sense includes, but is wider than, the kind of 'implicit definition' mentioned in Chapter III.

Some philosophers who use a restricted sense of definition and say that "probability" is indefinable would propose to leave the matter at that. This seems to me simply to ignore the problem of probability, which is that of elucidating the function in our thought of probability concepts. But most of these philosophers would agree with me in taking this to be the problem, in which case it is merely a matter of how the problem should be described. They might call it "finding the place of probability on our language-map", whereas I would call it defining the meaning of "probability" in scientific probability statements or analysing probability as it is used within a science. But these are two ways of describing the same philosophical activity. For myself I think it perfectly proper to describe the statement of the rules according to which probability words are used as defining these words, in a wide sense of definition; but I have no quarrel with those who would describe such a process in another way. Those with whom I disagree are those who would use the fact that probability statements are not translatable into statements of another form as an excuse for saying that there is nothing further to say about them; this is to commit the *trahison des philosophes*—that of saying there is no problem at exactly the point where the problem becomes most important and difficult.

THE ARBITRARY CONSTANT IN THE RULE OF REJECTION

Let us leave this question of the nature of philosophical analysis and return to the analysis of probability as used within a science. What seem to me to be good reasons have been given both for describing the mental operation upon the probability statements as rejection, and for the form of rule of rejection which has been adopted. But there is one element which remains arbitrary and for the choice of which no reason has been given—the value of the small positive number k in terms of which the k-rule-of-rejection has been stated. Since the occurrence of this number is an essential feature in the rule governing the use of probability concepts, some detailed discussion of its function is necessary.

There is first a comparatively minor point. The Bienaymé-Tchebichef theorem of class-ratio arithmetic, which provides the grounds for the k-rule-of-rejection which has been given, does not state the exact value of the proportion of n-fold selections having class-ratios outside the interval $[p - \sqrt{(pq/nk)},\ p + \sqrt{(pq/nk)}]$, but

only states that this proportion is less than k. For any set of particular values of p, n, k, this proportion can be exactly calculated (though the calculation is exceedingly tedious); and as n increases without limit the proportion can be proved to approach a value given by a fairly simple mathematical function of k.* All the exact values, and the limiting value, for the proportion are very much less than k; the Bienaymé-Tchebichef inequality is, in the mathematical sense, a very weak theorem, and the k-rule-of-rejection grounded upon it a much weaker rule of rejection than could be grounded upon the results of class-ratio arithmetic. Since as regards our purpose in explaining the logic of probability statements what is wanted is some rule of rejection, it does not matter if a stronger one is equally defensible. The k-rule given has been grounded upon the Bienaymé-Tchebichef inequality for the excellent reason that the inequality is simple and lends itself easily to numerical illustrations, and that its proof, though mathematically elegant, uses no more advanced techniques than those of school algebra, so that all the mathematics used in this book is above board and open to inspection. A person who would prefer to use a k-rule grounded on an equality rather than an inequality in class-ratio arithmetic could take the rule in the form: Reject the hypothesis if the α-ratio in a set of n observations lies outside the interval $[p - \sqrt{(pq/nh)},\ p + \sqrt{(pq/nh)}]$, where h is the number (dependent upon n, p, k) which is such that the proportion of n-fold selections having an α-ratio outside this interval is *equal to* k.†
Since h is always greater than k, the interval

$$\left[p - \sqrt{\frac{pq}{nh}},\ p + \sqrt{\frac{pq}{nh}} \right]$$

will always be smaller than the interval

$$\left[p - \sqrt{\frac{pq}{nk}},\ p + \sqrt{\frac{pq}{nk}} \right];$$

* $1 - \sqrt{\left(\dfrac{2}{\pi}\right)} \displaystyle\int_0^{1/\sqrt{k}} e^{-\frac{1}{2}z^2}\, dz.$

† Since, for given n, there are only a finite number of possible values which can be taken by the proportion of n-fold selections having an α-ratio lying within any given range, it will be impossible for such a proportion to be *exactly* equal to k, unless k is one of these possible values. To be strictly correct, "takes the greatest possible value equal to or less than k" must be substituted for "is equal to k" in the statement of the strong k-rule. This mathematical subtlety will be ignored in our subsequent discussion.

so any hypothesis rejected by the first rule will be rejected by the second, though the second rule will reject hypotheses which the first will fail to reject. But the first k-rule is a rule for rejection only; it requires us to reject a probability statement if certain empirical conditions are fulfilled. If these conditions are not fulfilled, it does not require us to reject the probability statement. But it does not prohibit us from rejecting a statement if we choose to use a stronger rule of rejection. The second k-rule-of-rejection, given above, is the strongest k-rule which is justifiable by class-ratio arithmetic. It might therefore appear proper to state it as "Reject the probability statement if *and only if*, etc.", and to take it as prohibiting the rejection of the probability statement if the empirical conditions requiring rejection are not fulfilled. But this would be a confusing way of stating the rule, since a probability statement which has failed to be rejected by one set of observations or at any one stage may well be rejected by another set of observations or at a later stage; so the prohibition of rejections would have to be provisional, and liable to subsequent cancellation. Moreover, a statistical hypothesis which passed direct observational tests might well come to be rejected because of its failure to fit into a more inclusive scientific deductive system with statistical hypotheses at a yet higher level. So the right way to state the strong k-rule is to state it in the same form as the weak k-rule (*if*, not *if and only if*), and to add as a footnote to the rule, but not as a part of the rule itself, that it is the strongest k-rule which is justified by considerations derived from class-ratio arithmetic.

The intervals and ellipses into which the observed class-ratios are required to fall in order that the probability hypothesis should fail to be rejected by the observational test correspond to the 'regions of acceptance' of contemporary statistical literature. In this literature these regions are used not with the explicit purpose of determining thereby the meaning of probability statements but—assuming this meaning is known—of providing a test for their truth as well as for their falsity. The regions are therefore supposed to be drawn in such a way that, on the basis of a set of n observations, not only must a probability hypothesis be rejected if the point (r, p) falls outside the region, but the hypothesis must be accepted if this point falls within the region (hence the name "region of acceptance"). The dichotomy of the square by the ellipse (or other closed figure) into two regions—the region on or within the closed figure and the region outside—must then be taken to correspond to a dichotomy of probability hypotheses into those to be accepted and those to be rejected. It is therefore essential to keep the closed figure to exactly the size justified by class-ratio arithmetic; since otherwise hypotheses would come to be accepted without due warrant. Consequently figures based upon the exact proportions of selective classes having α-ratios lying in certain intervals are required.* But we are not here concerned with criteria for the truth of probability statements, only with criteria for their falsity; so the statisticians' 'regions of acceptance' become for us merely 'regions of failure to reject'; and since we are interested in their outsides (the 'regions of rejection' or 'critical regions', as they are sometimes called) rather than in their insides, it does not affect the logic of our argument if the ellipses or other closed figures are drawn larger than theoretically they need be. All that happens is that we are using smaller regions of rejection, and consequently a weaker k-rule-of-rejection, than we need to do in order to ground the rule (as we wish to do) upon theorems of class-ratio arithmetic. Since any probability hypothesis rejected by a weak k-rule would be rejected by a stronger k-rule, this weakness is irrelevant to our problem of determining the meaning of probability statements by a k-rule of rejection; the arbitrariness inherent in grounding the k-rule to be used upon a fairly simple inequality of class-ratio arithmetic rather than upon the exact equality is, as it were, a stylistic arbitrariness consequent upon my method of exposition.

THE ESSENTIAL ARBITRARINESS INVOLVED

The existence, however, in the k-rule-of-rejection of an undetermined constant k is an arbitrariness which is one of the essential characteristics of probability statements. For the choice of the constant k cannot be determined by considerations either internal to the science in question or internal to the logic of this science. This is because the rejections imposed by the k-rule are never definitive rejections; we may always be mistaken in rejecting a

* Such figures will be found in C. J. Clopper and E. S. Pearson, *Biometrika*, vol. 26 (1934), pp. 404 ff. and in M. G. Kendall, *The Advanced Theory of Statistics*, vol. 2 (London, 1946), p. 68.

probability hypothesis, and the constant k is, as we have seen, an upper bound to the probability that we are mistakenly rejecting a true hypothesis. So our choice of the constant k will depend upon how important it is for us to avoid making a mistake in rejecting a hypothesis, and this will depend upon how useful the hypothesis would be to us if we were not to reject it. If the use of the hypothesis to us is purely intellectual (e.g. in the way of organizing hypotheses at a lower level into a more comprehensive deductive system than that in which they could otherwise be arranged), we shall wish to reject the hypothesis unless it fits the observed facts pretty closely. That is to say, we shall wish to reject the hypothesis if there is strong evidence against it even if there is quite a moderate probability of our rejecting it when it is true. So k will be taken as high as $\frac{1}{20}$ or even $\frac{1}{10}$, and we shall expect by choosing a high value of k mistakenly to reject the hypothesis in 5% or in 10% of the possible occasions for rejection. But suppose that the hypothesis, if it were true, would be of the greatest practical use to us. It might, for example, be a hypothesis which, if it were true, would enable a treatment to be developed for a disease for which no other treatment was known. We should then not wish to reject the hypothesis unless the evidence against it was overwhelming; we should not wish to reject it unless the probability was minute that we should be rejecting it if it was true; and we should choose k to be a very small number, e.g. $\frac{1}{100}$ or $\frac{1}{1000}$. So the choice of k will depend, to a large extent, upon how important it is to us not to reject a hypothesis which in fact is true; that is, how important it is to us to avoid mistaken rejections.

This arbitrariness, from the logical point of view, of the choice of the number k is an essential feature of the way in which we attach meanings to statements asserting probabilities which are neither 1 nor 0. When p is either 1 or 0, i.e. when the statement is a universal generalization, the k-rule-of-rejection will reject the hypothesis whenever one contrary instance is found, irrespective of the value of k (as has been shown on p. 165). So k as used in this case is, as logicians say, 'vacuous'; it could have any different value without affecting the result. Moreover, the probability that we shall be making a mistake in thus rejecting the hypothesis will be less than k, whatever positive number k may be. But if a number which is restricted to being positive or zero is less than every positive number, it must be zero; so the probability that we shall be mistaken in our rejection is o, and we shall never be mistaken. Thus in these two cases the rejection is a definitive rejection, and whether or not a hypothesis is rejected by the k-rule depends in no way upon the value of k. So the arbitrariness of k is unimportant.

In all other cases the decision whether or not to reject a probability hypothesis according to the k-rule may depend upon the value of k. On the basis of an observed α-ratio of r in a set of n observations the probability ellipse corresponding to k_1 (the b_1-ellipse, where $b_1 = nk_1$) lies inside the probability ellipse corresponding to k_2 (the b_2-ellipse, where $b_2 = nk_2$) if k_1 is greater than k_2. A person using the k_1-rule-of-rejection will reject the probability hypothesis if the point whose coordinates are (r, p) falls outside the smaller ellipse; a person using the k_2-rule will reject the hypothesis if this point falls outside the larger ellipse. If the point falls in the lunar region between the two ellipses, the hypothesis will be rejected by the first person but not by the second. There may be, as we have seen, excellent reasons for the first person to use the k_1-rule and the second to use the k_2-rule; the second person may have an economic or hedonic interest in the hypothesis being true which the first has not. But such considerations would seem to be irrelevant to the reasonableness of disbelieving the hypothesis. Logically the choice of k is arbitrary; there is, as a matter of pure logic, no reason to take one value of k rather than another; yet what value of k is taken may have a great influence upon whether or not a probability hypothesis is to be rejected. The fact which the recent subtle work of statistical mathematicians has brought into clear light,* that the principles according to which we reject probability hypotheses contain factors which are arbitrary from the logical point of view, and that the choice of these arbitrary factors must be justified by extra-logical considerations—by considerations of the purposes for which we propose to use the hypothesis—will become clearer when we consider in the next chapter the principles according to which we choose between alternative statistical hypotheses. Indeed, the question of the reasons for the acceptance of any hypothesis, whether statistical or universal, will

---

* Especially J. Neyman and E. S. Pearson, in *Proceedings of the Cambridge Philosophical Society*, vol. 29 (1933), pp. 492 ff., and A. Wald, in *Annals of Mathematical Statistics*, vol. 10 (1939), pp. 299 ff.

be found, in the last resort, to involve teleological considerations. The ultimate justification for any scientific belief will depend upon the main purpose for which we think scientifically—that of predicting and thereby controlling the future. The peculiarity of statistical reasoning is that it presupposes also at an early stage of the argument judgments as to what sort of future we want. In considering the rationale of such thinking we cannot avoid ethics breaking into inductive logic.*

In this chapter we are not concerned with the reasonableness of scientific thinking, whether by the use of statistical or of universal hypotheses; we are concerned with what determines the meaning of probability statements. Does the unavoidable arbitrariness of the k-rule-of-rejection vitiate its use for defining (in the wide sense of definition) probability statements in terms of empirical rejection tests? I do not think that it does—for the following reason.

We have seen that, if one person follows a k_1-rule-of-rejection while another follows a k_2-rule, where k_1 is greater than k_2, the first person may reject a statistical hypothesis which the second may not reject. But this was on the assumption that each person was basing his rejection test upon the same α-ratio found in the same n observations. Suppose now that each bases his rejection test upon exactly the same discovered α-ratio r but that the first has found this α-ratio r in a set of n_1 observations, the second has found the same α-ratio r in a set of n_2 observations.† Now the product of n and k is what determines the value of b for the probability ellipse, the falling outside of which of the point (r, p) serves to reject the hypothesis. So if $n_2 k_2 = n_1 k_1$ the b_2-ellipse used by the second person will coincide with the b_1-ellipse used by the first person, and the second person will reject the hypothesis whenever the first rejects it. The point (r, p) will for the first person represent the combination of p with an α-ratio of r found in a set of n_1 observations, for the second person the combination of p with an

---

* The general question of the relationships between ethics and inductive logic is discussed in my Henriette Hertz Annual Philosophical Lecture to the British Academy, *Moral Principles and Inductive Policies* (1950).

† Of course any two persons examining different sets of instances, whether of the same or of a different number of instances, would be very unlikely indeed to find *exactly* the same α-ratio r in the two sets; so the assumption is not a 'realistic' one. But this does not affect its use in a discussion of the *meaning* of probability statements, which is the sole concern of this chapter.

α-ratio of r found in a set of n_2 observations. If these α-ratios are the same and if $n_2 = n_1(k_1/k_2)$, the strength of the two observational rejection tests is the same for the two persons in the sense that a hypothesis rejected by the one test is rejected by the other, and vice versa.

Thus the second person using the smaller value k_2 of k in his k-rule-of-rejection can compensate for the fact that, on the basis of the same number of observations, he will fail to reject hypotheses which the first person, using the larger value k_1 of k, will reject by taking a greater number of observations, since the strength of the rejection test depends upon the product nk. The probability of making a mistaken rejection by using the k_2-rule is, of course, less than k_2; so the second person, by taking $n_2 = n_1(k_1/k_2)$, obtains a test of the same strength as that of the first person without sacrificing his advantage over him in having a lower upper-bound to the probability of a mistaken rejection. But this is not relevant to the point at issue here, which is that two different k-rules-of-rejection will reject the same hypotheses for the same α-ratio r if the number n_1 of observations upon which the k_1-rejections are based and the number n_2 of observations upon which the k_2-rejections are based are related by the equation $n_1 k_1 = n_2 k_2$.

This fact, though it cannot be said to reduce the arbitrariness of the choice of k, does reduce the element of privacy in the choice of k. For the result of a rejection test using a k-rule with one value of k will (for the same observed α-ratio) be the same as the result of a test using another value of k provided that the numbers of observations upon which the two tests are based are suitably related.

The fact that a k_1-rejection-test is of equal strength to a k_2-rejection-test if $n_1 k_1 = n_2 k_2$ has a consequence that, whatever value of k is chosen, however small, the k-rejection-test can be made of any assigned strength by a suitable choice of the number n of observations upon which it is to be based. In particular, the test can be made as strong as any other test, however strong, by taking n large enough. In terms of our Chinese-box probability ellipses if any one of these ellipses is taken, however narrow (i.e. however closely approximating to the diagonal straight line $r = p$), then for every value of k, however small, there is a number n which is such that the k-rule-of-rejection will reject a statistical hypothesis if on the basis of n or more observations an α-ratio r is found which is such that the point (r, p) falls outside this ellipse.* Thus any one of the Chinese-box systems of ellipses will serve as the rejection ellipse for a k-rejection-test however small k may be, if n is chosen large enough.

This leads to the following important consequence. Imagine a series of rejection tests in accordance with the same k-rule-of-rejection applied to a series of sets of observations of different members of β, the first set consisting of m observations, the second of $2m$, and so on.† Suppose that in the first set of observations the number of members of α is s, in the second set this number is $2s$, and so on; so that the α-ratio in each set of observations is the same fraction $r = s/m$. Now let us use this series of rejection tests to test the statistical hypothesis asserting that the probability of a member of β being a member of α is p, where p differs from r by a positive amount d. *At some point in the series of tests the statistical hypothesis will be rejected, and every subsequent test in the series will also reject it.* This is the case no matter what positive value is taken by k, and no matter what positive value is taken by d.

To prove this, draw the probability ellipse passing through the point whose r-coordinate is $p - d$ and whose p-coordinate is p. There is one and only one such ellipse passing through this point, and this ellipse also passes through the point $(p + d, p)$. The value of b for this ellipse will be pq/d^2.‡ Any probability ellipse with a larger value of b will lie inside this ellipse. So if n is greater than pq/d^2k, the probability ellipse whose value of b is nk will lie inside this ellipse, and the points $(p - d, p)$, $(p + d, p)$ will lie outside it. But one or other of these points represents the combination of a probability p with an observation of an α-ratio r differing from p by d; so the hypothesis will be rejected by any of the k-rejection tests based upon a number n of observations which is greater than pq/d^2k. Since the number of observations in the successive tests is m, $2m$, $3m$, ..., the lth and every succeeding test will reject the hypothesis, where l is any integer greater than pq/md^2k.

Thus whatever value be taken for k in a k-rule-of-rejection, the rule will always reject a statistical hypothesis if, in a sufficiently large number of observations, an α-ratio is found which differs

---

* $n = b/k$, where b is the constant of the ellipse.
† These sets need not be exclusive.　　　　‡ Since $\sqrt{(pq/b)} = d$.

from the value of a probability asserted by the hypothesis by more than any specified amount, however small. If two persons start with different rules of rejection, they may not agree in rejecting a statistical hypothesis on the basis of a small number of observations yielding the same α-ratio to both; but as they increase their number of observations, a point will always be reached at and after which they will agree in rejecting a hypothesis if its probability differs from the α-ratio they have found by more than a specified amount. So the arbitrariness of the choice of k for the rule of rejection has only a limited effect; a sufficiently large number of observations will make it not matter which value of k has been chosen out of a range of values of k whose lower bound is some positive number.

Let H_p be the hypothesis that the probability of a member of β being a member of α is p. We can now state precisely the 'invariance' property of different k-rules with regard to rejecting the hypothesis H_p when the α-ratios in all the sets of observations concerned differ from p by not less than a positive number d. Suppose k_1 and n_1 are any two non-negative numbers which are such that $k_1 n_1 = pq/d^2$. Then the property holding of H_p of being rejected by a k-test based upon a set of n observations will be invariant with respect to every value of k which is greater than k_1 and also with respect to every value of n which is greater than n_1.

In the case of a universal hypothesis, when $p = 0$ or $p = 1$, the restriction of the lower bounds k_1 of the value of k and n_1 of the value of n is given by the equation $k_1 n_1 = 0$, a solution of which is $k_1 = 0$, $n_1 = 0$. So the property holding of $H_{p=0}$ and of $H_{p=1}$ of being rejected by a k-test based upon a set of n observations will be invariant with respect to every k which is greater than 0 and with respect to every n which is greater than 0—as we have already seen (p. 165). So the invariance in the case of a statistical hypothesis differs from the invariance in the case of a universal hypothesis only in the fact that in the former invariance the lower bounds of k and of n are not zero, but are positive numbers which depend upon one another.

All this is really only stating in elaborate language the two fairly obvious facts that, if a k-rule-of-rejection with a given value of k is applied to a series of increasing sets of observations the α-ratio in each of which differs from p by more than some fixed positive number, then a point will come in the series of tests when the statistical hypothesis H_p will be rejected, and will continue to be rejected, in accordance with the k-rule-of-rejection. Thus although a logically arbitrary choice of one value rather than another for k will affect whether or not a hypothesis is rejected on the basis of a given number of observations, it will not affect whether the hypothesis will be rejected or not if the number of observations is increased sufficiently. The value of k is relevant to the result of using a k-rule-of-rejection in a particular case, but it is not relevant to the result of using the rule in a series of cases having increasing values of n. And for considering the definition of probability statements in terms of a rejection procedure, there is no objection to explaining the fact that the procedure looks on the face of it to be an arbitrary procedure making use of an arbitrary constant k by showing that 'in the long run' it is not an arbitrary procedure, since the value of k is irrelevant if we take n to be large enough. How large n must be for a given k will be settled by the lower bound d of the deviations from p of the α-ratios in the sets of observations, since nk must be greater than pq/d^2.

But has not this escape from the arbitrariness in the choice of k been attained by introducing another number equally arbitrary— namely, this lower bound d of the deviations of the α-ratios from p? The answer to this question is that the arbitrariness of d can be treated in the same way as the arbitrariness of k. If d_1 is smaller than d_2, then pq/d_1^2 is greater than pq/d_2^2, and the number of observations required in order that the exact value of k should be irrelevant will have to be greater. But for every positive value d_1 of d and for every positive value k_1 of k, there will be a value of n (namely, $n_1 = pq/k_1 d_1^2$) which is such that hypothesis H_p will be rejected by the k-rule if the α-ratio found in a set of n observations deviates from p by more than d_1. Thus the rejection test is invariant for all values of d which are not less than the value d_1 which, together with k_1, fixes the value of n_1. So the effect of choosing a smaller d_1 is like that of choosing a smaller k_1 in that an increased number of observations are required for invariance for both d and k to come into effect.*

* Qualitatively, not quantitatively, like, since n_1 depends upon the inverse square of d_1.

The numbers k_1, d_1, n_1 are related by the three arithmetically equivalent equations

$$d_1 = \sqrt{\frac{pq}{n_1 k_1}}, \quad k_1 = \frac{pq}{n_1 d_1^2}, \quad n_1 = \frac{pq}{k_1 d_1^2}.$$

In the definition of probability by means of a k-rule-of-rejection we took n and k as fixed in advance, and hence used the first of these equations in the statement of the k-rule-of-rejection. And no use was made at any point in the exposition of what would happen to this equation when the value of n was increased without limit. In order to explain the provisional nature of the rejection it was necessary to consider hypersets whose n_1 members were sets of n observations each, and hyperhypersets whose n_2 members were hypersets of sets of observations, and so on; but no use was made at any point of any property depending upon the size of any of these numbers; indeed, everything that has been said would be true (though not particularly interesting) if any or all of these numbers were 1. It was not until, having defined probability by means of a rejection procedure, we came to consider the nature of the arbitrary constant k involved in such a procedure that we examined what would happen with series of increasing sets of observations, and made use of the third equation $n_1 = p_1 q_1 / k_1 d_1^2$ to fix a value of n above which the rejection procedure would be invariant with respect of values of k and d which were not less than k_1 and d_1. Thus it was only at this point that we made use of values of n which were not fixed in advance independently of k and of d, but determined as to their lower bound by the lower bounds of k and of d. In this consideration no use was made of the second equation $k_1 = p_1 q_1 / n_1 d_1^2$, which determines k_1 in terms of n_1 and of d_1. If the Law of Great Numbers is taken (as it was on p. 143) as a name for the proposition that, for fixed p and d, $1 - k$ tends to 1 as limit as n increases without limit, no use has been made of the Law of Great Numbers in the definition of probability given in this chapter.

RELATION TO THE LAW OF GREAT NUMBERS

Of course the form of the Bienaymé-Tchebichef inequality which gives the reason for using for the definition of probability a rule of rejection in terms of the interval $[p - \sqrt{(pq/nk)}, \; p + \sqrt{(pq/nk)}]$ is mathematically equivalent to the form of the inequality which expressed in terms of probabilities, states that the probability that the α-ratio in a set of n observations lies outside the interval $[p-d,\ p+d]$ is not greater than pq/nd^2, from which the Law of Great Numbers immediately follows. But the definition of probability has not been given in terms of a diminishing k which is a function of n and which tends to zero as limit as n increases without limit. The value of k has always been taken as a fixed constant, and the interval with which the k-rule is concerned is given in terms of the independent numbers which are values of p, n and k. The rationale of the definition has been in terms of the possible cancellation of a rejection, not in terms of the probability of a mistaken rejection tending to zero. And the connexion between rejection and falsity has been shown by noticing that, if the proportion of possible mistakes is less than k, whatever positive number k may be, there can be no possibility of mistake. In the positive exposition the identification of 'having a very low probability' with 'being practically impossible', and of 'having a very high probability' with 'being practically certain', has never been made. The whole exposition has been in terms of a complicated and prima facie arbitrary rule which makes use of the notion of a rejection which may have to be cancelled. The way in which the number of observations enters into the rationale of the rejection procedure is merely that, the smaller k is chosen to be, the larger n must be taken to be for the rejection to be of the same strength.

The effect of increasing n throws light upon a question which may occur to a mathematically knowledgeable reader. The question may be put in this form: "You have given as a rule of rejection the rule to reject the hypothesis H_p if the α-ratio in a set of n observations lies outside the interval $[p-\sqrt{(pq/nk)}, p+\sqrt{(pq/nk)}]$. You have justified this rule by quoting the theorem of class-ratio arithmetic to the effect that the proportion in a hyperset of those sets of n observations whose α-ratio lies outside this interval is some number which is less than k (a number which class-ratio arithmetic will enable to be exactly calculated in any particular case and to be approximated to when n is large). But you have not given any reason for your choice of the interval $[p-\sqrt{(pq/nk)},\ p+\sqrt{(pq/nk)}]$ rather than any other interval surrounding p which has the required property that the proportion in a hyperset of those sets of n

observations whose α-ratio lies outside this interval is less than k. For there are an unlimited number of such intervals besides the ones which include the interval $[p - \sqrt{(pq/nk)}, p + \sqrt{(pq/nk)}]$. Why have you taken this interval as your interval of reference instead of one of these others? In terms of your probability ellipses, why have you taken one of these ellipses to divide the fundamental square into a region of k-rejection and a region of k-non-rejection instead of some other equally good method of division? In short, why have you not used some equally good alternative rejection rule?"

The reply to these questions is that, although there are many other intervals besides the interval $[p - \sqrt{(pq/nk)}, p + \sqrt{(pq/nk)}]$ which have the required property, any hypothesis rejected by a rejection rule on account of lying outside one of these other intervals will be rejected by my rejection rule on account of lying outside my interval $[p - \sqrt{(pq/nk)}, p + \sqrt{(pq/nk)}]$ for some value of n. This follows from the fact that the system of intervals $[p - \sqrt{(pq/nk)}, p + \sqrt{(pq/nk)}]$ forms a Chinese-box system for increasing n, the intervals in the system converging to the point p as n increases without limit. And the corresponding system of probability ellipses forms a similar Chinese-box system converging to the straight line $p = r$. So any interval whatever which includes the point p, and any suitable region* whatever which includes that part of the straight line $p = r$ falling within the fundamental square, will include all the intervals in the Chinese-box system of intervals $[p - \sqrt{(pq/nk)}, p + \sqrt{(pq/nk)}]$, or all the ellipses in the Chinese-box system of probability ellipses, from some value of n onwards. Consequently every point falling outside the new interval, or falling outside the new suitable region, will fall outside all these intervals, or all these ellipses, for sufficiently large values of n; and hypotheses rejected by the new rule will be rejected by the old rule for sufficiently large values of n.

Moreover, if, instead of considering intervals which are such that the proportion of sets which have α-ratios falling outside the interval is *less than* k, we consider instead intervals which are such that this proportion is *equal to* k, we find that all Chinese-box

---

* 'Suitable region' because the region must also fulfil the condition that every line $p = p_1$, for $0 < p_1 < 1$, cuts the boundary of the region not more than twice within the fundamental square.

systems, each of which converges to a point,* are equivalent, in a sense of "equivalent" appropriate to their use in rules of rejection. For if I_n (for $n = 1, 2, \ldots$) is an interval of the Chinese-box system of intervals $[p - \sqrt{(pq/nh)}, p + \sqrt{(pq/nh)}]$, where h is a number such that the proportion of sets lying outside each interval is equal to k, and J_m (for $m = 1, 2, \ldots$) an interval of any other point-convergent Chinese-box system of intervals with this same property, then not only for every m will there be an n which is such that the Chinese-box subsystem starting with I_n will fall wholly within J_m but also vice versa, i.e. for every n there will be an m such that the Chinese-box subsystem starting with J_m will fall wholly within I_n.† Thus the two Chinese-box systems will, as it were, interlock; and every hypothesis rejected by a rule which makes use of one system will be rejected by a rule which makes use of the other, though the number of observations required for rejection in one system will usually be different from the number required for rejection in the other. Since if two of these Chinese-box systems both interlock with the system using the intervals

$$\left[p - \sqrt{\frac{pq}{nh}}, \ p + \sqrt{\frac{pq}{nh}} \right]$$

they interlock with one another, all such Chinese-box systems are equivalent in the sense that rejections in the one will go with rejections in the other.

Any rejection rule will therefore serve our purpose of giving a meaning to probability statements if it is of the form: Reject H_p if the α-ratio in a set of n observations falls outside the nth interval in a system of intervals I_n which is such that the proportion of n-fold sets having an α-ratio lying outside the nth interval in this system is equal to k. Owing to the equivalence of all rejection rules of this form, it seems reasonable to choose a simple rule; and this accounts for my choice of the $[p - \sqrt{(pq/nh)}, p + \sqrt{(pq/nh)}]$ system of intervals upon which to base the rule. The rule has been further simplified by taking h as equal to k, and by justifying the

* A Chinese-box system of intervals converges to a point (is 'point-convergent') if the non-decreasing sequence of the left end-points of the intervals and the non-increasing sequence of the right end-points of the intervals both tend to the same point as n increases without limit.

† Both point-convergent Chinese-box systems converge to the same point whose value is p.

rule on the ground that it makes the probability of a mistaken
rejection less than instead of equal to k. By taking it in this
simplified form it has been possible to justify it with the use of
only elementary mathematics.

The equivalence of all Chinese-box systems explains the fact that
it is not incorrect to express the rule of rejection in the imprecise
form of a rule to reject a statistical hypothesis if an observed α-ratio
deviates from the value predicted by the hypothesis by more than
a certain amount, this required certain amount being smaller the
larger the number of observations. It is not wrong, but it is
imprecise. It was necessary for the philosophical discussion of
this chapter to expound the rule in a precise form, subsequently
explaining that any other precise form satisfying certain conditions
would have served our purposes equally well.

WHY THE REJECTION RULE IS IN TERMS OF INTERVALS

There is one further criticism that can be made against the form
of my definition which must be considered. The k-rule-of-rejection
that has been given corresponds to the rule used by most statisticians in their 'significance tests', namely, to reject a hypothesis
if the 'sample point' corresponding to a set of observations falls
into the 'critical region', the probability of a point falling into
which is below the 'significance level'; though they would not use
such a test (as I have done) as itself giving the definition of probability. This procedure has been criticized by Harold Jeffreys on the
ground that its reason for rejecting a hypothesis is not only the
improbability, given the truth of the hypothesis, of finding the
results which have been observed, but is also the improbability,
given the truth of the hypothesis, of finding results which have
not been observed and which, given the truth of the hypothesis,
are more improbable than those which have been observed. Thus
"a hypothesis that may be true may be rejected because it has not
predicted observable results that have not occurred. This seems
a remarkable procedure. On the face of it the fact that such events
have not occurred might more reasonably be taken as evidence
for the law, not against it."* This very plausible criticism leads
one to ask: Should not the proper criterion for the rejection of a
statistical hypothesis be sought in the proportion of sets having the

* *Theory of Probability* (Oxford, 1939), p. 316.

actual α-ratio observed, and not in the proportion of sets having an α-ratio lying in a range within which the observed α-ratio falls?

The difficulty about using the actual α-ratio as a criterion for the definition of probability is that, whereas class-ratio arithmetic proves that the proportion of sets having an α-ratio lying in the interval $[p-d,\, p+d]$ is an increasing function of n (it is not less than $1 - pq/nd^2$, by the Bienaymé-Tchebichef inequality), the same arithmetic proves that the proportion of sets having any particular α-ratio whatever is a decreasing function of n, which tends to zero as n increases without limit. Thus a set which has exactly p as its α-ratio is in no different position from a set having any other α-ratio as the number of observations increases.* It is therefore necessary to collect together sets with different α-ratios in order to secure classes whose proportions do not tend to zero as n increases without limit; and the natural way to collect the sets into such classes is by means of intervals of class-ratios. Once we have done this collection the class-ratio arithmetic developed in this book shows that, with increasing n, a larger and larger proportion of the total number of possible sets have α-ratios lying within any given interval including p, and that a smaller and smaller proportion of the possible sets have α-ratios lying within any given interval which does not include p. As n increases, the sets become more and more concentrated in any given interval surrounding p, and more and more sparse in any other interval or group of intervals. Since their sparseness in the region outside an interval surrounding p (a sparseness proved by the Bienaymé-Tchebichef inequality) entails their sparseness in any intervals included in this region, to use a rejection rule based upon sparseness in this region is weaker than to use one based upon sparseness in an interval included in this region—the interval, for example, which comprises only a small neighbourhood of the α-ratio observed.

Jeffreys's argument is that it is fallacious to "reject a hypothesis on the basis of observations that have not occurred".† But the set of observations with an observed α-ratio has got to be compared with something; since it is the only thing that has been observed, nothing we compare it with will have been observed.

---

* The proportion of such sets tends to zero more slowly than does the proportion of sets having any other α-ratio. Nevertheless, it tends to zero.

† *Theory of Probability*, p. 319.

Jeffreys suggests that it would be preferable to compare it with the "most probable result",* which is a set of n observations having the α-ratio p_1, where np_1 is the integral part of $(n+1)p$;† but this most probable result has not occurred; had it occurred a 'significance test' rule would not wish to reject the hypothesis. The procedure of dividing the interval $[0, 1]$ of possible values of the α-ratio into a subinterval surrounding p and the rest of the interval, and of comparing the number of sets with α-ratios lying in the two parts of this dichotomy, would seem preferable to any other method of comparison, if we take into account the fact that increasing the value of n concentrates more and more of the possible sets into the selected subinterval surrounding p.‡

The strongest reason, however, for this dichotomy and for taking into account in the rejection rule all the α-ratios within an interval $[p-d, p+d]$ surrounding p, arises when we come to consider not the hypothesis that the probability parameter has a specific value p, but the hypothesis that the probability parameter has some value lying in a specific interval $[p_1, p_2]$. Now the statistical hypotheses with which we are concerned in empirical tests (except when the hypotheses appear as lower-level hypotheses deducible from other statistical hypotheses) never in fact attribute a specific value p to a probability parameter, but state that the probability has a specific value p within a certain degree of approximation d, i.e. that it lies within the interval $[p-d, p+d]$. So for these hypotheses there will be no one 'most probable' α-ratio, but a 'most probable' range of α-ratios—the interval $[p-d, p+d]$; and any plausible rejection test will have to be in terms of the number of sets whose α-ratio lies somewhere within a range. A logical extension of the k-rule-of-rejection proposed in this chapter would reject the hypothesis that p lies in the interval $[p_1, p_2]$ if the α-ratio in n observations falls

* Op. cit. p. 317.

† If $(n+1)p$ is itself an integer, α-ratios of $(n+1)p/n$ and of p both yield 'most probable' results.

‡ The difference between the criterion suggested by Jeffreys—that of rejecting the hypothesis if the ratio of the number in the hyperset of sets of n observations having the observed α-ratio to the number in the hyperset of sets of n observations having the 'most probable' α-ratio is less than k—and my k-rule criterion diminishes as n increases. The alternative criterion may well be preferable in some inductive problems; but we are concerned here with the use of a rejection rule for the definition of probability, with which (it must be remembered) Jeffreys is not concerned.

outside the interval $[p_1 - \sqrt{\{p_1(1-p_1)/nk\}}, p_2 + \sqrt{\{p_2(1-p_2)/nk\}}]$. As n increases without limit, this interval contracts towards $[p_1, p_2]$, the interval mentioned in the hypothesis. The case of what mathematical statisticians call a 'simple' hypothesis attributing a unique value to the probability parameter may then be considered as a degenerate case of this more general procedure—the case in which $p_1 = p_2$ and the interval $[p_1, p_2]$ becomes the degenerate interval consisting of the single point p_1.

PROBABILITY IN THE CONTEXT OF GAMES OF CHANCE

Though probability statements used in connexion with games of chance are not scientific statistical hypotheses, one feature of them throws light upon one important fact about the way in which statistical hypotheses are used within a scientific system; so it will be convenient briefly to discuss the meaning of such probability statements.

Let us take, as a typical example, the statement that the probability of throwing a five with a die is $\frac{1}{6}$. In certain contexts (e.g. expositions of elementary probability theory) this statement will be taken not as an empirical statement referring to an actual die or to a class of actual dice but as a statement about an 'ideal' die used to illustrate a theorem in pure probability theory—perhaps the theorem that, if there are m exclusive and exhaustive alternatives, the probabilities of each of which are equal, the probability of any one of these alternatives is $1/m$. In this case the statement expresses a logically necessary proposition which logically follows from the definition of an ideal die as a machine for yielding equal probabilities to six alternatives; the statement is about a die only in the same 'vacuous' sense as the statement that two apples and two other apples make altogether four apples is about apples. This analysis holds also of the statement which appears to refer to an actual die ("This die") or to a class of actual dice ("These dice") but is taken as meaning that, if this die is an *ideal* die (or if these dice are *ideal* dice), the probability of throwing a five with the die (or with each of the dice) is $\frac{1}{6}$.

Leaving aside these cases in which the statement expresses a logically necessary proposition, let us pass to the interesting cases in which the statement expresses a logically contingent proposition which attributes an empirical property to an actual die or to a class of actual dice. If the statement is to be taken about a class of dice, e.g. the class of all the dice on this table, or the class of all the dice manufactured in a certain factory in a certain year, it will express the general proposition attributing the empirical property to every member of a restricted class. There are very few contexts in which the statement would be understood as expressing a general proposition about all dice, past, present and future, since such a statement would certainly be false if taken as expressing a contingent proposition about all actual dice; but if so understood the statement would attribute an empirical property to every member of an actual class.

What is this empirical property of throwing a five with a probability of $\frac{1}{6}$? The statement that a particular die has this property means, as C. S. Peirce has put it, that "the die has a certain 'would-be'; and to say that a die has a 'would-be' is to say that it has a property, quite analogous to any *habit* that a man might have". Peirce goes on to say that "just as it would be necessary, in order to define a man's habit, to describe how it would lead him to behave and upon what sort of occasion—albeit this statement would by no means imply that the habit *consists* in that action— so to define the die's 'would-be', it is necessary to say how it would lead the die to behave on an occasion that would bring out the full consequence of the 'would-be'; and this statement will not of itself imply that the 'would-be' of the die *consists* in such behavior".*

This striking description of Peirce's corresponds to saying, in the terminology used in this book, that the property of throwing a five with a probability of $\frac{1}{6}$ is a theoretical concept occurring in the hypothesis expressed by the sentence "The probability of throwing a five with this die is $\frac{1}{6}$", and that the method of testing this hypothesis is by observing how the die falls on occasions on which it is thrown, though the hypothesis is not identical with any set of observable propositions and the 'would-be' property of the die is not a logical construction out of observable properties.

* *Collected Papers of Charles Sanders Peirce*, vol. 2 (Cambridge, Mass., 1932), §2·664; *The Philosophy of Peirce*, ed. J. Buchler (London, 1940), p. 169. Peirce wrote these notes in 1910. The property of a die to which Peirce is referring is that of throwing a number divisible by 3 with a probability of $\frac{1}{3}$, but his remarks are equally applicable to the simpler property I have taken as an example.

The only way in which the hypothesis that the probability of throwing a five with this die is $\frac{1}{6}$ differs from a scientific hypothesis is that it is insufficiently general. It is, of course, a general proposition, since it is not about one particular throw of the die, but is about throws of the die generally; but the proposition is limited to being about this particular die. Even the proposition, understood as a contingent proposition, that all actual dice have this 'would-be' property is not sufficiently general to be a proposition of science. But the way of functioning of a probability statement about this particular die is the same as that of a scientific statistical hypothesis; the statement about the die is given its meaning by its place in a statistical deductive system in which it is testable by experience and rejectable on the basis of experience in accordance with a k-rule-of-rejection. The experience with which it is to be confronted in these tests is to be provided by observations of the proportions of throws of five to be found in sets of throws of the die. There is thus no logical difference between the way in which meaning is attached to probability statements within a science and the way in which it is attached to logically contingent probability statements occurring in connexion with games of chance.

There are many people, I believe, who, though they might be satisfied with the account of probability in scientific statistical hypotheses given in this chapter—an account in terms of the empirical criteria for their rejection—would feel an uneasiness about applying the same account to the probabilities in games of chance. This is due, I believe, to the fact that the reasons why we think of the scientific statistical hypothesis in the first instance are usually quite different from those which lead us to think of the game-of-chance probability statement. In the case of all the scientific hypotheses of the social sciences and for a great many of those in the biological and physical sciences, the numerical value which we assign to the probability parameter in the hypothesis is suggested to us by the class-ratio, or the range of class-ratios, found in sets of observed cases. No one, for example, would have even put up for consideration the hypothesis that 51% of births are births of boys if there were not a great amount of statistical evidence of sets of births with male-birth ratios of approximately this amount. But in the case of games of chance the method by which the numerical value is in the first instance assigned to the probability in the probability statement is usually quite different. If we are presented with a die which we have not seen before and are asked to ascribe a probability to throwing five with it, we shall be prepared to assign a number without having any data before us as to how many times fives have appeared when it was thrown in the past. All we require is to verify that it really has six faces, that only one of these faces has five dots on it, and that the die looks and feels symmetrical. We are then prepared to ascribe equal probabilities to the six alternative ways in which the die can fall, and from the equality of the probabilities of the six exhaustive and exclusive alternatives we can deduce, by pure probability theory, that the probability of any one of these alternatives (e.g. that of throwing a five) is $\frac{1}{6}$.

But although we are prepared in cases like that of the die to assign a value to the probability parameter without having any evidence of a statistical frequency, the assignment is always a tentative one to be abandoned if it does not sufficiently fit the class-ratios in sets of observations. However symmetrical the die may look and feel to our senses, and however symmetrical it may be according to any mechanical test, and however satisfied we may be that it is being thrown in a fair manner, yet we shall reject the hypothesis that the probability of its throwing a five is $\frac{1}{6}$ if the proportion of fives in a sufficiently large set of throws deviates sufficiently from $\frac{1}{6}$. The fact that the die has satisfied the prior test for unweightedness will have the effect of making us choose a small value for k, so that we shall only reject the hypothesis if, given the truth of the hypothesis, it is very improbable that we should have found the five-ratio we did find in our observed set of throws. But reject the hypothesis we shall, if the statistical evidence becomes strong enough. For if we decline to reject the hypothesis however much statistical evidence we have against it, on the ground that the die really is unbiased and is only appearing to be biased, we are treating the probability statement not as expressing a contingent hypothesis about a real die, but as expressing a logically necessary proposition about an ideal, unbiased die.

The same situation arises with a statistical hypothesis in an advanced science where the statistical hypothesis is itself deducible from a higher-level hypothesis which is strongly supported by evidence which is not direct evidence for the lower-level statistical hypothesis. Here we may well require very strong direct statistical evidence before we shall be prepared to reject the lower-level hypothesis, for its rejection will require rejection of the strongly supported higher-level hypothesis of which it is a logical consequence. And sometimes the independent evidence for the higher-level hypothesis is the evidence in favour of the equality of the probabilities of a number of alternative hypotheses, the alternatives being exclusive and exhaustive, just as in many games of chance. Vice versa, there can quite well be games of chance in which we are as little prepared to make any guess as to the probabilities before we have examined statistical evidence as we are about the male-birth ratio. The games of chance in which we start by assuming equalities of probabilities are games which we have designed for this purpose, making use of the fact that there are a great many physical systems (e.g. falls of dice, rotations of pointers) in which we know that a very small difference in the cause of a motion, a difference too small to be controllable by human agency, produces a very great difference in the effect. The importance of systems of this nature in connexion with assignments of equal probabilities was pointed out by Henri Poincaré;* the reason for ascribing equal probabilities to the six possibilities of fall of the die is not the negative reason that we have no reason for supposing that the die will fall with one face up rather than any other, but is the positive reason that the die is a Poincaré-system in which an imperceptible change in the position of throwing or in the direction or velocity of spin with which the die is thrown makes all the difference to the face it lands upon.

If it is agreed that contingent probability statements used in connexion with games of chance are given a meaning by a rule of provisional rejection in the same way as are probability statements within a science, the comparison will emphasize a feature of the latter which may perhaps be neglected. The reasons for believing a statistical hypothesis need not only be statistical evidence; indeed, they need not be statistical evidence at all. Direct

---

* *Calcul des Probabilités* (Paris, 1896), pp. 127 ff.; *La Science et l'Hypothèse* (Paris, 1902), pp. 233 ff. (English translation, London, 1905, pp. 201 ff.); *Science et Méthode* (Paris, 1908), pp. 67 ff. (English translation, London, 1914, pp. 67 ff.).

evidence for a statistical hypothesis would have to be statistical; but a statistical hypothesis in science, like a proposition about the probabilities in a Poincaré-system game of chance, can occur in a deductive system as a logical consequence of higher-level hypotheses for which there is independent evidence. Nevertheless, however strong the evidence, direct or indirect, may be for the statistical hypothesis, whether in science or in a game of chance, a sufficient amount of contrary direct evidence of class-ratios in sets of observations will serve to cause us to reject the hypothesis. So there is no inconsistency in holding both that the meaning of a probability statement is given by its rejectability by appropriate statistical evidence and that it can be reasonably believed independently of statistical evidence.

The view of probability, in science and in games of chance, which I have been advocating may properly be termed a *frequency* theory of probability, since the criteria for the meaning of probability statements are to be found in their being rejected or failing to be rejected on the basis of the frequencies (i.e. class-ratios) in observed sets.* But if my theory is called a frequency theory it should be remembered both that my view does not identify probability statements with statements about observable frequencies— the relations between the two statements is the complicated one described in this chapter—and that reasonable belief in probability statements may be based upon quite other data than those concerned with frequencies which correspond to the probabilities.

THE 'THEORETICAL' CHARACTER OF STATISTICAL HYPOTHESES

A statistical hypothesis expressed by the probability statement that the probability of a member of β being a member of α is p has been identified with the proposition that the α-parameter of the $\{\alpha, \beta\}$-probability hyperclass is p, this proposition being subject to empirical test by an observed set of n instances of β being treated as being a selection from this hyperclass. This chapter has been principally concerned with explaining how this empirical test works by being able provisionally to reject the hypothesis, the provisional

---

* Some writers (e.g. Kneale) would restrict the title of frequency theory to theories of the form advocated by von Mises and Reichenbach, in which probabilities are identified with *limits* of observed frequencies.

character of this rejection being the distinguishing characteristic of statistical hypotheses.

But there would seem to be two puzzling questions which have not yet been cleared up. First, what determines the size of the member-classes of the probability hyperclass? The numbers in these classes have been referred to as $m_1, m_2, ..., m_n$; what determines the values of these n numbers? Secondly, if a set of n_1 instances is to be regarded as a selection from a probability hyperclass and a selection of n_2 instances is also to be regarded as a selection from a probability hyperclass, these two probability hyperclasses must be different, since the first has n_1 member-classes and the second n_2 member-classes. How then can the probability p be identified with the probability parameter of one of these hyperclasses rather than with that of the other?

Let us put the questions concretely with reference to the statistical hypothesis that the probability of a child being born a boy is $\frac{51}{100}$. If this proposition is identified with the proposition that the maleness-parameter of a {maleness, human birth}-probability hyperclass is $\frac{51}{100}$, and particular sets of births are identified with selections from this hyperclass, first, what determines the sizes of the classes which are members of the hyperclass one member out of each of which forms the selection which is a particular set of births, and secondly, how can a set of one hundred births and a set of one thousand births both be selections from the same hyperclass?

The answers to these questions appear when we realize the 'theoretical' character of the probability hyperclass concerned. Its function is to enable deductions to be made about the numbers of selections from it which have certain class-ratios, and for this purpose it makes no difference what are the numbers in its member-classes, for these numbers $m_1, m_2, ..., m_n$ appear in both the numerators and the denominators of the fractions in the calculation, and so cancel out. They are like the phase factors in the Schrödinger wave-equations of quantum mechanics, which disappear in the process of the deduction of observable consequences. As for the second question, it may be answered in either of two ways. One way is to take the probability statement as asserting that *every* {maleness, human birth}-probability hyperclass has a parameter p. The disadvantage of this way of regarding the matter is that it conceals the theoretical character of the probability hyperclasses concerned; to say that the probability of a child born being a boy is $\frac{51}{100}$ is not to make a remark about every actual hyperclass of maleness-equiproportional classes of human births, but is to propound a hypothesis about the theoretical concept of such a hyperclass. It was to emphasize the theoretical character of the probability hyperclasses concerned that they have usually been spoken of in the singular—in the form of supposing a set of instances to be a selection from *a* probability hyperclass. If the point be pressed that the hyperclasses must be different if the numbers in the selections are different, a second way of meeting this point is preferable—that of taking the probability statement as asserting that every probability hyperclass concerned, e.g. B_1 with n_1 members, B_2 with n_2 members, etc., is a subclass of a probability hyperclass B_0 whose probability parameter is p. For it follows from the definition of a probability hyperclass that every non-empty subclass of it is a probability hyperclass having the same probability parameter.

If the probability hyperclass B_0 is to include subclasses with any finite number of members, it will itself have an infinite number of members. But there is no logical objection to this, since the classes whose α-ratios are to be equal are the finite classes which are the members of the infinite probability hyperclass B_0, and nothing is said about proportions in this infinite hyperclass itself.* Indeed, the infinitude of B_0 emphasizes the theoretical character of the concept: to treat a set of n observations as an n-fold selection from n classes which are members—some of the members—of an infinite probability hyperclass calls attention to the fact that the n-fold 'selection' is not to be regarded as being made by a process of picking one member out of each class of some of an infinite class of classes before us, like an n-handed Briareus picking one ball out of n of the infinite number of bags before him; but is rather a relationship between the theoretical concept of probability hyperclass (and the probability parameter assigned to it) and the actually observable sets.

Why there is a temptation to take the 'selections' in a more literal sense than is appropriate to a system of relationships

---

* Since all the selections concerned are selections from finite hyperclasses which are subclasses of the infinite hyperclass, no multiplicative axiom (axiom of choice) is required to ensure the existence of the selections.

between sets of observable entities and theoretical concepts is because, for the sake of simplicity, the class-ratio arithmetic has been expounded in the form of mathematical theorems about selections from classes of classes. But the notion of selection from a class of classes is a special case of that of selections from relations,* and the class-ratio arithmetic holds of this more general case. It would be out of place here to expound the pure deductive system of selections from relations, especially as the mathematical considerations involved are exactly the same as those concerned in selections from classes of classes; but it is worth giving a model of this system to be compared with the model for the other system of a three-handed Briareus drawing from three bags each containing two balls (p. 132). The comparable model is that of the number of ways of drawing a system of straight lines joining each

LLL   LLR   LRL LRR   RLL RLR RRL   RRR

Fig. 8.

of three lower points to one, and only one, of two upper points (see Fig. 8). The three-letter symbols under each of the eight figures indicate respectively whether the first, the second and the third of the three lower points is joined to the left or to the right of the two upper points; expressing the eight ways of drawing the system of lines by these eight symbols shows their formal equivalence to the eight ways of drawing three balls one from each of three bags each containing two balls.

If we think of statistical hypotheses as highest-level hypotheses in a deductive system for which diagrams of this sort are models, we shall tend to express their relationships to propositions about class-ratios in sets of observations in some such language as that of the sets of observations being 'representatives' of the theoretical concepts or of being their 'progeny'. The words describing their relationships will be so obviously metaphorical that there will be little danger of our failing to recognize the theoretical character of

* A. N. Whitehead and Bertrand Russell, *Principia Mathematica*, vol. 1, Part 11, Section D.

the concepts to which the observed entities are related by these relationships.

It is, however, verbally much lengthier to discuss probability deductive systems in the language of selecting from relations than in the language of selecting from classes of classes, and this is the reason why the latter procedure has been preferred. But we must then be careful not to ask questions (e.g. as to the number of members in the classes of a probability hyperclass) which have no parallel in the alternative form of discussion, and not to take the language of selecting as presupposing a similar status of the set of things selected and of the members of the hyperclass from which the set is 'selected'. The selection language must be regarded as *as-if* language useful for enabling the mathematics to be expressed in a convenient form.

CONCLUSION

The peculiarity of probability statements (other than those assigning probabilities of 0 or of 1), according to the view expounded in this chapter, is that the empirical criterion for rejecting them is never definitive, but is always provisional, so that rejection by the criterion is capable of being cancelled by further experience. A probability statement cannot therefore be identified with any statement of frequencies in any set of actual instances. But, as has been said, my view can properly be called a "frequency view of probability" in that the meaning of the probability statements are all given in terms of their rejectability (albeit provisional rejectability) by observable frequencies. After all that has been said in this chapter there should be little danger of misunderstanding in speaking, as will be done in the next chapter, of probability statements as being equivalent, *by and large*, to assertions of frequencies (i.e. class-ratios) in sets of observations, the qualification 'by and large' being intended to refer to the whole gamut of logical and epistemological considerations developed in this chapter.

CHAPTER VII

THE CHOICE BETWEEN STATISTICAL HYPOTHESES

Two of the problems of the logic of scientific hypotheses are the same whether or not the hypothesis in question is a statistical one. The problem of the meaning of a statistical hypothesis has been treated in the last two chapters on lines suggested by the logic of universal hypotheses. Just as the statement of a universal hypothesis was given a meaning in terms of the observations on the basis of which it was to be rejected, and definitively rejected, so the statement of a statistical hypothesis has been given a meaning in terms of the observations on the basis of which it was to be rejected, though provisionally and never definitively. Thus the solutions of this problem in the two cases are in terms of rejection; in both cases it is a rejection procedure which gives meaning to the statements of the hypotheses. Until this problem had been solved no further progress could be made. The second problem—that of the ultimate justification for accepting a scientific hypothesis—is also similar, indeed, it is the same, for the two types of hypotheses. But before treating of this I wish to discuss a third problem which arises for statistical hypotheses in a different way from that in which it arises for universal hypotheses—the problem of the reason for preferring one statistical hypothesis to another.

Of course questions arise as to whether evidence supports one universal hypothesis (e.g. Every B is A) more than it supports a rival one (e.g. Every C is A); and adherents of a non-empirical theory of probability who have attempted to produce a formal logic of induction (Harold Jeffreys, J. M. Keynes, Rudolf Carnap) have included the rationale of the choice between rival hypotheses within its scope. None of these writers, it seems to me, have produced criteria for making such a choice which are both plausible and definite enough to be useful. If the standpoint of this book is correct, there are no such general criteria for making such choices; the extent to which a hypothesis would fit into the body of scientific knowledge is highly relevant to its claim for incorporation into such a body of knowledge, and it is difficult to see how a formal logic of 'credibility' or 'acceptability' or 'confirmation' can help in the matter, beyond pointing out such obvious truisms as that if a hypothesis is supported by good evidence, any lower-level hypothesis which is a consequence of it is also supported by good evidence and may well also be supported by other good evidence.

The situation is different in the case of statistical hypotheses.* Every statistical hypothesis (of the sort considered in this book)† asserts that a specified probability has a definite value, or one out of a definite range of values; and the question immediately arises, in considering it, why that particular value (or range of values) rather than any other? To consider the hypothesis that the probability is 51% that a birth is the birth of a boy is well-nigh impossible without also comparing it with the rival hypotheses that this probability is 50% or is 52%, or is any other specific percentage. Indeed, in developing the logic of the meaning of probability statements, I have had the greatest difficulty in avoiding bringing in such a comparison, and I may well have laid myself open to criticism for this avoidance. But whether or not I am right in my contention that an account of the meaning of probability statements can be given in isolation, there is no question but that any grounds for *believing* a probability statement must be grounds for believing in that statement rather than in another one which ascribes a different probability to the same proposition on the same evidence. Moreover, many of the probability statements (statistical hypotheses) with which we are concerned in science—and almost all of those in the social sciences—have little indirect support; the evidence for them is entirely, or almost entirely, the direct evidence of the agreement of their consequences with observed facts. It is easy to see that finding a black swan in Australia makes the hypothesis that all European swans are white preferable to the hypothesis that all swans are white. But there is no obvious way of

---

* When the rival to a universal hypothesis is not another universal hypothesis but a statistical hypothesis, the universal hypothesis is being regarded as a statistical hypothesis ascribing a probability of 1.

† More complicated statistical hypotheses asserting that a conjunction or an alternation of a set of specified and related probabilities have a set of assigned values, and statistical hypotheses concerned with continuous probability distributions. are not being considered.

seeing whether the fact that I draw (replacing after each draw) eight black balls in twenty draws from a bag which I know contains ten balls which are either six black and four white or are three black and seven white, makes the first of these hypotheses preferable to the second or vice versa.

To put the matter another way: For universal hypotheses the logic of the relation of the direct evidence to the hypothesis is either very simple or involves only mathematical difficulties; if the evidence follows, either immediately or by a complicated mathematical deduction, from the hypothesis, the hypothesis is supported by the evidence; and, if not, it is not so supported. So far as direct evidence is concerned all that is required to justify preferring one hypothesis to another is that the evidence for the preferred hypothesis includes all and more than all the evidence for the non-preferred hypothesis. Given the same direct evidence the choice between one universal hypothesis and another depends upon the indirect evidence. But for alternative statistical hypotheses which ascribe different probabilities to the same proposition on the same evidence the choice between them may depend upon this direct evidence itself. Just as the rejection of a statistical hypothesis by observation is not a matter of deductive logic but is a matter of a special rule of rejection, so the choice between alternative statistical hypotheses is not a matter of deductive logic but is a matter which requires first choosing a policy before making such choices. And it is by no means obvious what policy is best, nor indeed what is meant by calling a policy a 'best' policy for this purpose.

Since the pioneer work of Sir R. A. Fisher from 1912 onwards mathematical statisticians have developed various such policies. These have been devised for different types of choice, and have been called by different names. Policies devised for deciding whether or not the hypothesis that the value of a certain probability differs from a certain number should be preferred to the hypothesis that it is equal to it have been called "significance tests"; policies devised for preferring one value (or range of values) to any other value (or range of values) have been called "tests of estimation" ("point estimation" or "interval estimation" respectively). Many of these tests were first put forward as *ad hoc* tests for particular problems arising in applied statistics, and were presented without much attempt at rational justification. It is only in the last twenty years that general principles governing such policies have been evolved by Jerzy Neyman, E. S. Pearson and other mathematical statisticians. Recently, Abraham Wald has worked out a completely general 'decision procedure' for what he calls "statistical inference".* All these writers have developed their methods with an eye upon the practical application to cases in which the statistical hypotheses are not singular probability statements but are statements about functional laws about continuous probabilities (such as that the probability that an Englishman has height x is a certain function of x in which the parameters take certain values); consequently a great deal of complicated mathematics is involved which makes it difficult to distinguish the logical wood from the mathematical trees. The logical skeleton of the procedures can be seen, however, in the simplest cases; since Wald's treatment raises a point of great philosophical interest I shall in this chapter expound what is essentially Wald's 'minimax' policy in connexion with an example which I have constructed to be as simple as possible.†

But before doing so there is one important point which must be made. The mathematical statisticians whom I have mentioned—Fisher, Neyman, Pearson, Wald—are quite properly not concerned with the philosophical question of the justification of induction which has been the question of supreme interest to philosophers. Consequently they frequently assume that if a criterion fails to reject a statistical hypothesis, then the statistical hypothesis is to be accepted. This is to suppose that the scientist is confronted with two and only two possibilities—that of rejecting the hypothesis and that of accepting it; and that the business of the statistical mathematician is to provide him with a policy for telling him which of these two courses he should follow when his observed evidence is

---

* Wald sketched his method in a brochure *On the Principles of Statistical Inference* (Notre Dame, 1942), Chapter VI, and developed it in a series of articles in *Annals of Mathematical Statistics*, vol. 10 (1939), pp. 299ff.; vol. 18 (1947), pp. 549ff.; vol. 20 (1949), pp. 165ff.; *Annals of Mathematics*, second series, vol. 46 (1945), pp. 265ff.; *Econometrica*, vol. 15 (1947), pp. 279ff. [The method is expounded systematically in Wald's book *Statistical Decision Functions* (New York, 1950), which I did not see until after this chapter was completed.]

† Though several writers on logic refer to Fisher's 'maximum likelihood' method, I know of only two works on logic, C. W. Churchman, *Theory of Experimental Inference* (New York, 1948) and Rudolf Carnap, *Logical Foundations of Probability*, which refer to Wald's work—or indeed to Neyman and Pearson's work which is as old as 1933.

of certain sorts. Statistical mathematicians frequently do this by a geometrical representation in which the possible observation values are represented as points in a 'sample space', which is to be divided into a 'region of acceptance' and a 'region of rejection'; and consider that their problem is to find a principle which will justify this dichotomy. All the scientist then has to do is to see in which of these two regions the point which corresponds to his set of observations falls, and to accept or to reject the hypothesis accordingly.

The philosopher will, of course, immediately point out that to fail to reject a hypothesis is an entirely different matter from accepting it, and that even if a rejection policy rejected every alternative hypothesis except one, this would not imply that that one should be accepted, except on the assumption that one of the alternative hypotheses must be true. But this need not be the case; it need not, for example, be the case either that 50% or less of births are male births or that more than 50% of births are male births; for there may quite well be no statistical hypothesis at all about the proportion of male births which is true. An estimation procedure of a mathematical statistician is really an answer to the hypothetical question: Assuming that there is a statistical law to the effect that the probability of a B being an A has the value p, what is the best estimate of this value? (Or, on this assumption, what is the best estimate of a range within which this value lies?)

There can be no objection to the mathematical statistician making this assumption and describing a region in his sample space where his principle instructs him to refrain from rejecting the hypothesis as the region of 'acceptance' of the hypothesis, provided that it is clear to all that 'acceptance' in this context means non-rejection. However, to change this negative sort of acceptance into a more positive sort requires following the general inductive policy of accepting hypotheses which have been confirmed and have not been refuted; and the justification for this policy raises the philosophical question of the fundamental justification of induction which will be the theme of the next chapter.

The simple case which we shall study in some detail is that of the policy for choosing, on the basis of experience, between two alternative statistical hypotheses. The general assumption behind this simple case is that it is to be supposed that one or other, but not both, of the two hypotheses is true, so that a policy for rejecting one will *ipso facto* be a policy for accepting the other and vice versa. With the reasons for accepting this general assumption this chapter will not be concerned; it will be concerned with the choice of a policy for preferring one statistical hypothesis to another, it being granted both that some statistical hypothesis is true and that the two hypotheses in question are the only claimants for the position of the true hypothesis.

'UTILITIES' AND 'MATHEMATICAL EXPECTATIONS'

The policy to be developed in this chapter for choosing between two alternative statistical hypotheses, which is based upon Wald's 'minimax' policy, depends essentially upon regarding belief in a hypothesis as having desirable effects if the hypothesis believed is true and undesirable effects if the hypothesis believed is false. The choice is to be determined by a comparison of the desirable effects of believing one hypothesis if it is true and the undesirable effects of believing it if it is false with the desirable and undesirable effects of believing the other hypothesis if it is true or false. Thus notions of measurable *value* cannot be avoided; the criteria used will be in terms of the arithmetical relationships of four values— the gain or the loss obtained by choosing one hypothesis if it is true or false respectively, the gain or the loss obtained by choosing the other hypothesis if it is true or false respectively. Different assignments of these four values will make a difference to the choices. But though numbers have to be assigned to these gains or losses in order to provide a basis for what will be called the "prudential" policy, it is irrelevant to the theory to which feature of the effects these numbers are to be assigned. The gains and losses considered may be quantities of happiness, or quantities of eudaemonia, or of joy, or of absolute goodness; there is no need for the logician in devising a choice-policy to settle any ethical question as to the nature of the ultimately desirable things. What the logician requires is that there should be some measurable criterion of the amount the scientist stands to gain by accepting a hypothesis which in fact is true and stands to lose by accepting a hypothesis which in fact is false. Just as parts of economics depend upon the assumption of a measurable utility which serves to determine economic choice, so this part of inductive logic depends upon the assumption of a measurable value—which it is most non-committal also to call "utility"—which serves to determine the choice between rival hypotheses. To the extent that these assumptions do not hold, the pure theories of utility-economics and of 'prudential' statistical inference (which are impure or mixed deductive systems in the sense of Chapter II) will not enable deductions of practical value to be made. But if the assumptions hold approximately, or hold approximately in a limited field, approximate criteria can be obtained which, *faute de mieux*, will be of great practical value.

The way in which the value elements—the 'utilities'—enter into our theory of choice between hypotheses is not direct, but is by way of the notion of 'mathematical expectation'. Suppose that p is the probability of a member of β being a member of α and a is the utility gained if the member of β is a member of α and b that gained if the member of β is not a member of α. Then if the utility is averaged over a large number of members of β, by and large this average utility will be

$$ap + b(1-p).$$

This by-and-large average will be called, in accordance with customary usage, the "mathematical expectation of the aggregate utility"—or shortly, the "mathematical expectation" when, as in this chapter, it will always be the mathematical expectation of a utility. Mathematical expectations, like the probabilities by which they are defined, refer not to particular members of a class β, but to every member of β; the traditional name "mathematical expectation" is an unfortunate one, since it carries the suggestion of a subjective act of expecting a utility on a particular occasion of a particular member of β being or not being a member of α, whereas what is concerned is a by-and-large average over a large number of occasions. When the mathematical expectation is negative it is equivalent to the notion of risk, the term used by Wald, who worked in terms of negative mathematical expectations. Since I shall be more cheerful in my treatment than Wald, and shall allow for gains as well as for losses, I prefer to use the term "mathematical expectation" instead of "risk" in spite of its misleading associations.

DETAILED DISCUSSION OF A SIMPLE CASE

Let us suppose that we have to choose between two statistical hypotheses H_1 and H_2. H_1 asserts that the probability of a member of a class β being also a member of a class α is p_1, H_2 that this same probability is p_2. We will exclude the case in which either of the hypotheses is a universal hypothesis; that is, we will suppose that $0 < p_1 < 1$ and that $0 < p_2 < 1$. If $p_1 = p_2$ the two hypotheses are identical and no choice is possible; so we will suppose that one or the other is the greater, and for convenience take p_1 to be the greater. The hypotheses are then logically exclusive; they cannot both be true. Suppose that we know that they are also exhaustive, i.e. that they cannot both be false. This knowledge will not be a piece of logical knowledge; it will have been obtained presumably as the result of empirical knowledge, e.g. that the bag from which we are drawing balls either contains one proportion p_1 of white balls or contains another proportion p_2 of white balls, or that the plants we are examining either all have a gene-ratio yielding a probability p_1 of exhibiting a certain characteristic or all have a gene-ratio yielding a probability p_2 of exhibiting that characteristic, the other possibilities not being permitted in the Mendelian deductive system which is being accepted. Suppose further that, besides this knowledge that the probability has only these two possible values, the only relevant knowledge we possess is that one set of n members of β has been examined and has been found to contain exactly s members of α. Then if we have to make the choice between preferring the hypothesis H_1 to H_2 and preferring H_2 to H_1, how are we to decide? That is to say, what is a rational justification for such a decision?

Instead of discussing this question in general I shall discuss it in a simple numerical case. Suppose $p_1 = \frac{3}{5} = 0 \cdot 6$ and $p_2 = \frac{3}{10} = 0 \cdot 3$. These values have been selected, after many thought-experiments, so that the geometrical representation I shall use should be of a suitably visible shape, and also so that the values of the probabilities derived from p_1 and p_2 which will be required should be multiples of $\frac{1}{100}$ and thus can be conveniently expressed as percentages or as decimals to two decimal places. Suppose further that the number n of the instances of β which we have examined is two. This may seem to make the whole problem ridiculous;

how, it may be asked, can any rational decision be made on the basis of so small a sample? But, in fact, a rational choice can be made; the whole logic of the procedure is shown in this very simple case, and I do not want my readers to have to do more arithmetic to follow me than is strictly necessary. Nothing in the general principles to be propounded depends either upon the probabilities of the hypotheses being such simple fractions or upon the number in the examined sample being so small. (A larger sample, of twenty instances, will be mentioned when the question of the size of the sample becomes relevant.)

On the basis of knowledge of a sample of two there are eight possible ways of deciding between the two hypotheses—eight alternative 'decision rules' or 'preference strategies'—according to whether H_1 or H_2 is to be preferred if the sample contains 0 or contains 1 or contains 2 members of α. It will be convenient to enumerate them in the following order, s being the number of members of α found in the sample:

T_1.   If $s=0$ or 1 or 2, prefer H_1 (i.e. always prefer H_1).

T_2.   If $s=1$ or 2, prefer H_1; if $s=0$, prefer H_2.

T_3.   If $s=2$, prefer H_1; if $s=0$ or 1, prefer H_2.

T_4.   If $s=0$ or 1 or 2, prefer H_2 (i.e. always prefer H_2).

T_5.   If $s=0$, prefer H_1; if $s=1$ or 2, prefer H_2.

T_6.   If $s=0$ or 1, prefer H_1; if $s=2$, prefer H_2.

T_7.   If $s=0$ or 2, prefer H_1; if $s=1$, prefer H_2.

T_8.   If $s=1$, prefer H_1; if $s=0$ or 2, prefer H_2.

Since each of these strategies is a complete strategy, i.e. deals with all the three possible ratios that may be found in the sample, the problem for us is to select a best strategy on rationally defensible grounds. There may, of course, be two or more 'optimum' strategies which are equally good and better than any other. We shall see that in our example this will not happen (except in extreme cases), so that an optimum strategy will be the best. The problem is therefore: What is a rational policy for selecting optimum strategies?

We must first calculate the probability that each of the eight strategies prescribes a correct choice of each of the two hypotheses, i.e. the probability that, if the hypothesis is true, the strategy in question will prescribe that that hypothesis should be chosen. For example, if H_1 is true and we follow strategy T_1, which prescribes that H_1 should be chosen irrespective of what is found in the sample, this strategy will always prescribe a correct choice; its probability of correct choice, given the truth of H_1, is 100%. But if H_1 is false, H_2 being the true hypothesis, this strategy will never prescribe a true choice, since it leads us always to select the false hypothesis; its probability of correct choice, given the truth of H_2, is 0%. The strategy T_2 will sometimes prescribe a correct choice if H_1 is true and H_2 false, and sometimes do so if H_2 is true and H_1 false. Its probability of correct choice, given the truth of H_2, is the probability, given H_2, of a set of two members of β comprising no members of α, which is $^2C_0 p_2^0 (1-p_2)^2$, i.e. $1.1.(\frac{7}{10})^2$, which is 49%. And its probability of correct choice, given H_1, is the probability, given H_1, of a set of two members of β comprising either one or two members of α, which is the sum of the probability of the set's comprising one member of α and of the probability of its comprising two members of α. The first of these probabilities is $^2C_1 p_1^1 (1-p_1)^1$, i.e. $2.\frac{3}{5}.\frac{2}{5}$, which is 48%; the second of these probabilities is $^2C_2 p_1^2 (1-p_1)^0$, i.e. $1.(\frac{3}{5})^2.1$, which is 36%; so that the sum is 84%, the probability, given H_1, of correct choice for strategy T_2.

The formula giving the probability of correct choice for each strategy is $\Sigma\,^2C_s p^s (1-p)^{2-s}$, where $p=\frac{3}{5}$ or $\frac{3}{10}$ according as the probability of correct choice is taken on the basis of H_1 or of H_2 being true, and where the sum is taken over those values of s for which the strategy prescribes that H_1 or H_2 (as the case may be) should be chosen. The pair of probabilities of correct choice x_i, y_i for each of the eight strategies are given in Table V (expressed as percentages) as calculated from this formula.

TABLE V. *Probabilities of correct choice*

Strategy T_i	Probability, given H_1, of correct choice (%) x_i	Probability, given H_2, of correct choice (%) y_i
T_1 (H_1 for $s=0$, 1, 2)	100	0
T_2 (H_1 for $s=1$, 2; H_2 for $s=0$)	84	49
T_3 (H_1 for $s=2$; H_2 for $s=0$, 1)	36	91
T_4 (H_2 for $s=0$, 1, 2)	0	100
T_5 (H_1 for $s=0$; H_2 for $s=1$, 2)	16	51
T_6 (H_1 for $s=0$, 1; H_2 for $s=2$)	64	9
T_7 (H_1 for $s=0$, 2; H_2 for $s=1$)	52	42
T_8 (H_1 for $s=1$; H_2 for $s=0$, 2)	48	58

We shall also require the probability that each of the eight strategies prescribes an incorrect choice on each of the two hypotheses, i.e. the probability that, if the hypothesis is true, the strategy in question will prescribe that that hypothesis should not be chosen. Since choosing a hypothesis and not choosing a hypothesis are, under the conditions of our problem, exclusive and exhaustive alternatives, the probability of incorrect choice for a strategy T_i if H_1 is true or if H_2 is true is $1 - x_i$ or $1 - y_i$ respectively, where x_i, y_i are the probabilities of correct choice for the strategy given the truth of the first or of the second hypothesis. Thus we can calculate from Table V, for example, the probability of incorrect choice for strategy T_3 given hypothesis H_1 as being $1 - \frac{36}{100}$, i.e. 64%.

It is at this point that considerations of value must be brought in. Suppose that if H_1 is true I gain a units of value (utility, happiness, eudaemonia, joy, or what you will) by choosing H_1—and of course acting on my choice; but that if H_1 is false I lose b units of value by choosing H_1. Suppose that I gain c units by choosing H_2 if H_2 is true, and lose d units by choosing H_2 if H_2 is false. Since gains and losses have been separately allowed for, none of the numbers a, b, c, d will be negative. Except for this restriction they can be any rational numbers whatever, integers or fractions.* Mathematical statisticians who have made use of measures of value in discussing statistical inference have allowed only for losses; for example, Wald's 'weight function' is a non-negative function measuring "the relative importance of the error committed by accepting" one hypothesis when another is true.† To allow only for losses is equivalent to taking $a = c = 0$. Since it seems unnecessarily gloomy to treat of statistical inference in terms of minimizing expectations of losses rather than in terms of maximizing expectations of gains, and since to allow only for losses or only for gains involves restricting the generality of the policy and neglecting an interesting possibility which will arise, my exposition will use all four quantities. Of course it will only be their ratios that will be relevant, since the scale of measurement

* The restriction to rational numbers is only required for the argument of pp. 233 ff. But since the mathematicizing of the theory of value has not advanced to the stage of requiring irrational numbers as measurements of value, it seemed best to exclude these from the beginning.

† *On the Principles of Statistical Inference*, p. 40.

of value is irrelevant; in fact, it will be shown that it is two numerical relationships between the four numbers (the four utility coefficients or utility values) which affect the choice of strategy.

These four quantities determine the mathematical expectations of each of the eight strategies given the truth of each of the two hypotheses. Suppose I use strategy T_i. If H_1 is true the probability of my correctly choosing H_1 is x_i, and the mathematical expectation of gain arising from choosing H_1 when it is true is accordingly ax_i. But the probability of incorrectly choosing H_2 when in fact H_1 is true is $(1 - x_i)$; so the mathematical expectation of loss arising from choosing H_2 when it is not true is $d(1 - x_i)$. So the net mathematical expectation yielded by following strategy T_i when H_1 is true is $ax_i - d(1 - x_i)$. If, on the other hand, H_2 is true, the mathematical expectation of gain arising from correctly choosing H_2 when it is true is cy_i, and the mathematical expectation of loss arising from incorrectly choosing H_1, which is false, is $b(1 - y_i)$, with a net mathematical expectation of $cy_i - b(1 - y_i)$. Calling the mathematical expectations for the strategy T_i, X_i for the case in which H_1 is true, and Y_i for the case in which H_2 is true, we have

$$X_i = ax_i - d(1 - x_i) = (a + d)\,x_i - d,$$
$$Y_i = cy_i - b(1 - y_i) = (b + c)\,y_i - b.$$

X_i depends only upon the values of x_i, of a and of d; Y_i only upon the values of y_i, of b and of c. For given values of a, b, c, d the pair of mathematical expectations can be calculated from the table of probabilities of correct choice, e.g.

$$X_2 = (a + d)\tfrac{84}{100} - d, \quad Y_2 = (b + c)\tfrac{49}{100} - b.$$

In the simplest case in which $b = d = 0$ and $a = c$, the case in which we stand to lose nothing if we choose a hypothesis which is false and stand to gain the same amount if we choose a hypothesis which is true, the sixteen mathematical expectations are proportional to the corresponding probabilities of correct choice; since their absolute values are irrelevant, they can then be taken as equal to these corresponding probabilities.

The essential point is that, by this method of calculation, a pair of numbers X_i, Y_i are found for each strategy T_i, the first measuring the mathematical expectation, given the truth of H_1, yielded by making the choice prescribed by T_i, the second measuring the mathematical expectation, given the truth of H_2, yielded by making this choice. It is the relations between these sixteen numbers that will form the basis for the 'prudential' policy for selecting an optimum strategy.

THE 'PRUDENTIAL' POLICY

The rational policy for selecting optimum strategies which I propose is the following. *First step*: from each pair of numbers giving the two mathematical expectations for a strategy, select that one which is the lesser; if these numbers are equal, select that number. We now have eight numbers, one for each strategy. *Second step*: select the greatest of these numbers; the strategy to which this number corresponds is to be regarded as an optimum strategy, and is to be the one employed.

It may be the case that there is no unique number which is the greatest of these numbers, but that two or more of them are equal and greater than any other. In this case a *third step* is required: select as optimum strategy that one among the strategies corresponding to these equal maxima for which the other mathematical expectation is the greatest; if two or more of them yield equal maxima for this other mathematical expectation, all of these will be equally optimum strategies.*

To illustrate this policy for selecting strategies let us look at Table V of probabilities of correct choice (p. 205), and take the numbers as measuring the corresponding mathematical expectations, which, as we have seen, is the case if $b=d=0$, $a=c$. The first step in following the policy is to look at the numbers along each row, and select the minimum (or a minimum if the two numbers are equal).† Thus we get 0% from the T_1 row, 49% from the T_2 row, 36% from the T_3 row, and so on down to 48% from the T_8 row. The second step is to find the maximum of these eight numbers, which is 49%, and to take as optimum strategy the strategy in whose row this maximum is found—which in our case is T_2.

---

* This, in fact, never happens with mathematical expectations derived from probabilities of correct choice for alternative hypotheses, unless $a=b=c=d=0$. So, except in this case, there will always be a unique optimum strategy if we follow the prudential policy.

† "Minimum" and "maximum" will always be used in senses which do not imply uniqueness, so that, for example, 50 will be said to be a minimum (and also a maximum) of the pair of numbers 50, 50.

If the 48% in the T_8 row had been 49%, there would have been two equal maxima. Then the third step would be required of comparing the other numbers in the two rows. The other number for the T_8 row is 58%, which is smaller than the other number for the T_2 row, 84%; so the strategy T_2 would still be the one to be taken.

When no complicating equalities occur, the policy may be shortly expressed as that of selecting that strategy the minimum of whose pair of mathematical expectations is the maximum for the possible strategies, or more briefly still, as the policy of *maximizing the minimum mathematical expectation*. Wald describes his similar policy inversely as that of minimizing the maximum risk and calls it the "minimax method". Analogously I could call my policy, expressed positively, the "maximin policy" and the strategies under it "maximin strategies"; but I prefer to call the strategies, and the policy prescribing them, "prudential", since their characteristic feature lies in manifesting the cardinal virtue of prudence.

For what will be the result of following the prudential policy? It will be to protect ourselves as much as possible against the vagaries of Nature. If we select a strategy different from the prudentially optimum one, we may reap a greater benefit, but we may just as well do much worse. In the case in which $b = d = 0$, $a = c$, so that the mathematical expectations correspond to the numbers in Table V, if we had selected strategy T_3 instead of strategy T_2, we should better our prospects if H_2 were to be true (from 49 to 91%), but we should worsen our prospects if H_1 were to be true (from 84 to 36%). Since the condition of our problem is that we have no basis whatever for preferring one of the hypotheses to the other except that provided by observation of the sample of two, to select T_3 instead of T_2 would be to trust to luck. It would be a lucky chance, and not due to forethought on our part, if we scored more by using T_3 than by using T_2; the selection of T_3 on a particular occasion might be fortunate, but it would remain imprudent. By employing T_2, and thus choosing hypothesis H_1 if we find either one or two α-specimens in our twofold β-sample, and hypothesis H_2 only if we find no α-specimens in this β-sample, we are obtaining the maximum advantage which is consistent with our complete ignorance of what Nature is doing, except for the information which Nature has provided for us in the observed sample.

It would be rash to maintain that to employ the prudential policy in choosing between hypotheses would be to use that policy which it would in all cases be the most reasonable one to use. The criteria for determining which action is the most reasonable one in particular circumstances are very various, and prudence is only one out of these criteria. But if we agree that to be prudent is one way of being reasonable, a policy which is the most prudent one will be the most rational policy in circumstances in which to act most reasonably is to act most prudently. The logic of prudence, provided by the theory of 'the prudential policy', will therefore be part—how important a part it is not for a logician to say—of the rationale of rational action.

RELATION TO VON NEUMANN'S THEORY OF GAMES

It is very easy to slip, as I have done, into anthropomorphic language in describing the prudential policy and to speak as if we were trying our hardest to secure the greatest advantage while a personified Nature was trying its hardest to contrive that we should not succeed. But, in fact, this metaphor is an excellent one in many ways; it is highly profitable to treat our problem on the analogy of how it would be reasonable to behave were we playing a game with Nature as opponent. The logic of games has been the subject of remarkable work by John von Neumann, who has worked out a complete theory of the rational policy for playing a general 'zero-sum two-person game', which is the analogue for the problem with which we are concerned.* A zero-sum two-person game is one in which there are two players playing against one another, and in which the gain or loss of each of the players is equal to the loss or gain respectively of the other. Von Neumann has shown that every such game, however complicated (e.g. chess), and however much chance may enter in (e.g. two-handed poker), is theoretically reducible to a 'normalized form' in which each player has one and only one move, and has to make that move in

* Von Neumann's solution of the general zero-sum two-person game, first published in *Mathematische Annalen*, vol. 100 (1928), pp. 295 ff., is expounded in John von Neumann and Oskar Morgenstern, *Theory of Games and Economic Behavior* (Princeton, 1944), Chapters II–IV. I shall refer to this book as *NM*.

complete ignorance as to how his opponent will make his move. The 'outcome' of the game for player A is a function of two independent variables, one variable being the move made by player A, the other being the move made by player B; it is the amount which A will receive if he and his opponent make these moves. Since the game is 'zero-sum', the outcome of the game for player B is equal to minus the outcome for player A. Calling the outcome for player A the "score", we see that A's object in choosing his move is to maximize the score, B's object in choosing his move is to minimize the score. The score itself is a resultant of the two moves, one of which is controlled by A, the other of which is controlled by B.

If there are l possible moves for player A and m possible moves for player B, there are lm possible scores (some of which may be the same number); and the game can be represented by a rectangular table with the lm scores arranged in l rows and in m columns. These lm scores are fixed by the rules of the game. Then the play permitted to A may be represented as his picking out a row, the play permitted to B as his picking out a column, in the rectangular table, these pickings-out to be made quite independently of one another. The score in the game is then the number appearing in the row picked out by A and in the column picked out by B. A will therefore try to pick his row so as to maximize, and B to pick his column so as to minimize, the resultant score. The characteristic, as von Neumann says, of this "peculiar tug-of-war" is that it "is not a tug-of-war. The two players have opposite interests, but the means by which they have to promote them are not in opposition to each other."*

For example, consider this rectangular form of representation for a game in which A has eight possible moves (i.e. rows to choose from) and B two possible moves (i.e. columns to choose from). Suppose the rules of the game determine the numbers in the rectangular table as being those in Table VI. It will be clear from examining this table (and von Neumann has proved it rigorously in the general case for any finite number of rows and columns) that if A chooses his row according to the policy of choosing that row which contains the maximum of the minima for each row— that is, if he chooses the second row, whose minimum -2 is

* *NM*, §14.1.3.

greater than the minimum in any other row—the score which he will secure will be not less than this number (-2 in our case) whatever column his opponent B chooses. But this will not be the case were A to choose any other row. Were he, for example, to choose the first row he would win more if B were to choose the first column, but he would lose more if B were to choose the second. Did he know that B would choose the first column, it would be sensible for him to choose the first row (and indeed stupid of him to choose any other row), but such knowledge is excluded by the conditions of the game. Thus by making the move of choosing the second row, A will protect himself against scoring less than -2 in a situation in which he cannot protect himself against scoring less than any other number; he will be maximizing his minimum gain.

Table VI

B's range of choice

A's range of choice		
100		-100
68		-2
-28		82
-100		100
-68		2
28		-82
4		-16
-4		16

To put the matter another way. The conditions of the game (i.e. the numbers in the rectangular table) settle that A cannot lose more than 100 nor gain more than 100 whatever moves A or B make. But the conditions equally determine that, if A plays prudentially by choosing the row which contains the maximum of the row-minima, he cannot lose more than 2. The effect of a prudential policy of play is to cut off the lower part of the range of his possible scores; a prudential policy usually also has the effect of cutting off an upper part also, but this cannot be helped if A's object is to maximize his score irrespective of how B may play.

This prudential policy for playing a zero-sum two-person game is, so far as the method for selection is concerned, exactly the same as the prudential policy expounded for selecting a strategy for preferring one statistical hypothesis to another. The prudential policy for this choice corresponds to the prudential policy for playing a zero-sum two-person game in which player B (corresponding to Nature) has a choice of only two possible moves and player A (corresponding to the scientist) has a choice of 2^{n+1} possible moves, where n is the number in the observed sample. The game given by the example of Table VI corresponds to the problem of the choice between our hypotheses H_1 (with $p_1 = \frac{3}{5}$) and H_2 (with $p_2 = \frac{3}{10}$) on the basis of observation of two instances for the case in which $a = b = c = d$, i.e. when we stand to gain the same amount if we choose that hypothesis which is true and stand to lose an equal amount if we choose that hypothesis which is false. In general we can treat every problem of choice between m alternative statistical hypotheses H_1, H_2, ..., H_m on the basis of the observation of n instances on the analogy of a zero-sum two-person game in which player A (the scientist) has the choice of m^{n+1} possible moves and player B (Nature) the choice of m possible moves, and we may use any of von Neumann's mathematical methods or results which may be of service to us.

However, it behoves a philosopher to be cautious in pressing analogies; for he must see that there is no relevant material difference concealed beneath a formal similarity. Now there are two obvious differences between choosing a hypothesis according to the prudential policy and playing a game prudentially. The first difference is more apparent than real. In the game we play for monetary or other stakes; in the choice of hypotheses we aim only at securing mathematical expectations. But to secure a mathematical expectation of a possible benefit is to secure a benefit; a mathematical expectation is itself a benefit. Indeed mathematical expectations are frequently marketable commodities; lottery tickets, which are titles (in the legal sense) to mathematical expectations, are bought and sold, and can perfectly well be used as stakes in games, where their use would make the outcome of the game exactly comparable in this respect with the outcome of a choice between hypotheses.

But the second difference appears at first sight to be so great as to prevent us from making any serious use of the analogy. In a zero-sum two-person game the object of player A is to maximize the score in spite of the best efforts of player B to minimize it; the interests of player B are directly opposed to those of player A, and he will try to defeat A's strategy. The prima facie most engaging feature in the prudential policy for playing such a game is that, by using it, *A* can always secure not less than a certain score however much *B* knows of how *A* will play. To present the prudential policy in this way makes it look as if the intentions or amount of knowledge of *B* was relevant to the question as to how *A* should play. Were these considerations relevant, the analogy of a zero-sum two-person game for the problem of the choice between statistical hypotheses would be grossly misleading—unless we were willing to suppose a permanent desire on the part of Nature to defeat our aims (or at least to defeat the aims of scientists), and also perhaps be prepared to allow for the possibility of Nature's knowing beforehand which hypothesis we should choose and of adjusting its laws so as to defeat our aims as much as possible. Wald, in the article in which he first remarked upon the similarity of his theory of statistical inference to von Neumann's Theory of Games, seemed to think that such a supposition was necessary.* This, I think, is a misunderstanding. In order for it to be rational for *A* to play according to the prudential policy, he must wish to score as much as possible and he must be unable to know beforehand how his opponent will play. But he need make no assumptions whatever as to his opponent *B*'s intentions, if any, or knowledge, if any. The prudential policy allows for *B*'s playing purely at random or for his trying to secure that *A* shall gain as much as possible just as much as it allows for *B*'s finding out beforehand how *A* is going to play and choosing his move so that *A* shall gain as little as possible. The case for the prudential policy is simply the fact, which depends only upon arithmetical properties of maxima and minima, that to make the prudential move (or one of the prudential moves, if there are more than one) secures,

---

* I quote from his article, altering some of his terms to those I have used: "Of course we cannot say that Nature wants to [minimize the score]. However, if the [scientist] is in complete ignorance as to Nature's choice, it is perhaps not unreasonable to base the theory of a proper choice of [a strategy for deciding between statistical hypotheses] on the assumption that Nature wants to [minimize the score]. Under this assumption a problem of statistical inference becomes identical with a zero-sum two-person game" (*Annals of Mathematics*, second series, vol. 46 (1945), p. 279; words in square brackets my changes). [In his book *Statistical Decision Functions*, p. 27, Wald added to a passage in similar terms the qualification that "even if one is not willing to take this attitude, the theory of games remains of fundamental importance for the problem of statistical decisions", since the mathematics of the former can be transferred to the latter.]

however the opponent plays, a score not less than could be secured, however the opponent plays, by any non-prudential move.*

To put the matter precisely. Let v be the maximum of the minima for each row in the rectangular table. Then if A chooses the row containing v, or one of the rows containing v, if more than one do so, A will secure a score not less than v whatever column B chooses. If A chooses any row not containing v, it will be false that A will secure a score not less than v whatever be the column chosen by B, i.e. there will be a choice of column by B which will make A score less than v. Stating it in this form makes it clear that what is in question is merely all the possibilities open to B; how B discriminates among these possibilities is irrelevant to the case for the prudential policy.

Thus to regard the principles governing the choice between statistical hypotheses as equivalent to the principles of the best way for a scientist to play a game against Nature does not attribute will or knowledge to Nature, if the policy for playing the game is one, like the prudential policy, which is determined by all the possibilities open to the opponent. So we shall unhesitatingly use the analogy of a game played by the scientist against Nature whenever the analogy is useful.

GEOMETRICAL REPRESENTATION FOR THE PRUDENTIAL POLICY

The prudential policy for choosing between the two statistical hypotheses H_1 and H_2 is based upon a previous assignment of numerical values to a, b, c, d—the gains or losses occasioned by choosing one or other of these hypotheses when it is true or when it is false. In order to see the exact way in which these values affect the strategies of choice we must examine how the strategy which is the prudential strategy changes with changes in the values of a, b, c, d. To do this it is convenient to make use of a simple geometrical representation of the possible strategies which will enable us to solve the maximum-minimum problem geometrically and which will have the incidental advantage of enabling us to

---

* Though von Neumann talks in various places about the intentions and the knowledge of the opponent, he makes it quite clear that no hypothesis as to the 'rationality' of the opponent enters into the definitive form of his solution. See *NM*, §§ 15.8.3, 17.8.2.

discover intuitively the possibility of a modification of the prudential strategy which will, in many cases, make us even more
independent of Nature's vagaries.

Represent the eight possible strategies for choosing between
H_1 and H_2 on the basis of a set of two observations by eight points
whose x-coordinates and y-coordinates are the x_i's and y_i's
respectively given in Table V of probabilities of correct choice
(p. 205). All these eight points lie on or inside the unit square in
Fig. 10 (p. 254 opp.). The point representing strategy T_1 is (1, 0),
the bottom right-hand corner of the square; the point representing
T_2 is (0·84, 0·49); and so on. In Fig. 10 the point representing
T_i ($i = 1$, 2, ..., 8) is marked as T_i; it will cause no confusion to
use the same symbol both for the strategy and for the point
representing it. The six strategy-points T_1, T_2, T_3, T_4, T_5, T_6 will
be seen to be the vertices of a convex hexagon with its first and
fourth vertices T_1 and T_4 at the bottom right-hand corner and at
the top left-hand corner of the square respectively. All the six
sides of the hexagon slope downwards from left to right. The other
two points T_7, T_8 lie inside the hexagon. The eight points are
centrally symmetrical about the centre of the square.*

Now draw the straight line represented by the linear equation

$$(a+d)\,x - d = (b+c)\,y - b,$$

where a, b, c, d are the utility coefficients, none of which are
negative. Call the equation the *utility equation* and the line the
utility line. It will be shown that the prudential policy for
selecting a strategy is represented by geometrical relations between
the strategy points and the utility line.

If a, b, c, d are all zero, the utility equation becomes the identity
$0 = 0$ and there is no utility line at all.

If $b+c=0$ (i.e. if $b=0$ and $c=0$) but $a+d$ is positive, the utility
equation reduces to $x=d/(a+d)$ and the utility line is vertical.
If $d=0$ it is the left-hand side of the square; if $a=0$ it is the right-
hand side; in other cases it divides the square into two regions.

If $a+d=0$, i.e. if $a=0$ and $d=0$, but $b+c$ is positive, the utility
equation reduces to $y=b/(b+c)$ and the utility line is horizontal.
If $b=0$ it is the bottom side of the square; if $c=0$ it is the top
side; in other cases it divides the square into two regions.

* No use will be made of this property in our discussion.

If $b+c$ and $a+d$ are both positive, the utility line slopes upwards from left to right (since its slope $(a+d)/(b+c)$ is positive). It intersects the vertical axis at the point $y=(b-d)/(b+c)$, which is lower than the top left-hand corner of the square unless $c=d=0$, when it passes through this corner. It intersects the horizontal axis at the point $x=(d-b)/(a+d)$, which is to the left of the bottom right-hand corner unless $a=b=0$, when it passes through this corner. Hence the utility line divides the square into two regions, except when $a=b=0$ or $c=d=0$.

Fig. 10 shows the parts, lying on or within the square, of eight utility lines—the left-hand side of the square for $b=c=d=0$, the top side of the square for $a=c=d=0$, the right-hand side of the square for $a=b=c=0$, the bottom side of the square for $a=b=d=0$, the upward diagonal U_1 for $a=c$, $b=d$; the line marked U_2 for $a+d=2(b+c)$, $b=d$; the line marked U_3 for $a+d=2(b+c)=4(d-b)$; and the line marked U_4 (which will not be used until p. 227) for $a+d=11(b+c)/26=55(b-d)/49$.

To simplify exposition we will exclude for the moment the exceptional cases in which the utility line is vertical or is horizontal or passes through one corner without cutting the square. We shall suppose therefore either that a and c are both positive or that b and d are both positive. This will ensure that the utility line will always slope upwards and will divide the square into two regions. Call the region which includes the top left-hand corner the *superior region*, that which includes the bottom right-hand corner the *inferior region* with respect to the utility line.

Since the utility line is given by the utility equation

$$(a+d)\,x-d=(b+c)\,y-b,$$

the superior region contains all the points (x, y) of the square for which

$$(a+d)\,x-d<(b+c)\,y-b,$$

and the inferior region contains all the points (x, y) of the square for which

$$(a+d)\,x-d>(b+c)\,y-b.$$

But the mathematical expectation, given the truth of H_1, yielded by using strategy T_i is

$$X_i=(a+d)\,x_i-d_i,$$

and the mathematical expectation, given the truth of H_2, yielded by using T_i is

$$Y_i = (b+c)\,y_i - b_i.$$

Hence X_i is less than Y_i if and only if the strategy-point T_i lies in the superior region, and X_i is greater than Y_i if and only if T_i lies in the inferior region. Thus the utility line serves to separate the strategies for which the mathematical expectation, given the truth of H_1, is less than the mathematical expectation, given the truth of H_2, from those for which the converse is true. So the minimum of each pair of mathematical expectations is X_i for those strategies T_i whose strategy-points (x_i, y_i) lie in the superior region, and is Y_j for those strategies T_j whose strategy-points (x_j, y_j) lie in the inferior region. The first step in applying the prudential policy—that of selecting the minimum in each row—has been effected by the division of the square into the two regions by the utility line. [In Fig. 10 the strategy-points T_3, T_4, T_5, T_8 fall into the superior region, and T_1, T_2, T_6, T_7 into the inferior region, with respect to both the utility lines U_1 and U_3; T_3, T_4, T_5 into the superior region, and T_1, T_2, T_6, T_7, T_8 into the inferior region, with respect to the utility line U_2.]

The second step in applying the prudential policy is to discover which of these minima is the greatest. Since

$$(a+d)\,x_i - d > (a+d)\,x_j - d$$

if and only if $x_i > x_j$, the greatest of the minima of the pairs of mathematical expectations for the strategies whose strategy-points lie in the superior region is $(a+d)\,x_1 - d$, where x_1 is the x-coordinate of the strategy-point (x_1, y_1) lying farthest to the right in the superior region. Similarly, since $(b+c)\,y_i - b > (b+c)\,y_j - b$ if and only if $y_i > y_j$, the greatest of the minima of the pairs of mathematical expectations for the strategies whose strategy-points lie in the inferior region is $(b+c)\,y_2 - b$, where y_2 is the y-coordinate of the strategy-point (x_2, y_2) lying highest in the inferior region. Thus the choice is restricted to two strategy-points—(x_1, y_1) in the superior region, (x_2, y_2) in the inferior one. [In Fig. 10 the strategy-points (x_1, y_1) and (x_2, y_2) are T_8 and T_2 respectively for both the utility lines U_1 and U_3; they are T_3 and T_8 respectively for the utility line U_2.] Now drop a perpendicular from (x_1, y_1) on to the horizontal axis; let (x_1, y_1') be the point at which this perpendicular cuts the strategy line. Then

$$(a+d)\,x_1 - d = (b+c)\,y_1' - b,$$

and
$$(a+d)\,x_1 - d > (b+c)\,y_2 - b$$

if and only if
$$(b+c)\,y_1' - b > (b+c)\,y_2 - b.$$

But
$$(b+c)\,y_1' - b > (b+c)\,y_2 - b$$

if and only if $y_1' > y_2$. Similarly

$$(a+d)\,x_1 - d < (b+c)\,y_2 - b$$

if and only if $y_1' < y_2$. So we only have to compare the height of the strategy-point (x_2, y_2) above the horizontal axis with the height above this axis of the point (x_1, y_1') where the perpendicular dropped upon this axis from the strategy point (x_1, y_1) cuts the utility line. If the former point is higher than the latter, the strategy whose strategy-point is (x_2, y_2) is to be selected; if the former point is lower than the latter, the strategy whose strategy-point is (x_1, y_1) is to be selected. [Applying this technique to Fig. 10 we find that, if the utility line is U_1, the strategy to be selected is T_2;* if the utility line is U_2, the strategy to be selected is T_3; if the utility line is U_3, the strategy to be selected is T_2.]

If the two points (x_2, y_2) and (x_1, y_1') are at an equal height from the horizontal axis, the third step is required of comparing $(a+d)\,x_2 - d$ with $(b+c)\,y_1 - b$. This can be done geometrically by dropping a perpendicular from (x_2, y_2) on to the horizontal axis, and extending this perpendicular upwards to meet the utility line at the point (x_2, y_2'). Then, since

$$(a+d)\,x_2 - d = (b+c)\,y_2' - b,$$

$(a+d)\,x_2 - d$ will be greater or less than $(b+c)\,y_1 - b$ according as y_2' is greater or is less than y_1.†

* In Fig. 10 the point (0·48, 0·48), where the perpendicular dropped from T_8 cuts U_1, is just perceptibly lower than T_2 (0·84, 0·49).

† The third step is not required in the case of any of the utility lines U_1, U_2, U_3 in Fig. 10. But it would have been required in the case of U_1 had the coordinates of T_8 been (0·49, 0·58) instead of (0·48, 0·58), for then the perpendicular dropped from T_8 would cut U_1 at the point (0·49, 0·49), whose height above the horizontal axis is equal to that of T_2 (0·84, 0·49). The third step would then select strategy T_2, since the perpendicular dropped from T_2 cuts U_1 at the point (0·84, 0·84), and 0·84 > 0·49.

We have not considered the case in which a strategy-point T_i actually lies on the utility line (as T_8 does on the utility line U_4 in Fig. 10). If this happens the two mathematical expectations are equal for this strategy, and each is a minimum for the pair. This minimum will be greater than the minimum of the pair for any other strategy if T_i stands to the right of the rightmost strategy-point in the superior region T_j and higher than the highest strategy-point in the inferior region T_k. If these conditions are fulfilled T_i is to be selected. If they are not fulfilled either T_j or T_k is to be taken instead, and the dropping-perpendicular method may be used to choose between them.

We are now in a position to examine how different assignments of utility values a, b, c, d affect the selection of the prudential strategy. If we start with one assigned set of utility values, it is simpler to find the maximum of the minima of each pair of mathematical expectations by constructing and examining a rectangular table of these mathematical expectations rather than by using the geometrical method. And, though the geometrical method can be modified to deal with the exceptional cases which were excluded when it was expounded, it is simpler to deal with them directly by means of rectangular tables, since the strategies selected do not depend upon the exact values of the utilities, but only upon which of them are, and which are not, zero.

If we construct the rectangular tables for these exceptional cases we find the results given in Table VII. In all these cases the choice of the hypothesis is entirely independent of the result of any observation upon the set of two (or, indeed, any number of) instances; if the utility values form one of the exceptional cases, there is no object in making any observation whatever, since the prudential choice of the hypothesis is settled beforehand. This may seem surprising; but the surprise will vanish if we look at the prudential strategies appropriate to the exceptional cases.

If $a = b = c = d = 0$, it matters nothing whatever to us whether either of the hypotheses is true or is false, and the prudential policy provides no criterion for preferring one to the other. The prudential policy, therefore, will not satisfy idle curiosity. This may seem shocking to intellectualist philosophers, but accords well with the healthy pragmatism of common sense. In the language of games it is a case of "No stake, no game."

If $b = c = 0$ but one at least of a and d is positive, we stand either to gain by choosing H_1 if it is true or to lose by choosing H_2 if it is false, i.e. by failing to choose H_1 if H_1 is true, whereas we stand to gain or lose nothing by choosing or by failing to choose H_2 if H_2 is true. If H_1 is true it is always to our advantage to choose H_1; but if H_2 is true it is never to our advantage to choose H_2. To choose H_1 is a case of "Heads you win, tails you don't lose"; so the prudential strategy of always choosing H_1 is in accord with common sense. Similarly, if $a = d = 0$ but one at least of b and c is positive, it is always to our advantage to choose H_2.

TABLE VII. *Summary of exceptional cases*

Exceptional case	Position of utility line on the square	The prudential strategy
$a = b = c = d = 0$	No utility line	All strategies equally prudential
$a > 0, b = c = d = 0$	Left-hand side	T_1, i.e. always prefer H_1
$b > 0, a = c = d = 0$	Top side	T_4, i.e. always prefer H_2
$c > 0, a = b = d = 0$	Bottom side	T_4, i.e. always prefer H_2
$d > 0, a = b = c = 0$	Right-hand side	T_1, i.e. always prefer H_1
$a > 0, b > 0, c = d = 0$	Through top left-hand corner	T_4, i.e. always prefer H_2
$c > 0, d > 0, a = b = 0$	Through bottom right-hand corner	T_1, i.e. always prefer H_1
$a > 0, d > 0, b = c = 0$	Vertical	T_1, i.e. always prefer H_1
$b > 0, c > 0, a = d = 0$	Horizontal	T_4, i.e. always prefer H_2

To say that at least one of a and d is positive is equivalent to saying that $a + d$ is positive. It will be convenient to give this quantity $a + d$ a name; I will call it in Stock Exchange parlance the value of the *option* if H_1 is true, or shortly the H_1-*true option*, since it is what I stand to benefit by choosing correctly rather than incorrectly if H_1 is true.* The prudential policy in the last three cases may then be expressed by saying: if only one option is positive choose the hypothesis with the positive option; if neither option is positive, it does not matter how you choose.

If $c = d = 0$ but a and b are both positive, we stand to gain or lose nothing by choosing H_2, whereas by choosing H_1 we stand to gain if H_1 is true and to lose if H_1 is false. The prudential strategy in this case is always to choose H_2; this seems paradoxical, but it

---

* The quantity $b + c$ will similarly be called the H_2-*true option*.

must be remembered that the purpose of the prudential policy is to play for safety, and if H_2 is always chosen no loss will ever be incurred.

Similarly, if $a = b = 0$ but c and d are both positive, the prudential policy is to protect ourselves against a possible loss by always choosing H_1.

This disposes of all the exceptional cases. In the other cases the prudential strategy appropriate to a particular assignment of utility values depends upon exactly what these utility values are; the geometrical method will enable us to see most easily how the appropriate prudential strategy varies with variations in a, b, c, d. To exclude the exceptional cases we will suppose that a and c are both positive or that b and d are both positive; this secures that the utility line will always intersect the square and will be neither vertical nor horizontal.

Up to now we have worked with four separate utility values entering separately into the utility equation

$$(a+d)\, x - d = (b+c)\, y - b.$$

But now that $b+c$ is restricted to being greater than zero it will be convenient in interpreting the results to divide through by $b+c$, and to use the equivalent equation

$$y = \frac{a+d}{b+c}\, x + \frac{b-d}{b+c}.$$

This form of expression shows clearly that, for variations of a, b, c, d, the utility line, and the prudential strategy determined by it, depend upon only two parameters $(a+d)/(b+c)$ and $(b-d)/(b+c)$.* $(a+d)/(b+c)$ is the ratio of the H_1-true option to the H_2-true option, $(b-d)/(b+c)$ is the ratio to the H_2-true option of $b-d$, which is the difference between the loss incurred by incorrectly choosing H_1 and that incurred by incorrectly choosing H_2; it will be called (somewhat arbitrarily) the *penalty-disparity*. The position of the utility line depends then upon two

---

* Had the equation been converted into other forms which are equivalent when $b+c > 0$ and $a+d > 0$, e.g.

$$1 - x = -\frac{b+c}{a+d}\, y + \frac{a+b}{a+d},$$

a different pair of parameters would have occurred, but they would be functions of $(a+d)/(b+c)$ and $(b-d)/(b+c)$.

numbers, namely the ratios of these three quantities: the H_1-true option, the H_2-true option, the penalty-disparity.

In Fig. 10 the utility line U_1—the upward diagonal of the square—represents the case in which the two options are equal, and the penalty-disparity is zero. The prudential strategy in this case is T_2, i.e. to prefer H_1 to H_2 if the observed set of two β-specimens contains either one or two α-specimens, but to prefer H_2 to H_1 if the observed set contains no α-specimens. Inspection of the method of calculating the prudential strategy will show that this strategy would be the prudential one for a set of two observations whenever $p_2 < \frac{1}{2} < p_1$ and $p_1 - \frac{1}{2} < \frac{1}{2} - p_2$.

Suppose now a, b, c, d to vary in such a way that the ratio of the H_1-true option to the H_2-true option increases to $2:1$ while the penalty-disparity remains zero. The utility line will rotate about the bottom left-hand corner of the square in a counter-clockwise direction to become the line U_2, and the prudential strategy then to be selected will be T_3, i.e. to prefer H_1 to H_2 if and only if there are two α-specimens in the observed set. The effect of the relative increase in the H_1-true option is to diminish the number of occasions upon which it would be prudential to choose H_1. This may appear surprising but it is comprehensible. For an increase, relatively, in the H_1-true option makes it relatively more risky to prefer H_1 to H_2; and the aim of the prudential policy is to avoid unnecessary riskiness.

Now suppose that a, b, c, d vary again in such a way that the ratio of the H_1-true option to the H_2-true option remains at $2:1$ while the penalty-disparity decreases from zero to a point where its ratio to the H_2-true option becomes $-\frac{1}{2}:1$. The utility line will move from left to right remaining parallel to itself (performing a 'parallel displacement') to become the line U_3, and the prudential strategy yielded will return to the original strategy T_2, i.e. to prefer H_1 to H_2 unless the observed set of two β-specimens contains no α-specimen. In this case the increase in the riskiness of choosing H_1 due to the relative increase in the H_1-true option is counterbalanced by an increase in the penalty incurred by incorrectly choosing H_2 as compared with that incurred by incorrectly choosing H_1.

The relations between the utility values in a simple form of these three cases are schematically shown in the following diagram, where b has been taken as equal to the unit of value and c as equal to $2b$. Thus b and $c = 2b$ are constant in the three figures:

	$b = 1$	$a = 2$	
U_1. $a + d = b + c$, $b = d$ (yielding strategy T_2)			
	$d = 1$	$c = 2$	
	$b = 1$		$a = 5$
U_2. $a + d = 2(b + c)$, $b = d$ (yielding strategy T_3)			
	$d = 1$	$c = 2$	
U_3. $a + d = 2(b + c)$, $b - d = -\frac{1}{2}(b + c)$ (yielding strategy T_2)	$b = 1$		$a = 3\frac{1}{2}$
	$d = 2\frac{1}{2}$	$c = 2$	

Since any straight line on a plane can be transformed into any other straight line on the same plane by combining (in either order) a rotation of it about a fixed point on the plane with a parallel displacement of it along the plane in a fixed direction, any utility line can be converted into any other by altering successively the ratios of the H_1-true option and of the penalty-disparity to the H_2-true option, as we have done in our example. Wald, and the other mathematical statisticians who have used in their work on statistical inference what I have called utility values, have only considered the losses which would be suffered by accepting hypotheses that were false. This is equivalent, in my terminology, to taking $a = c = 0$; and in Fig. 10 to considering only those utility lines which pass through the top right-hand corner of the square.* This restriction has, I think, no restricting effects upon the principles of statistical inference which they have worked out; and it is highly convenient when the number n of the set of observed instances becomes large, since the upper vertices of the polygon then recede towards the top and the right-hand sides of the square (five sides of the polygon for $n = 20$ are shown as dotted lines in Fig. 10). But in considering as generally as possible a very simple case, which we are doing here, it is important to notice that utility lines have a greater degree of freedom than is permitted by those

* The utility line reduces to $-d(1 - x) = -b(1 - y)$.

who would restrict them to an upward diagonal of the square and the lines obtained by rotating this diagonal about its end-points.

THE APPROPRIATE STRATEGIES

The prudential strategies appropriate to the three utility lines which were used as illustrations were T_2 and T_3. T_2 prescribes that the hypothesis H_1 should be chosen if and only if either one or two α-specimens are found in the twofold sample, T_3 that H_1 should be chosen if and only if two such specimens are found. Both these strategies prescribe the choice of a number between 0 and 1 inclusive—$\frac{1}{2}$ in the case of T_2, 1 in the case of T_3—which is such that H_1 is to be chosen if and only if the α-ratio in the observed set is not less than this number. T_1 similarly prescribes 0 as this number; and T_4, T_5, T_6 prescribe a certain number— 0, $\frac{1}{2}$, 1 respectively—which is such that H_2 is to be chosen if and only if the α-ratio is not less than this number. The six strategy-points lying on the hexagon therefore represent choices of H_2 or of H_1 (or of H_1 or of H_2) according as the observed α-ratio is or is not less than a certain number. They therefore represent choices which are prima facie plausible; we expect to base our choice of one or other hypothesis upon this observed ratio. But these six strategies are not the only logically possible ones. There are two— T_7 and T_8—which do not connect the choice of a hypothesis with the observed α-ratio's being greater or less than a fixed number. Strategy T_7 enjoins us to choose H_1 in the two extreme cases in which there are either no α-specimens or two α-specimens in the set of two instances and to choose H_2 in the intermediate case in which there is exactly one α-specimen in the set. Strategy T_8 prescribes the reverse procedure. Now these strategies are prima facie unplausible. If the prudential policy ever were to select one or other of these strategies, this would be a strong reason against accepting the prudential policy as a rational method for selecting optimum strategies.

It is easy to see that the prudential policy will never select T_7. For each of the probabilities of correct choice for T_7 (52 and 42%) is less than the corresponding probability of correct choice for T_2 (84 and 49%); so there can never be an advantage in selecting T_7 rather than T_2. To use Wald's terms, T_2 is 'uniformly better' than T_7, and so T_7 is not 'admissible'. For similar reasons neither of the strategies whose strategy-points are lower vertices of the hexagon—T_5 and T_6—are admissible.

But there is no strategy which is uniformly better than T_8, i.e. there is no strategy each of whose probabilities of correct choice is greater than or equal to the corresponding probability of correct choice for T_8 (48 and 58%). So T_8 is an admissible strategy; and there will be assignments of utility values which will make it the prudential strategy. Indeed, it only narrowly avoids being selected as the prudential strategy by the examples of utility lines that have been already taken. If the upward diagonal utility line U_1 is rotated counter-clockwise round the bottom left-hand corner of the square through the very small angle (less than two degrees) required to increase its slope from 1 to more than $\frac{49}{48}$ (that is, increasing $(a+d)/(b+c)$ from 1 to more than $\frac{49}{48}$, with b remaining equal to d), the strategy which is prudential changes from being T_2 to being T_8. And it continues to be T_8 while the line continues to rotate until its slope becomes greater than $\frac{58}{36}$, when the prudential strategy becomes T_3. There would be a similar change of prudential strategy from T_2 through T_8 to T_3 if the line U_1 (or, for that matter, U_3) were moved parallel to itself to the left.

The dotted staircase line in Fig. 10 has been drawn to show how the strategy prescribed by the prudential policy depends upon the position of the utility line and hence upon the assignment of the utility values a, b, c, d. To start from the right: any utility line which intersects either the side of the hexagon stretching from T_1 (1, 0) to T_2 (0·84, 0·49) or the horizontal dotted line running left from T_2 up to and including the point (0·48, 0·49) determines T_2 as the prudential strategy; any utility line which intersects either the vertical dotted line from (0·48, 0·49) to T_8 (0·48, 0·58) or the horizontal dotted line from T_8 up to and including (0·36, 0·58) determines T_8 as the prudential strategy; any utility line which intersects either the vertical dotted line running upwards from (0·36, 0·58) to T_3 (0·36, 0·91) or the side of the hexagon stretching from T_3 to T_4 (0, 1) determines T_3 as the prudential strategy line. Since every utility line which cuts the square and which is neither vertical nor horizontal intersects the broken line T_1—T_2—staircase-line—T_3—T_4 once and once only, every assignment of utility values a, b, c, d with a and c both positive or b and d both positive selects one and one only out of the three strategies T_2, T_3 or T_8.* All this follows from the reasoning upon which the geometrical method for selecting strategies has been granted.

So we are in the unfortunate position, that, according to the prudential policy, an assignment of utility values which gives rise to a utility line which crosses the square between the points (0·36, 0·58) and (0·48, 0·49), e.g. U_4, will make prudential the strategy of choosing H_2 if no α-specimens are found in a set of two β-specimens, of choosing H_1 if one α-specimen is found, and of choosing H_2 if two α-specimens are found. A strategy prescribing such an arbitrary picking and choosing cannot lay claim to be a reasonable one. If to follow the prudential policy prescribes the use of such a 'freak' strategy, so much the worse for the claim of the prudential policy to be a rational one.

It may perhaps be thought that the difficulty has arisen because we have taken a very small number (2) for the number n of observed instances. Perhaps it has only been possible for T_8 to qualify as an admissible strategy because the strategy-points T_2 and T_3 are so far apart. As n is increased the strategy-points which are the upper vertices of the polygon will crowd together, and will perhaps make all freak strategies inadmissible.

Now it is indeed true that, as n increases, the strategy-points along the top of the polygon will lie closer together. Towards the top right-hand corner of Fig. 10 the strategy-points P_8, P_9, P_{10}, P_{11}, P_{12}, P_{13} have been put in for six strategies when $n = 20$ (the remaining strategy-points along the top of the polygon lying so close either to the top side or to the right-hand side of the square as to be unrepresentable on a small-scale diagram).† But, for any given n, there are 2^{n+1} strategy-points of which $2n + 2$ will lie on the $(2n + 2)$-sided polygon which includes all the strategy-points, and hence $2^{n+1} - (2n + 2)$ of them will be points representing freak strategies. Though most of these freak strategies will be inadmissible, there will remain some near enough to the top of the $(2n + 2)$-sided polygon to be admissible, and these will determine prudential strategies for appropriate sets of utility values. Moreover, even if the difficulty did not arise with a larger n, we are

* All the admissible strategies (i.e. T_1, T_2, T_3, T_4, T_8) lie on this broken line.
† Their probabilities of correct choice (the coordinates of the strategy-points) are given in the footnote to p. 240.

asking for a rational procedure for the case $n = 2$; it would be very surprising if the prudential policy were the reasonable one for some utility lines, but not for all. So even if there were no admissible freak strategy-points for large n, their existence for a small n would still be paradoxical.

INTRODUCTION OF RANDOMIZED STRATEGIES

There is, however, a most ingenious way of getting out of the difficulty by a method which makes all freak strategies inadmissible. This is done by considering a new sort of possible strategy—a sort which was not considered when the eight possible strategies for $n = 2$ were enumerated. The idea of the new sort of strategy comes from von Neumann's Theory of Games; it has been used by Wald in his most recent work on statistical inference. But since I am introducing it to resolve the freak-strategy paradox, with which neither von Neumann nor Wald were concerned, I shall expound it independently.

The best way to do so is to indulge in a little wishful thinking. Consider any one particular utility line which determines T_8 as the prudential strategy. Since the paradox arises for every such utility line, I will take the case that will make my exposition simplest, which is the utility line U_4 which passes through T_8 and the point R on the line $T_2 T_3$ where $T_2 T_3$ is cut by the upward diagonal U_1.* Now suppose that there were to be a possible new strategy for which R was the strategy-point. Let us call it, as well as its strategy-point, R. The pair of mathematical expectations for this new strategy R would be equal to one another, as they are for T_8, since R, like T_8, would lie on the utility line U_4. Since both the mathematical expectations for R would be greater than the corresponding mathematical expectations for T_8, R would be 'uniformly better' than T_8, and T_8 would cease to be an admissible strategy. The prudential strategy determined by the utility line would thus become this new strategy R. So if there were to be this new strategy, and if it was not itself a freak strategy, the paradox would be resolved.

Now let us step back a little and consider how exactly the prudential policy works. Each of the eight old strategies prescribes

* U_4 is given by $\dfrac{a+d}{b+c} = \dfrac{11}{26}$, $\dfrac{b-d}{b+c} = \dfrac{49}{130}$.

the choice of one or other of the hypotheses under certain observable conditions. The old strategy which was the prudential strategy was that one among the eight whose pair of mathematical expectations included the maximum of the minima of the pairs of mathematical expectations for each of the eight strategies. The selection of the prudential strategy was determined by these sixteen mathematical expectations, and the case for the reasonableness of the prudential policy was derived from the reasonableness of basing our actions upon the general principle that we should try to maximize the mathematical expectations of the utilities of their consequences. This reasonableness does not consist in the fact that any one particular action done in accordance with this general principle will produce consequences having greater utility than an action done in accordance with a different principle, but in the statistical fact, i.e. in the truth of the statistical hypothesis, that a number of actions done in accordance with the general principle will, by and large, yield more utility than a number of actions done in accordance with a different principle. The reasonableness of a selected strategy by the prudential policy must equally be a statistical reasonableness. The case for selecting T_2 rather than T_3 as the optimum strategy when the utility line is the upward diagonal U_1 is that, if we use strategy T_2 in making a number of choices between H_1 and H_2, between H_1' and H_2', etc., on the bases of twofold samples (where H_1, H_1', etc., assert that the values of the probabilities with which they are concerned are all equal to $\frac{3}{5}$, and H_2, H_2', etc., assert that the values of the probabilities with which they are concerned are equal to $\frac{3}{10}$, and the sets of utility values for each pair H_1 and H_2, H_1' and H_2', etc., are all such as to determine the upward diagonal as the utility line), we shall, by and large, behave more prudently than if we used strategy T_3 in making these choices.

But the fact that the reason for using a prudential strategy is a statistical reason in terms of the by-and-large effect of using the strategy a number of times opens the possibility that it might be reasonable *not always to employ the same strategy but sometimes to use one strategy and sometimes another*. For example, suppose we decide sometimes to use strategy T_2 and sometimes to use strategy T_3 and to settle when to use T_2 and when to use T_3 by shuffling and drawing a card from a pack of eighteen cards containing eleven spades and seven hearts, and using T_2 or T_3 according as a spade or a heart is drawn. The probability, given the truth of H_1, of correct choice for this 'mixed' strategy is obtained by multiplying the probability, given H_1, of correct choice for T_2 (0·84) by the probability of drawing a spade (which is $\frac{11}{18}$), by multiplying the probability, given H_1, of correct choice for T_3 (0·36) by the probability of drawing a heart (which is $\frac{7}{18}$), and then adding these two products together. The result is 0·6533. A similar calculation will give the probability, given the truth of H_2, of correct choice for this mixed strategy as $\frac{11}{18} . 0·49 + \frac{7}{18} . 0·91 = 0·6533$. The strategy-point representing the mixed strategy is the point (0·6533, 0·6533), which is the R of our wishful thinking—the point where the line T_2T_3 cuts the upward diagonal U_1. So if we permit ourselves to use mixed strategies, the paradox is resolved; R becomes a strategy-point, and its presence makes T_8 inadmissible as a prudential strategy.

Strategies obtained by using a combination of 'basic' strategies in connexion with a suitable 'chance-machine' arranged so that the probability of any one of the basic strategies being used on any occasion is a pre-assigned number, were introduced by von Neumann in his theory of zero-sum two-person games, and called by him "mixed strategies" as opposed to the "pure strategies" out of which they are compounded. I shall follow Wald and the other mathematical statisticians who have used them in calling them *randomized strategies*, randomization being the technique, widely used in biological experiments, of deliberately arranging things in such a way that the probabilities of encountering them will be definite pre-assigned numbers. The notion of arranging to select one thing out of many 'at random' is common knowledge; it is to arrange so that it is as probable that any one thing will be selected as any other, and it can be done by such devices as shuffling and drawing from a pack of cards. To select one thing out of many 'randomizedly' is the generic notion of which selecting 'at random' is the special case in which the probabilities to be pre-assigned in the randomization are all equal. Anyone who has ever tossed a coin to tell him how to choose between alternative decisions—whether or not to get married, whether or not to take out a mackintosh when it is not raining at the moment, whether or not to finesse the Queen at whist—is using what Wald calls a "randomized decision function", though he would usually be hard put to justify his method of randomization.

To use, however, a randomized strategy in order to choose between two statistical hypotheses does seem a strange and, on the face of it, an irrational procedure. For it involves choosing sometimes one hypothesis and sometimes another on relevantly similar data. If it is ever proper to use strategy T_2 to choose H_1 on finding one α-specimen in a set of two β-specimens, how could it ever be proper to use strategy T_3 and to choose instead H_2 on finding on another occasion one α-specimen in a similar sample? To answer this puzzling question one has to realize what is involved in a strategy. To use the unrandomized (or, as I have called it, *basic*) strategy T_2 is not to choose one hypothesis on all occasions; it is to choose H_1 when the set of two β-specimens contains either one or two α-specimens and to choose H_2 when the set contains no α-specimens. The choice prescribed by a strategy (except in the case of the extreme strategies) is not a universal prescription for all cases; it is a prescription to choose differing hypotheses under differing conditions. All that is done in passing from basic to randomized strategies is to complicate these conditions. Since the randomized strategy R is compounded out of the basic strategies T_2 and T_3, it will prescribe the same choice of hypothesis as do T_2 and T_3 in the cases in which they agree in their prescriptions; so it will only be necessary to complicate the condition in the one case in which they disagree—when exactly one α-specimen is found in the set of two β-specimens. Disentangling what is prescribed by the randomized strategy R, we find that R prescribes that, when the observed set contains two α-specimens, hypothesis H_1 is to be chosen; that, when it contains no α-specimen, H_2 is to be chosen; that, when it contains one α-specimen and a spade is drawn from a shuffled pack of eighteen cards containing eleven spades and seven hearts, H_1 is to be chosen; and that, when it contains one α-specimen and a heart is drawn from such a pack, H_2 is to be chosen. The prescriptions of R are conditional as are the prescriptions of T_2; the only difference is that the condition of finding one α-specimen in the sample is, in the case of R, to be divided into two subconditions—one of these being the conjunction of the finding of one α-specimen with the drawing of a spade, the other being the conjunction of the finding of one α-specimen with the drawing of a heart. Thus no novel and irrational principle seems to me to be involved in introducing randomized strategies.

The 'chance-machine' which I have used in my exposition of the randomization procedure is drawing from a shuffled pack of suitably selected cards. Of course any other suitably designed chance-machine would have served the same purpose, e.g. a roulette wheel or teetotum or a table of 'random numbers'.* All that is required is some procedure for giving a specified probability p' for using one of the basic strategies out of which the randomized strategy is compounded and for giving the probability $1 - p'$ for using the other basic strategy. It may be asked whether it is necessary to use such a chance-machine at all; would it not do as well, in our example, to choose H_1 eleven times and H_2 seven times out of every eighteen times that the observed set of two β-specimens contained one α-specimen? But this would not serve the required purpose. It is essential that the probability p' for using one of the basic strategies should be the same on each occasion, and that it should be independent of how many times that basic strategy has been used on previous occasions. It is also essential that p' (the *mixing probability*, as I shall call it) should be independent of the number of times the various α-ratios have been found in the observed sets; otherwise it could not be multiplied into the probability of finding that α-ratio in order to give the probability of the conjunction required in the calculation of the probability of correct choice. These two independences can only be secured by making use of a chance-machine whose working is unrelated to the truth of the statistical hypotheses under consideration.

Randomized strategies for choosing between statistical hypotheses, though peculiar at first sight, are thus logically quite respectable; and there is no reason why they should not be used when there is an advantage in doing so. And, apart from resolving the paradox of the existence of freak strategies, there is a very great advantage in using them. Instead of having only eight possible strategies to play with, by using randomized strategies we have an unlimited number of possible strategies at our disposal.

In the example of a randomized strategy given on pp. 229f. the mixing probabilities $\frac{11}{18}$ and $\frac{7}{18}$ have been carefully selected so that

* The 'chance-machine' need not be strictly speaking a machine (i.e. an artefact) if there is some natural process at hand (e.g. the emission of electrons from a radioactive substance) which gives exactly the right probabilities.

the strategy-point representing the randomized strategy should fall on the intersection of the side T_2T_3 of the hexagon with the upward diagonal U_1.* By choosing any other pair of fractions p', $1 - p'$, another randomized strategy with a strategy-point on the line T_2T_3 would have been constructed.† So, by selecting p' suitably a randomized strategy can be constructed whose strategy-point is any rational point on the line, i.e. any point whose co-ordinates are rational numbers.‡ There are an unlimited number of such rational points, and they are, in mathematical language, 'everywhere dense', i.e. every point on the line has an unlimited number of them within any neighbourhood, however small. Since we have assumed that the utility values are rational numbers, every utility line which crosses T_2T_3 will cut it at a point representing a possible randomized strategy.§

In an exactly similar way a randomized strategy can be compounded out of the basic strategies T_1 and T_2 in such a way that any rational point on the side of the hexagon T_1T_2 is its strategy-point, and a randomized strategy can be compounded out of the basic strategies T_3 and T_4 with any rational point on the side T_3T_4 as its strategy-point.‖ Since every utility line crossing the square cuts the broken line $T_1T_2T_3T_4$ at one and only one point,¶ every such utility line will determine a unique strategy-point, which will represent a randomized strategy unless it is either T_2 or T_3. Every point lying below the broken line $T_1T_2T_3T_4$ will then become inadmissible as a strategy-point, and the paradox of freak strategies

---

* These mixing probabilities are p' and $1 - p'$, where p' is given by the equation $84p' + 36(1 - p') = 49p' + 91(1 - p')$.

† $p' = 1 - p' = \frac{1}{2}$ gives the randomized strategy-point falling on the intersection of T_2T_3 with the utility line U_3.

‡ Irrational randomized strategy-points can easily be introduced, if required, by a Dedekind-cut construction. They would be required if the utility values were irrational.

§ This depends upon the fact that mathematical expectations are all linear functions, so that no trouble will arise like that of the square root which originated the Pythagorean discovery of irrational numbers. For the important part played by the linearity of mathematical expectation see *NM*, § 17.7.2.

‖ Every rational point on the broken line $T_4T_5T_6T_1$ can similarly be a randomized strategy-point. And we can construct randomized strategy-points by compounding three basic strategies (using a chance-machine giving pre-assigned mixing probabilities p', p'', $1 - p' - p''$) in such a way that every rational point within the hexagon can be a randomized strategy-point. But since all these randomized strategies will be inadmissible, they are not required for my argument.

¶ The converse is not true. In Fig. 10, U_1 and U_4 both cut T_2T_3 at the point R.

is resolved. All the admissible strategy-points will lie on the line $T_1 T_2 T_3 T_4$, and every rational point on this line will be an admissible strategy-point.

Let us now take the prudential policy in an extended sense to cover the selection of a prudential strategy from among all possible strategies, basic or randomized. If we exclude the exceptional cases in which either the prudential strategy does not exist or it is T_1 or T_4, we shall find by the geometrical argument of p. 220 that the strategy R_i whose strategy-point is the point where a utility line U_i intersects the broken line $T_1 T_2 T_3 T_4$ is the prudential strategy corresponding to U_i, i.e. corresponding to an assignment of utility values a_i, b_i, c_i, d_i which determine U_i. For, since every utility line (for the non-exceptional cases) slopes upwards from left to right and the three parts of the broken line $T_1 T_2 T_3 T_4$ all slope downwards from left to right, the strategy-point R_i lies to the right of every strategy-point in the utility line's superior region and is higher than every strategy-point in the utility line's inferior region. Since the strategy-point representing the prudential strategy R_i lies on the utility line U_i, the pair of mathematical expectations for R_i are equal; if R_i is a randomized strategy each of these two mathematical expectations will be greater than the minimum of the pair of mathematical expectations for each of the basic strategies out of which R_i is compounded. Thus, by following the prudential policy, the least we can gain however Nature may behave if we employ randomized strategies is never less than the least we can gain if we restrict ourselves to basic strategies; it is, in fact, more, except when the prudential strategy is itself a basic strategy, which happens only when the set of utility values is one of the exceptional cases or when the utility line determined by them passes either through T_2 or through T_3.

The geometrical method for discovering the strategy, randomized or basic, which is that strategy which is the prudential one for a given assignment of utility values, now becomes extremely simple. The method is merely to draw the utility line determined by this assignment of utility values, and to find where this line cuts the broken line $T_1 T_2 T_3 T_4$. If R', the point where it cuts it, is either T_2 or T_3, the basic strategy T_2 or T_3 respectively is the prudential strategy. If R' lies in the part $T_2 T_3$, the prudential strategy is the randomized stategy obtained by compounding the use of strategy

T_2 with a probability p' and the use of strategy T_3 with a probability $1 - p'$, where p' is the ratio of the length of $R'T_3$ to the length of T_2T_3. So all we have to do to find the mixing probability p' is to arrange our chance-machine (pack of cards, teetotum, etc.) so as to give us this probability. Similarly, if R' lies in the part T_1T_2 or in the part T_3T_4, the randomized strategy is obtained by compounding strategy T_1 and strategy T_2, or strategy T_3 and strategy T_4 respectively, with probabilities p' and $1 - p'$, where p' is the ratio of the length of $R'T_2$ to the length of T_1T_2, or the ratio of the length of $R'T_4$ to the length of T_3T_4, respectively.*

The extension of the prudential policy to include the use of randomized strategies has thus two great virtues:

(1) The fact that any rational point on the broken line $T_1T_2T_3T_4$ may be regarded as a possible strategy-point renders inadmissible all strategies whose strategy-points lie below this line, and thus resolves the freak-strategy paradox. There are such freak strategies, but there are no assignments of utility values which will make any of them the prudential strategy. To put the matter positively, the strategies represented by the four strategy-points T_1, T_2, T_3, T_4 are fundamental in the sense that every strategy which is the prudential strategy for some assignment of utility values is either one of these four strategies or is a randomized strategy obtained either by compounding T_1 and T_2 or by compounding T_2 and T_3 or by compounding T_3 and T_4 with appropriate mixing probabilities.

(2) The pair of mathematical expectations for every prudential strategy are equal, and are greater than the minimum of the two mathematical expectations for the strategy which would have been the prudential one if the use of randomized strategies had not been allowed (unless, of course, this strategy is one which would remain prudential even if randomized strategies were permitted).

---

* The coordinates x', y' of the prudential strategy-point R' and the mixing probability p' are given algebraically by solving the three simultaneous equations

$$(a+d)\, x' - d = (b+c)\, y' - b \quad \text{[the utility line]},$$

$$\frac{x_j - x'}{x_j - x_i} = \frac{y_j - y'}{y_j - y_i} = p' \qquad \text{[the mixing probability]},$$

where x_i, y_i; x_j, y_j are the coordinates of the permissible basic strategy-points nearest the utility line on either side. The pair of (equal) mathematical expectations for R' are $(a+d)\, x' - d$ and $(b+c)\, y' - d$.

RANDOMIZED STRATEGIES IN THE THEORY OF GAMES

Randomized strategies were introduced by von Neumann into his theory of the general zero-sum two-person game to meet a difficulty which arose in producing a 'solution' of the game which would be, in a clear sense, the 'best way of playing' the game.* In a game where randomized strategies are not used, for one player A to follow the prudential policy of making his move in such a way as to maximize the minimum score he can make is not obviously the best way for him to play. It is the way in which he will obtain the maximum advantage irrespective of whatever move his opponent B makes, and hence it is clearly the best policy for him to pursue if his opponent knows beforehand how he will play and will choose his own move accordingly. But it is not obviously the best way for him to play if he is able to 'find out' his opponent's move quite as much as, or more than, his opponent is able to 'find out' his own.† There seems to be no possibility of a clear-cut answer as to the best way to play the game, since each player will properly take into account how much he can guess, or infer, of what his opponent's move will be and how much he can guess, or infer, that his own move would be guessed or inferred by his opponent.

The use of randomized strategies is a device by which each player can protect himself against having his move being 'found out' by his opponent. For the player will not himself know what his own move will be; so it will be impossible for him to disclose it, directly or indirectly, to his opponent. Von Neumann has proved formally, without any reference to the knowledge or lack of knowledge of the players, that, if randomized strategies are permitted, to use such a strategy devised in accordance with prudential policy is, in a clear sense, the 'best way of playing' for each player.

The reason for introducing randomized strategies into the solution of our inductive problem is different. It is not a matter of protecting ourselves against Nature 'finding out' beforehand which

* See *NM*, §14.2.

† In an important class of cases—those which von Neumann calls "strictly determined games"—neither player would obtain any advantage by finding out what his opponent's move would be; so to follow the prudential policy is in a clear sense the 'best way of playing' for each player. But the games which are the analogues of choices between hypotheses are never 'strictly determined'— in von Neumann's original sense of this term (*NM*, §14).

of the two statistical hypotheses we intend to choose; for if we were to personify Nature to that extent, there could be little reason for supposing that Nature would not also have foreknowledge of the mixing probabilities which our chance-machine would provide. Our reasons for introducing randomized strategies are, first, that to do so disposes of the freak-strategy paradox, and, secondly, that it increases the maximum of the minimum mathematical expectation. By using randomized strategies the prudential policy can improve what it has to offer without sacrificing its characteristic feature of prudence.

Thinking of the matter in terms of the Theory of Games as a game against Nature immediately raises the question: What happens if Nature also uses a randomized strategy? What happens if H_1 is the true hypothesis on $\frac{7}{15}$ths of the occasions and H_2 is the true hypothesis on $\frac{8}{15}$ths of the occasions on which we choose between them on the basis of the observed number of α-specimens in a set of two β-specimens, these fractions being the mixing probabilities with which Nature would have to compound H_1 with H_2 in order to behave prudentially?* Stated in this form the question is, of course, nonsense. A general hypothesis, even if it is a statistical hypothesis, cannot be sometimes true and sometimes false. But the question can be restated as a question about a third hypothesis H_3, and can be most conveniently expressed by reference to the model of a bag. Suppose that there are in a large bag fifteen small bags each containing ten balls. In seven of the small bags six of the balls are black and four are white; in the remaining eight small bags three of the balls are black and seven are white. Then H_3 is to be the hypothesis that the probabilities of there being o or 1 or 2 α-specimens in a set of two β-specimens are the same as the probabilities of o or 1 or 2 black balls being drawn when, first, one of the small bags is drawn from the large bag and, secondly, two balls are drawn from this small bag (the first ball drawn being replaced before the second drawing). The hypothesis H_3 thus specified is a statistical hypothesis which is in itself a perfectly proper one; but it has been excluded by the conditions of our problem, which were that one or other of H_1 and H_2 should

* The mixing probabilities are q' and $1-q'$, where q' is given by the equation $84q' + 49(1-q') = 36q' + 91(1-q')$. Cf. the mixing probabilities p' and $1-p'$ of p. 233 footnote.

be the true hypothesis, any third alternative being excluded. Thus, to speak in the language of games, while it is not cheating for the scientist to use a randomized strategy formed by compounding two or more of his basic strategies, it would be cheating (i.e. it would not be in accordance with the rules of the game) for Nature to compound randomizedly the two hypotheses H_1 and H_2 into a third hypothesis H_3.

The fact that the conditions of the problem permit of only two hypotheses being considered need not prevent us from considering hypothetical mixtures of them if such consideration would assist us in calculating the prudential strategies. The mixing probabilities of the hypotheses would then be theoretical concepts in a deductive system in which would be deduced consequences in which these concepts did not occur. Wald found it convenient to introduce hypothetical probability distributions of his probability parameters (these probability distributions correspond to the mixing probabilities for our statistical hypotheses) in order to prove the existence of minimax strategies, since he could prove that the minimax strategy was one which minimized the average risk relative to a least favourable mixture of the hypotheses.*

Regarded from the standpoint of the Theory of Games the situation would seem to be that to consider Nature as randomizedly mixing her hypotheses may be an excellent method for finding out how to treat her when she does not mix them. But for the scientist randomizedly to mix his basic strategies is a technique by which he definitely increases his mathematical expectations. By deliberately using artificial probabilities he can improve his prospects of discriminating between the natural probabilities presented to him.

Von Neumann was not able, except in the case in which each player is limited to only two moves,† to give a method for calculating the mixing probabilities with which the basic strategies would have to be compounded in order to form the prudential randomized strategy. My geometrical method is equivalent to a

---

* *Annals of Mathematics*, second series, vol. 46 (1945), p. 267. In his original article Wald made it perfectly clear that his reason for introducing hypothetical probability distributions (which he sometimes called "*a priori* probability distributions") into his treatment was "simply that [they] prove to be useful in deducing certain theorems and in the calculation of the best system of regions of acceptance" (*Annals of Mathematical Statistics*, vol. 10 (1939), p. 302).

† *NM*, §18.2.

Theory-of-Games method for calculating the mixing probabilities if one player B is limited to only two basic moves, the other player A being permitted any finite number of basic moves. The important special property of this case, which is the one corresponding to the choice between two hypotheses, is that, if A uses the randomized strategy which is the prudential one, his score will be the same whichever of the two possible moves is made by B.*

THE EFFECT OF AN INCREASE IN THE NUMBER OF INSTANCES

We have examined in considerable detail the prudential policy for preferring hypothesis H_1 to hypothesis H_2 when the evidence to be observed is the number of α-specimens in a set of two β-specimens, and have found that (except for the exceptional cases of assignments of utility values) the prudential strategy is that represented by the point where the utility line cuts a three-sided broken line running downwards from the top left-hand corner to the bottom right-hand corner of the square in our diagram. What will happen when the evidence is of a sample with a larger number of β-specimens than two?

Although in order to simplify our exposition a number, n, of instances as small as two was taken, nothing in the general argument depends upon the value of n. So we may in general say that the prudential policy for preferring H_1 to H_2 on the basis of observation of the number of α-specimens in an n-fold β-sample is (except for the exceptional cases of assignments of utility values, which can be treated exactly as before) to select that strategy—the 'prudential' strategy—represented by the point where the utility line cuts the $(n+1)$-segmented line formed by joining the points $(x_1, y_1), (x_2, y_2), ..., (x_{n+2}, y_{n+2})$, where (x_j, y_j) represents the basic strategy of choosing H_1 if $s = j-1, j, j+1, ..., n$ and of choosing H_2 if $s = 0, 1, 2, ..., j-2$, and x_j, y_j are the probabilities, given the

---

* It would be inappropriate to set out in detail here the straightforward generalization of my geometrical method required in order to apply it to the general case of the zero-sum two-person game with one player restricted to only two moves. The mixing probabilities for A's prudential strategy are, in the general case, determined by the intersection of the upward diagonal with one of the admissible sides of the strategy-point polygon (if all admissible sides lie wholly above or wholly below this diagonal, the game is 'strictly determined' in von Neumann's original sense): the mixing probabilities for B's prudential strategy are determined by the angle between this upward diagonal and this side of the polygon.

truth of H_1 and of H_2 respectively, of correct choice for the basic strategy. (The calculation of these pairs of probabilities is only a matter of complicated arithmetic.) Unless the utility line passes through one of the strategy-points T_2, T_3, ..., T_{n+1}, the prudential strategy will be a randomized strategy compounded out of the two basic strategies whose strategy-points lie nearest on each side to the strategy-point of the prudential randomized strategy. For the prudential strategy R_i the pair of mathematical expectations $(a+d)\,x_i - d$, $(b+c)\,y_i - b$ will be equal. All this follows by similar reasoning to that used for the case $n = 2$.

If the $(n+1)$-segmented line, all of whose rational points are permissible strategy-points, is drawn for $n = 2, 3, 4, ...$, it will be found that each of these lines lies above all its predecessors in the series. Towards the right-hand top corner of Fig. 10 will be seen five dotted lines joining six points, which represent six—P_8 to P_{13}—of the twenty-two basic strategies for $n = 20$.* The remaining basic strategy-points lie too near the right-hand side or the top side of the square to be conveniently shown on the figure. The prudential strategy appropriate to the utility line which is the upward diagonal U_1 will be a randomized strategy compounded out of P_{10} and P_{11} with appropriate mixing probabilities and with the intersection of U_1 with $P_{10}P_{11}$ as its strategy-point. Since this strategy-point lies farther up the upward diagonal U_1 than R, the prudential strategy-point when $n = 2$, its pair of equal mathematical expecta-

---

* The coordinates of these six strategy-points, calculated from the Binomial Distribution Table given by M. G. Kendall, *The Advanced Theory of Statistics*, vol. 1 (London, 1943), p. 119, are:

Strategy		Probability, given H_1, of correct choice (x)	Probability, given H_2, of correct choice (y)
P_8.	(H_1 for $s = 7, 8, ..., 20$; H_2 for $s = 0, 1, ..., 6$)	0·9935	0·6079
P_9.	(H_1 for $s = 8, 9, ..., 20$; H_2 for $s = 0, 1, ..., 7$)	0·9789	0·7722
P_{10}.	(H_1 for $s = 9, 10, ..., 20$; H_2 for $s = 0, 1, ..., 8$)	0·9434	0·8866
P_{11}.	(H_1 for $s = 10, 11, ..., 20$; H_2 for $s = 0, 1, ..., 9$)	0·8724	0·9520
P_{12}.	(H_1 for $s = 11, 12, ..., 20$; H_2 for $s = 0, 1, ..., 10$)	0·7553	0·9828
P_{13}.	(H_1 for $s = 12, 13, ..., 20$; H_2 for $s = 0, 1, ..., 11$)	0·5956	0·9948

tions will both be greater than the pair of equal mathematical expectations for the strategy R. The effect of increasing n from two to twenty is to increase these mathematical expectations.

Up to this point we have limited ourselves to the case in which the hypothesis H_1 states that the probability of a β-specimen being an α-specimen has the value $p_1 = \frac{3}{6}$ and the hypothesis H_2 states that this probability has the value $p_2 = \frac{3}{10}$. But these numbers also were taken merely for the sake of illustration and to simplify the arithmetic. Nothing in the general argument requires anything more than that $0 < p_2 < p_1 < 1$; so the whole argument can be applied to the general case. If p_1 and p_2 are close together (i.e. if $p_1 - p_2$ is small), it can be shown that the $(n+1)$-segmented line will be lower for a given n than if p_1 and p_2 are farther apart. Thus, for a given n and a given utility line, the pair of equal mathematical expectations for the prudential strategy will be less than they would have been had $p_1 - p_2$ been greater.

Let us now examine what happens to the prudential policy for choosing between any two statistical hypotheses H_{p_1}, H_{p_2} which assert that the probability of a β-specimen being an α-specimen is p_1 or is p_2 respectively (where $0 < p_2 < p_1 < 1$), when the number n of observed β-specimens upon which the choice is to be based increases without limit. It can be shown that, as n increases without limit, the point P where the $(n+1)$-segmented line crosses the upward diagonal approaches more and more closely to the right-hand top corner of the square, i.e. its coordinates x and y, which are equal, both tend to 1 as n increases without limit.* Thus the parts of the $(n+1)$-segmented line above the upward diagonal approach more and more closely to the top side of the square, the parts below this diagonal more and more closely to the right-hand side of the square.

* As n increases without limit, x, which is the probability, given H_{p_1}, of correct choice, tends to

$$F_x(n) = \sqrt{\left(\frac{n}{2\pi p_1(1-p_1)}\right)} \int_{f(n)}^{\infty} \exp\left[-\frac{n(z-p_1)^2}{2p_1(1-p_1)}\right] dz;$$

and y, which is the probability, given H_{p_2}, of correct choice, tends to

$$F_y(n) = \sqrt{\left(\frac{n}{2\pi p_2(1-p_2)}\right)} \int_{-\infty}^{f(n)} \exp\left[-\frac{n(z-p_2)^2}{2p_2(1-p_2)}\right] dz;$$

where $f(n)$ is a number, dependent upon n, which is such that $F_x(n) = F_y(n)$. Since $p_2 < p_1$, $p_2 < f(n) < p_1$ for every n; and $F_x(n)$, $F_y(n)$ both tend to 1 as n increases without limit.

Suppose that a and c are both positive or b and d are both positive, so that the utility line $(a+d)\,x-d=(b+c)\,y-b$ cuts the square without being either vertical or horizontal. If the utility line cuts the top side of the square, which it does if $a>c$, the pair of equal mathematical expectations for the prudential strategy appropriate to the utility line tend to c as n increases without limit. If the utility line cuts the right-hand side of the square, which it does if $c>a$, the pair of equal mathematical expectations for the prudential strategy appropriate to the utility line tend to a as n increases without limit. If the utility line passes through the top right-hand corner of the square, which it does if $a=c$, the pair of equal mathematical expectations tend to the value common to a and to c.

The interpretation of this result is that, as n, the number in the set of observed instances, increases without limit, the pair of equal mathematical expectations obtained by following the prudential policy in the selection of a strategy for choosing between the two hypotheses H_{p_1} and H_{p_2} tend to the minimum of a and c, i.e. to the minimum of the gains to be obtained by choosing H_{p_1} when it is true and by choosing H_{p_2} when it is true. If the minimum of a and of c is positive, so is the other one; and the result holds quite irrespective of the values of b and of d. If the minimum of a and of c is zero, it is necessary that both b and d should take positive values (in order to ensure that the utility line should cut the square and be neither vertical nor horizontal); but the result holds irrespective of what positive value is taken by b and what by d.

This result seems to me of the greatest interest. I have expounded it in terms of the use of randomized strategies, since in this case the prudential strategy-point lies on the utility line and moves upwards and to the right along it as n increases. But the result holds equally if randomized strategies are not used. For as n increases, the two basic strategies nearest on either side to the utility line, which are those between which the selection has to be made in the second step of the non-randomizing prudential procedure (explained on p. 218), get closer and closer to this utility line; and the four mathematical expectations, a pair for each of these basic strategies, all approach one another, the limiting value which they all approach when n increases without limit being the value represented by the point on the utility line where it is cut by the line joining the two basic strategy-points. This value is that of the pair of equal mathematical expectations which would have been obtained had randomized strategies been used; so the result that, by employing prudential policy on the basis of a set of n observations, the H_{p_1}-true mathematical expectation and the H_{p_2}-true mathematical expectation both tend to the minimum of a and c as n increases without limit irrespectively of the values of b and of d (except that these must both be positive if the minimum of a and c is zero) holds whether or not the use of randomized strategies is permitted.*

To speak of the behaviour of the pair of mathematical expectations as n increases without limit is a convenient way of speaking of their values when n is any number, however large. This way of speaking may perhaps be confusing to an incautious reader, since it might lead him to suppose that, after having chosen one or other of the hypotheses in accordance with prudential policy on the basis of the observed number of α-specimens in an n-fold β-sample, we then increase the number in the observed sample to be $n+1$, and consider how our choice of hypothesis should be affected by the addition to our knowledge provided by the new observation. But this is a mathematically more complicated problem than the one with which we are concerned: it will not be discussed here, since it raises no new logical problems.† What we are concerned with in this chapter is purely with what statisticians call "fixed samples"; when the effect of increasing the number n of a set of instances is spoken of, we are not comparing some property of that set of instances with the same property of a larger set of instances including that set, we are comparing some property of that set with the same property of a larger but quite separate set of instances. Each set of instances, that is to say, is to be considered as inde-

---

* The introduction of randomized strategies was called for by the fact that the strategy points T_1, T_2, ..., T_{n+2} are isolated points on the broken line $T_1 T_2 ... T_{n+2}$. If continuous probabilities are being used in such a way that the basic strategy-points are continuously distributed along a line (usually a curved line) joining the top left-hand corner of the square to its bottom right-hand corner, there will be less occasion for the use of randomized strategies. Continuous probabilities are not discussed in this book.

† Wald's ingenious technique of 'sequential analysis' is a way for treating this problem. Wald expounded it in his *Sequential Analysis* (New York, 1947), and extended his general minimax policy to include it in *Econometrica*, vol. 15 (1947), pp. 279ff. [See also his *Statistical Decision Functions*.]

pendent of any other set considered; the probability of there being s α-specimens in one n-fold β-sample is independent of this probability in any other β-sample we consider; none of our probabilities are 'connected'. So to speak of increasing the number n of the set is merely a short way of speaking of considering an independent set with a larger number of instances; the result we have established may therefore be expressed by saying that the larger the set of examined instances upon which the choice between the hypotheses is to be made, the nearer will the pair of equal mathematical expectations for that randomized strategy which is the prudential one approach to the minimum benefit to be received by a correct choice of hypothesis.

The point of this important result may be brought out in several ways. One way is to compare the mathematical expectations for the prudential strategy based upon a set of n observations with the mathematical expectations for the two extreme strategies of choosing H_{p_1} or H_{p_2} whatever may be the α-ratio disclosed by the observation. We arrive at the following comparative table, ϵ_n being a small number which tends to o as n increases without limit, and min$[a, c]$ denoting the minimum of a and c:

	Mathematical expectation, given H_{p_1}	Mathematical expectation, given H_{p_2}
Strategy T_1 (always choose H_{p_1})	a	$-b$
Prudential strategy	min$[a, c] - \epsilon_n$	min$[a, c] - \epsilon_n$
Strategy T_{n+2} (always choose H_{p_2})	$-d$	c

Since the mathematical expectation, given the truth of H_{p_1}, is the greatest among all possible strategies for the strategy T_1 which chooses H_{p_1} irrespective of the evidence, and the mathematical expectation, given the truth of H_{p_2}, is greatest amongst all possible strategies for the strategy T_{n+2} which chooses H_{p_2} irrespective of the evidence, the effect of using the prudential strategy when n is large is to secure a pair of equal mathematical expectations which approximate to the minimum of the maximum of the mathematical expectations, given the truth of each hypothesis. So, by following the policy of aiming at maximizing over all strategies the minimum of the pair of mathematical expectations for each strategy, when n is large we very nearly secure the minimum over the two hypotheses of the maximum of the mathematical expectations given the truth of H_{p_1} and the maximum of the mathematical expectations given the truth of H_{p_2}.*

If $a=c$, i.e. if the same gain is to be obtained by choosing either hypothesis if it is true, the prudential strategy will, for a large n, yield a pair of mathematical expectations which approximate to this gain. If $a=c=0$, the case considered by Wald, the prudential strategy yields a pair of mathematical expectations which approximate to zero; in Wald's language, the maximum risk minimized by the prudential policy tends to zero as n increases without limit.

In all these cases the values of b and of d, i.e. the losses to be incurred by choosing H_{p_1} when it is false or H_{p_2} when it is false, enter less and less into the mathematical expectations for the prudential strategy as n increases. So another way of expressing the result is to say that, when n is large, the mathematical expectations for the prudential strategy are approximately independent of b and of d; the values of b and of d are highly relevant in determining which is the prudential strategy for choosing between the hypotheses, but they scarcely enter at all into the values of the mathematical expectations yielded by that strategy which is the prudential one. What we stand to gain by using the prudential strategy when n is large depends very little upon how much we should stand to lose if the hypothesis we should choose in accordance with this strategy were to be false. What the values of b and d do affect is the point on the broken line $P_1 P_2 \ldots P_{n+2}$ cut by the utility line, and thus what is the prudential strategy to be selected. In Wald's case in which $a=c=0$, the prudential strategy is settled entirely by the ratio of d to b, which determines the slope of the utility line

$$-d(1-x) = -b(1-y)$$

passing through the top right-hand corner of the square.

This behaviour of the prudential strategy as the number n of observations upon which it is based becomes large is a strong additional reason for accepting the prudential policy as a rational policy for choosing between two statistical hypotheses. The strategy selected by the prudential policy depends upon the

---

* If it is thought that this result is too good to be true, and that the scientist cannot have such an advantage in his game with Nature, it must be remembered that, under the conditions of the problem, whereas Nature is limited to two possible moves, the scientist has the great advantage of having 2^{n+1} possible moves at his disposal, even if he does not use randomized strategies.

position of the utility line across the square, and thus upon relationships between the values of the utilities. If it be objected that this has the effect of making the choice of the hypothesis on the basis of the observed number of α-specimens in a β-sample to depend upon extra-logical considerations of value, and is thus to make inductive logic subsidiary to economics or hedonics or ethics, the reply can be made that it is surely better to make the choice of hypotheses depend upon the advantages or disadvantages to be obtained by believing a true hypothesis, which, however difficult they may be to measure, are at any rate genuine factors in our life, than to make the choice dependent upon such obscure notions as the prior probabilities of the hypotheses. And the theorem that, as n increases, the pair of mathematical expectations for the prudential strategy tend to the minimum of the gains to be obtained by accepting one or the other of the hypotheses if it is true irrespective of the amounts of the losses to be incurred by accepting one or the other of the hypotheses if it is false, seems to me to be a theorem about limiting values which shows the advantages to be gained by increasing the number of examined instances much better than any attempt to fit to the facts a calculus using symbols for probabilities which are not to be interpreted in terms of observable frequencies.

CONFIDENCE INTERVALS

I have presented the case for regarding the prudential policy as the best principle for deciding between statistical hypotheses by examining in detail how it works in a very simple case. It would be out of place to go into the elaborate mathematics required to apply it to cases of statistical inference more complicated than those of choices between two statistical hypotheses. But to show how the prudential policy is related to some other methods of statistical inference, I will indicate very briefly its relation to the technique of *confidence intervals* for interval estimation.

This ingenious method, devised by Jerzy Neyman,* may be introduced as follows. Suppose that the probability ellipses of Chapter V are redrawn as closed figures (call them *quasi-ellipses*) which are such that, for a fixed number n in a sample, the prob-

* *Philosophical Transactions of the Royal Society*, series A, vol. 236 (1937) pp. 333 ff.

ability of the sample point (r, p)—the coordinate r being the α-ratio in the n-fold sample, the coordinate p being the parameter of a statistical hypotheses—falling outside one of these quasi-ellipses is k, different values of k determining different quasi-ellipses. The only difference between these quasi-ellipses and the probability ellipses of Chapter V is that, whereas the probability of the sample point (r, p) falling outside the probability ellipse is less than k, the probability of its falling outside the quasi-ellipse is to be exactly k.*
The quasi-ellipses will form a Chinese-box system converging to

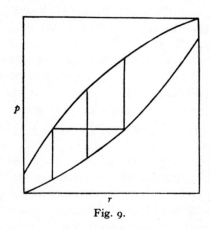

p

r

Fig. 9.

the diagonal $r = p$ just as the probability ellipses do; the whole exposition of Chapter V would have been given in terms of these quasi-ellipses instead of in terms of ellipses had not much more difficult mathematics been required to do so. The quasi-ellipses would then have been used in Chapter VI as methods for representing the intervals of non-rejection for the various values of p; the quasi-ellipses would have been considered as cutting off

* Strictly speaking, the finite number of values which r can take, for any given n and any given k, will make it impossible for us to draw a *convex* figure which will be such that, for all values of p, the probability of the sample point falling outside the figure is *exactly k*. (See C. J. Clopper and E. S. Pearson, *Biometrika*, vol. 26 (1934), p. 405, for a correct figure with saw-toothed sides.) Since, when n is large, the appropriate saw-toothed figure approximates to a convex figure with both top and bottom sides sloping upwards, this mathematical subtlety (like the similar subtlety mentioned on p. 169 footnote) will be ignored in our subsequent discussion.

appropriate lengths on lines representing the different values of p drawn horizontally across them.

Neyman's idea, brilliant in its simplicity, was to consider *vertical* sections of these quasi-ellipses instead of horizontal sections; and to examine what is represented by the lengths cut off on these vertical lines by a quasi-ellipse. This is equivalent to transforming the equation for the quasi-ellipse from the form $r = f_1(p)$ for the bottom side of the quasi-ellipse, $r = f_2(p)$ for the top side of the quasi-ellipse into the form $p = g_1(r)$ for the bottom side, $p = g_2(r)$ for the top side.* Since the proposition that the sample point (r, p) falls within one particular quasi-ellipse Q is logically equivalent to the proposition that the vertical interval $[g_1(r), g_2(r)]$ includes the value p, for any p the probability that this interval includes the value p is equal to the probability that the sample point falls within the quasi-ellipse Q, which is $1 - k$, where k is the k-value determining the quasi-ellipse Q. Hence, whatever be the value p of the probability parameter, the probability is $1 - k$ that this value p is included in the interval $[g_1(r), g_2(r)]$, where r is the α-ratio in a sample of n β-specimens (see fig. 9).

This, of course, is a theorem in the pure probability deductive system. It can be applied to the inductive problem of the estimation by means of an interval of a probability parameter p by taking as the strategy to be used that of choosing as the interval which is to include the probability parameter the vertical interval $[g_1(r), g_2(r)]$, where r is the α-ratio in the observed set of n β-instances. This strategy prescribes the choice of the probability parameters as follows: Given an observed α-ratio r, construct the vertical interval $[g_1(r), g_2(r)]$, and take the probability parameter as being included in this interval. This strategy, which estimates the probability parameter, not as being situated at a certain point, but as being situated within a certain interval, is Neyman's 'confidence interval' strategy. The justification he gives for it is that the probability of correctly assigning a probability parameter p to a confidence interval $[g_1(r), g_2(r)]$ when this interval does in fact include it, is $1 - k$ (which for the same quasi-ellipse can be made as near 1 as we please by taking n large enough), so that in a large number of uses of the confidence-interval strategy (based upon

---

* This transformation is always possible, since $f_1(p)$ and $f_2(p)$ are both increasing functions of p with $f_1(p) > f_2(p)$ for $0 \leqslant p \leqslant 1$.

samples with different values of their α-ratios) by and large the proportion of correct assignments of the probability parameter to the including confidence interval will be $1 - k$.

This confidence-interval strategy can be brought under the prudential policy for choosing between hypotheses if we take for each of these hypotheses the proposition that the probability parameter p is included in the interval $[g_1(r), g_2(r)]$, where r is one of the possible values for the α-ratio in the n-fold sample. Since there are $n + 1$ such possible values, there are $n + 1$ of these hypotheses. They are not necessarily exclusive; there may well be two or more values of r which determine intervals $[g_1(r), g_2(r)]$ that overlap, so that a particular probability parameter p may be included in all of them. So the confidence-interval strategy is not properly speaking a strategy for choosing between alternative hypotheses, i.e. for accepting one alternative and rejecting all the other alternatives; it is a strategy for accepting one hypothesis without either accepting or rejecting those of its alternatives which are logically compatible with it.* But on the assumption that the amount I gain by accepting one of these hypotheses if it is true is the same as the amount which I stand to gain by accepting any other of the hypotheses if it were true, and if the amount b which I stand to lose by accepting one of these hypotheses if it is false is the same as the amount I stand to lose by accepting any other of the hypotheses if it were false, then the mathematical expectation, given that the probability parameter p lies in the confidence interval $[g_1(r), g_2(r)]$ determined by the observed α-ratio r, is

$$a(1 - k) - bk = a - (a + b) k.$$

This mathematical expectation given by the confidence-interval strategy is the same whatever be the value of the probability parameter p. The confidence-interval strategy therefore shares the important characteristic of the prudential strategies considered in this chapter—that of yielding a unique mathematical expectation which is independent of the vagaries of Nature. Whether or not this unique value is a maximum of the minima of mathematical expectations for alternative strategies depends upon which set of strategies is taken as the set of exclusive and exhaustive alter-

* If the intervals $[g_1(r_1), g_2(r_1)]$, $[g_1(r_2), g_2(r_2)]$ do not overlap or touch one another, it is logically impossible for p to be included in both of them.

CHOICE BETWEEN STATISTICAL HYPOTHESES

natives. If the set of alternative strategies are those of accepting different hypotheses out of the set of $n+1$ confidence-interval hypotheses when a particular α-ratio r is observed, e.g. instead of accepting in accordance with the confidence-interval strategy the hypothesis that p is included in the interval $[g_1(r),\ g_2(r)]$, to accept instead the hypothesis that p is included in the interval $[g_1(r'),\ g_2(r')]$ or the hypothesis that p is included in the interval $[g_1(r''),\ g_2(r'')]$ or ..., where r', r'', ... are possible values of the α-ratio such that no two of these intervals overlap, then the confidence-interval strategy is certainly the strategy which is the prudential one from among this set of alternatives.

The confidence intervals with 'confidence coefficient' $1-k$ were defined as vertical sections of a quasi-ellipse which was such that the probability of the sample point (r, p) falling outside the quasi-ellipse was to be k. But there is no unique quasi-ellipse having this property; there are a great number of different functions g_1, g_2 which are such that the probability that the interval $[g_1(r),\ g_2(r)]$ includes p is $1-k$. Mathematical statisticians have suggested various additional conditions to be satisfied by 'optimum' confidence intervals. Wald has shown that an interval surrounding in a particular way the maximum-likelihood estimate is a confidence interval with an important optimum property.* So if the prudential policy is accepted as the most rational policy, the use of confidence intervals can be justified on prudential considerations, and the maximum-likelihood strategy for point estimation can be justified as yielding a particularly desirable type of confidence interval. This would seem to provide a rationale for the maximum-likelihood method which is lacking if the remarkable properties of this method are merely pointed out in a disconnected manner, and which to my mind is a much more satisfactory justification than that provided for the method by an ascription of equal prior probabilities to all the possible hypotheses.

CONCLUSION

So much space has been devoted to developing the theory of prudential policy for preferring one statistical hypothesis to another because the theory is novel and important and has not

* That of being an 'asymptotically shortest unbiased confidence interval'; *On the Principles of Statistical Inference*, Chapter V.

been expounded before by a philosophical logician. I have preferred to give attention to advocating it rather than to discussing
and criticizing theories based upon non-empirical notions of
probability (Jeffreys, Keynes, Carnap) because, while such theories
have themselves been damagingly criticized (e.g. by Kneale), their
great strength lies in their claim to answer a question which
empirical theories of probability fail to answer—that of providing
a reason for choosing among statistical hypotheses. The non-
empirical theorists are quite right in recognizing that an analysis
in terms of observable frequencies alone is not sufficient to justify
preferring one hypothesis to another; and that the fact that the
observed frequency would be the most probable one, given the
truth of a certain hypothesis, is no reason in itself for accepting
that hypothesis rather than another. The merit of the view advocated in this chapter is that it recognizes frankly that another
consideration is involved besides that of the values of the observed
frequencies, and finds this in the relative advantages and disadvantages of acting on belief in a statistical hypothesis if it is
true or false respectively.

The fact that any theory of statistical inference will have to
take into account the relative importance of falling into the error
of rejecting a hypothesis which is true and of falling into the error
of accepting a hypothesis which is false has been finding general
acceptance among mathematical statisticians since Neyman and
Pearson distinguished the two types of error in 1933. It has not
been so clearly recognized that there is equally an evaluative
element in the meaning of probability statements themselves, and
that to say that it is 'practically certain' that the next 1000 births
in Cambridge will include the birth of at least one boy includes
a hedonic or ethical assessment. The arbitrariness, from a logical
point of view, of the value of k used in the k-rule-of-rejection
which determines the meaning of a probability statement has been
emphasized in the last chapter; if the thesis of that chapter is
accepted as a correct account of the way in which meaning is
attached to probability statements within a science, and thus of
the way in which statistical hypotheses function in scientific deductive systems, a theory of statistical inference in which extra-logical
elements occur is a natural development. If ethical considerations
cannot be excluded from the definition of probability, they can

scarcely be kept out of the reasons for preferring one probability statement to another.

A philosopher who might be prepared to admit ethical elements into inductive logic may well be inclined to object to the private character of the evaluative notions required, since a strategy which is the prudential one for the assignment of values to the 'utilities' by one person will not be the prudential strategy for another such assignment by another person. But, as in the case of the definition of probability, so in the case of statistical inference an increase in the number of instances will tend to diminish differences based upon different assignments of utility values. In the last chapter we saw that a k-rule-of-rejection based upon any one value k_1 of k is equivalent to a k-rule-of-rejection based upon any other value k_2 of k, however small, provided that the set of instances upon which this latter rule has to work is taken to be large enough. In this chapter we have seen that 'at infinity' the mathematical expectation yielded in choosing one hypothesis rather than another in accordance with the prudential policy is independent of the assignment of losses to be incurred by an erroneous choice. How the actual choice varies when the number of instances increases without limit is a mathematically complicated question which cannot be discussed here; Wald has shown that various statistical procedures can be justified 'at infinity' by prudential considerations which are independent of the assignment of utility values.*

It should be remarked, however, that this privacy is not peculiar to the theory of statistical inference here advocated. Logicians like Jeffreys and Keynes who use a non-empirical theory of probability have in their account of statistical inference to assign *prior probabilities* to the alternative hypotheses; these assignments, for all that they frequently base themselves upon judgments of equiprobability, are as private to each person as are our assignments of utilities. As in the case of some of our prudential strategies, the variation in these private assignments becomes less and less relevant as the number in the set of observed instances increases, so that 'at infinity' the values of the prior probability assigned to the alternative hypotheses become irrelevant. But the proof that these values are irrelevant 'in the limit' presupposes that they are relevant in the series of which the limit is the limit. So the fact that

* *On the Principles of Statistical Inference*, p. 46.

the prudential theory expounded in this chapter makes use of assessments of utility which may vary from person to person is no argument against the theory as contrasted with a theory making use of equally personal assignments of prior probability.

If one stands back a little from the problem, there would seem to be no reason to be surprised at the involvement of inductive logic in evaluative considerations. In a deductive chain of reasoning the conclusion is a logical consequence of the premisses: a person believing the premisses and recognizing the logical relationship between the premisses and the conclusion can give this logical-consequence relationship as his reason for believing the proposition which is the conclusion rather than any alternative proposition. In a case of statistical inference, where the premisses are propositions about observable class-ratios and the result of the inference is to be the choice of one hypothesis out of a set of alternatives, none of these hypotheses is a logical consequence of the premisses; so a person believing the premisses cannot give a logical-consequence relationship as his reason for choosing one hypothesis rather than another. He can produce logical relationships between probability statements, as has been done in this chapter, and as is done by logicians like Jeffreys and Keynes in their use of 'inverse probability' and by statisticians who use 'maximum likelihood'. But these logical relationships between probabilities, though highly relevant to the choice of the hypothesis, are not sufficient in themselves to justify the choice, in the way in which the logical-consequence relationship is sufficient to justify the choice of a deductive conclusion. What else can possibly serve to guide us in our choice? Jeffreys and Keynes propose as guides judgments of prior probabilities which are not subject to empirical test. The proposal of this chapter is to use as guides the relative advantages and disadvantages to be obtained by acting on beliefs in the alternative hypotheses, combining these with probabilities (used in an empirical sense based upon frequencies) to form 'mathematical expectations'. These mathematical expectations are, in spite of the name, the statistical properties of the by-and-large averages over a large number of occasions. The use of mathematical expectations to guide choices of courses of action is the foundation not only of gambling on games of chance but of all forms of insurance: to use mathematical expectations to guide choices of beliefs is to apply customary criteria to the actions to which the beliefs give rise. It is not necessary to define a belief solely in terms of the actions to which it gives rise* in order to be willing to take a feature of such actions as our guide in the choice of a belief. To base a preference for one scientific hypothesis to its alternatives upon preferences among the results of different courses of action is to make use of no new criterion which is not already involved in the principles of our active life; inductive behaviour is subject to the same teleological standards as any other behaviour; and it would seem less in accordance both with the principle of empiricism and with that of parsimony to justify the choice between scientific hypotheses by attributing non-empirical prior probabilities to these hypotheses than to make use of those evaluative criteria for courses of action which we employ every day in our ordinary life.

* As was done by Alexander Bain. I have discussed this 'propensity to action' view of belief in *Proceedings of the Aristotelian Society*, n.s., vol. 33 (1932–3), pp. 129 ff.

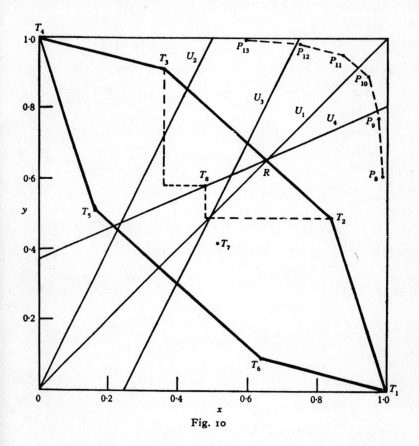

Fig. 10

CHAPTER VIII

THE JUSTIFICATION OF INDUCTION

Up to this point in this book there has been no serious consideration of the logic of the confirmation of a scientific hypothesis. The first few chapters were concerned to insist that scientific hypotheses enter into our thinking by virtue of the places they occupy in scientific deductive systems, the lowest-level propositions of which are empirical generalizations directly confrontable with experience. If the experience with which this system was to be confronted turned out contrary to one of the lowest-level propositions, this proposition and all the higher-level propositions of which it was a logical consequence would have to be rejected. If, as is usually the case, the rejected lowest-level proposition was a consequence of a conjunction of higher-level propositions, one at least of these higher-level hypotheses would have to be rejected, though the experience refuting the lower-level proposition would not indicate which this rejected higher-level proposition would have to be. The empirical criterion of rejection for a scientific hypothesis is so fundamental that it is most convenient to treat the meaning of universal sentences expressing empirical generalizations as being determined by the experiences which would refute them.* Consequently, the meanings of the theoretical terms occurring in the higher-level hypotheses of a scientific system which, as we have seen, are determined (although with a certain 'arbitrary choice') by the lower-level hypotheses which are their consequences are in the last resort determined (with the same 'arbitrary choice') by the empirical criteria for the rejection of the system as a whole. And in our discussion of probability statements which has occupied the last three chapters, we have emphasized the role of probability statements within a science as being hypotheses which function in a particular way in a scientific deductive system, this particular way of functioning being due to the fact that the probability statements are only provisionally and never definitively rejectable by experience. When further experience

* See above, p. 153 and footnote.

cancels a provisional rejection, this does not imply that experience instructs us to accept the statistical hypothesis; what experience instructs us to do is, negatively, not to reject the probability statement on that evidence. Whether or not that evidence is sufficient for us positively to accept the statistical hypothesis is quite another question—unless, of course, we are arguing on the assumption that a hypothesis has either to be accepted or to be rejected. An assumption which included this assumption was deliberately made in the last chapter, where the policy for the choice between statistical hypotheses was discussed in the simplest case in which only two hypotheses were candidates, so that to reject one was to accept the other. But nowhere up to this point have the policies governing the acceptance of a scientific hypothesis been discussed in themselves. Such a discussion cannot properly be postponed any longer.

We may by confusing non-rejection with acceptance deceive ourselves into supposing that a discussion of rejection, and a definition of the terms involved by means of the concept of rejectability, will have implicitly included a treatment of acceptance. And we may fall into the subtler error that, when there are a limited number of competing mutually exclusive hypotheses of the same type (e.g. that the number of electrons in the outer ring of a certain atom is less than or equal to 10, or that this number is greater than 10; that the probability of a particular penny falling heads upmost is less than $\frac{2}{5}$, or that this probability is greater than or equal to $\frac{2}{5}$ and less than $\frac{3}{4}$, or that this probability is greater than or equal to $\frac{3}{4}$), the rejection of all these hypotheses save one requires the acceptance of this one—forgetting that no hypothesis of this type may hold, or indeed that the phenomena in question may fall under no general law whatever. But, if we think about the matter, it will be clear that any use of the rejection of alternative hypotheses to support the acceptance of a non-rejected alternative must work on the assumption that some one of the alternatives is the true law of nature. Thus criteria for rejecting, or for failing to reject, a scientific hypothesis only provide partial criteria for accepting such a hypothesis—partial criteria to the extent that the rejection criteria do not reject it. Whether or not a hypothesis which is not rejected by experience is one which it is permissible to *accept* is a further question requiring for its answer reference to criteria independent of the empirical rejection criteria which, directly or indirectly, determine the meaning of the sentence expressing the hypothesis.

THE CONFIRMATION OF A SCIENTIFIC HYPOTHESIS

Statements expressing scientific hypotheses come to us with a well-defined meaning given by the class of observations which would refute them, or which would refute conjunctions of them with other scientific hypotheses. The philosophical problem of the confirmation of a scientific hypothesis is thus independent of that of determining the meaning of the statement expressing it. But it is not independent of the problem of determining the meaning of "reasonable belief" in a scientific hypothesis, nor is it independent of the meaning of a "valid inductive inference" from empirical evidence as premiss to scientific hypothesis as conclusion. We do not come to the discussion of how we obtain reasonable belief in a scientific hypothesis by means of a valid inductive inference already knowing what we mean by such a 'reasonable' belief and such a 'valid' inference; in stating the conditions which justify inferences we shall, *ipso facto*, be giving criteria which determine the meaning of the phrases "valid inference" and "reasonable belief" in the case of an inductively established hypothesis.

This is because induction is not a demonstrative form of inference like deduction.* In deduction the reasonableness of belief in the premisses as it were overflows to provide reasonableness for the belief in the conclusion. This happens because the conclusion is a logical consequence of the premisses, and cannot be false while the premisses are true. The empirical circumstances under which the premisses collectively would be true are included in the circumstances under which the conclusion would be true; so a belief that the circumstances are such as to make the premisses true involves, implicitly if not explicitly, a belief that the circumstances are such as to make the conclusion true.

---

* The word "induction" will be used throughout in the sense in which it is used by philosophers to cover the inference of an empirical generalization from its instances, or of a scientific hypothesis from empirical evidence for it. It will not be used in the Aristotelian sense of the logic text-books to cover an inference which may be demonstrative by virtue of having a major premiss which, in conjunction with the instances, entails the generalization which is the conclusion.

257

If the former belief is a reasonable one, the latter belief will also be reasonable.*

In inductive inference there is no such automatic overflowing of reasonableness. This is because there is no logical impossibility in the premisses being true and the inductive conclusion false. The circumstances which would make the premisses true are not included in the circumstances which would make the conclusion true; indeed, when the induction is an inference to a generalization from its instances, or to the whole of the highest-level hypotheses in a scientific theory from instances of the generalizations which are the lowest-level hypotheses in the theory, the reverse is the case, the premisses in the induction being logical consequences of its conclusion. So belief in the premisses does not in any way involve belief in the conclusion (except as a matter of psychological causation), and the reasonableness of belief in an inductive conclusion requires more justification than that of citing the reasonableness of belief in the premisses.

The tradition of logicians has been to try to justify induction by assimilating it to deduction. This assimilation has been attempted in two ways. One is that of supposing that in every induction there is a suppressed major premiss, the explicit recognition of which will make the inference a deductive one. But no plausible major premiss can be proposed which would make the inductive conclusion a logical consequence of the premisses taken together. So a second way of assimilating has been tried—that of supposing that the conclusion of an induction is not the scientific hypothesis itself but is instead a proposition assigning a number to the hypothesis in relation to the evidence, this number measuring the 'probability' in the sense of Kneale's acceptability or Carnap's degree of confirmation; and that the argument thus understood is a deduction in a deductive system of probability statements in this sense of probability. It is the great merit of Keynes's work to have shown conclusively that this second way of assimilating induction to deduction is not defensible unless it is combined with the first.†

* There is another sense of reasonable belief in the conclusion of an inference—a subjective sense—which depends upon a belief that the conclusion follows from the premisses, and not upon whether the conclusion does so follow. This subjective sense will be important in the discussion later in this chapter of the circularity of the justification of induction to be given.

† J. M. Keynes, *A Treatise on Probability* (London, 1921), Part III.

A deductive system of probability statements will only give the probability of an inductive hypothesis on the basis of its confirmation in a set of instances (its 'posterior probability') as an arithmetical product one of whose factors is its probability before it has been confronted with experience (its 'prior probability'); and a supreme major premiss is necessary to ensure that this prior probability should be greater than zero. Thus the justification of induction propounded by the modern assimilating school (J. M. Keynes, H. Jeffreys, C. D. Broad) requires both a well-developed theory of probability in the sense of degree of confirmation and a supreme major premiss to stand at the head of every inductive inference.

But such a supreme major premiss, whether in Mill's form of a principle of Uniformity of Nature or in Keynes's form of a principle of Limited Independent Variety, will have to be a logically contingent proposition in order to fulfil the function which is required of it. That it is reasonable to believe such a proposition can then only be justified by an inductive inference. The overwhelming objection to the assimilation of all induction to deduction is that this would require that we should reasonably believe a very general empirical major premiss, the reasonableness of belief in which would have to be justified by another inductive argument.*

INDUCTION AS THE USE OF INDUCTIVE POLICIES

The attempt to justify induction as quasi-deduction was based upon the philosophical principle that the right way to discover a justification for a particular valid inductive inference was minutely to examine that valid inference in the hope of a discovery of a suitable 'suppressed' major premiss, or of intermediate stages in the argument, which, when clearly discerned, would settle the question of validity. This is the correct procedure to use in establishing the validity of a deductive inference. Here, when it is not intuitively apparent that the conclusion is a logical consequence of

---

* Attempts have been made to escape from this objection by calling the major premiss a "postulate", by which is meant a proposition which is neither knowable directly nor is inferable, deductively or inductively, from directly knowable propositions, but whose truth has to be *postulated* in order to justify other knowledge or reasonable belief. To call an empirical generalization a *postulate*, however, does not make it any the less an empirical generalization; nor does re-naming a proposition provide a new method for getting to know it.

the premises, it is necessary either to disclose a suppressed assumption or to insert intermediate steps in the deduction (or both) in order to become satisfied with its validity. This line of approach has led to a dead-end in the case of induction. So let us instead stand back from the particular inductive inference whose validity we wish to establish and, instead of examining it minutely, compare its general features with those of other presumedly valid inductive inferences. What then becomes clear is that all the inductions whose validity we wish to establish are inferences in which the inductive conclusion has been derived from the empirical premisses in accordance with one or other of a limited number of inductive principles of inference.

These inductive principles are those discussed in books on inductive logic and scientific methodology; though they can be classified in various ways, they fall into two main types according as they rely upon the bulk or upon the variety of the evidence. There are, first, principles of induction by simple enumeration according to which an inductive hypothesis is to be treated as being well established if it has not been refuted by experience and has been confirmed by not fewer than n positive instances. (Different values of n will yield the different simple-enumeration principles.) There are, secondly, principles of elimination according to which an inductive hypothesis is to be taken to be well established if, while it has not been refuted by experience, alternative hypotheses have been so refuted. Mill's Methods of Agreement and of Difference fall into this class. An inductive inference proceeding according to an eliminative principle could be regarded as a deduction with a suppressed major premiss asserting that one of a set of hypotheses consisting of the one which has been confirmed and the alternatives which have been refuted is the true one. If the induction is justified in this way, the major premiss will have to be considered as having been established by a previous induction by simple enumeration. In fact, we are prepared to use eliminative principles of induction when we should not be prepared to regard as well established, or even as plausible, the proposition that the possible hypotheses are limited to those of which we know whether they are confirmed or refuted by experience. The eliminative principles we use are those according to which a hypothesis is treated as well established when, while it has not been refuted by experience, a certain number of the alternatives to it have been refuted. The different eliminative principles correspond to the different ways of selecting this certain number of alternatives.

The other inductive methods of the text-books follow one or other of these types of principle, or a combination of a principle of one type with one of another or with a deductive principle. One method for the establishment of a higher-level 'functional' hypothesis is that of induction by simple enumeration with lower-level laws regarded as being 'instances' of the higher-level functional hypothesis. The hypothetico-deductive method is that of deducing the hypothesis in question from higher-level hypotheses which have themselves been inductively established. The different methods employed for establishing a statistical hypothesis asserting that a certain probability parameter has an assigned value combine inductive principles of both types.

The various inductive principles can be expressed in terms of the construction of scientific deductive systems which this book has put in the forefront of its picture of scientific method. The simple-enumerative principles permit us to accept a system whose lowest-level hypotheses have been frequently confirmed but never refuted; the eliminative principles permit us to accept a system whose lowest-level hypotheses have not been refuted by experience while the lowest-level hypotheses of alternative systems have been so refuted.

Policies for establishing general hypotheses in accordance with inductive principles of inference on the basis of empirical data will be called "inductive policies". They all have the feature in common that they require a basis of experience to build upon; in this they differ from many non-inductive policies for establishing general hypotheses, e.g. that of deducing them from metaphysical premisses. It is not the function of this book to discuss in any detail the different inductive policies and their relations to one another, which can be expounded in different ways. It is sufficient for our purpose to recognize that there are a limited number of inductive policies which are explicitly used by scientists in establishing hypotheses on the basis of observations or which they would use in defending their supposedly established hypotheses against criticism. The first thesis of this chapter is that the justification of an inductive inference is to be given in terms of the principle of inductive inference used in the induction and not in terms of a relationship of logical consequence holding between the scientific hypothesis which is the conclusion of the inference and the premisses, even including in these premisses a 'suppressed' major premiss.

Many contemporary philosophers, in reaction against the attempt of the logical tradition to assimilate induction to deduction, feel happy in stopping at this point and in regarding the fact that an inductive inference has been made in accordance with a recognized principle of inference as being a complete justification for the induction. Some of them would support this view by citing the usage of such expressions as "established" or "well-grounded" as employed by scientists of inductive hypotheses. When a scientist says that the evidence is sufficiently good to *establish* a scientific hypothesis, he means that the evidence is such that the hypothesis could be inferred from it according to one of the recognized principles of scientific inference. His interest is in the adequacy of the evidence for this purpose, for this is what would be in dispute if a fellow scientist questioned the 'well-groundedness' of the hypothesis. He is not interested in the adequacy of his principle of inference, for he takes for granted that the use of this principle would be common ground to him and to any other scientist.

Thinking only of the considerations which are relevant to a scientist a philosopher will be quite right in taking the question as to whether or not an inductive inference is valid and the belief in a scientific hypothesis yielded by it is a reasonable belief as being a question as to whether or not the empirical evidence is such that the hypothesis can be obtained from it by following recognized 'rules of procedure'. But a philosopher cannot avoid saying something in answer to the question "What is the justification for using these recognized rules of procedure in making inductive inferences?" If he brushes this question aside, he is ceasing to be a philosopher.

Of course he may say, with Felix Kaufmann, that the question is a nonsense question in that "there is no ultimate justification for these rules [of scientific procedure]; we cannot go beyond them in discriminating between correct and incorrect scientific decisions".*

* *Methodology of the Social Sciences* (New York, 1944), p. 230. See also Chapter IV.

But to say this is to make every statement attributing validity to inductive inference an elliptical statement with an uneliminable reference to the rules of procedure used, and to forbid any comparison between the validity of an inference made according to one rule of procedure and that made according to another. But we frequently wish to choose between different rules of procedure in considering the same scientific problem; an example was discussed at length in the last chapter, where one procedure was preferred to alternative procedures for choosing between statistical hypotheses for reasons which, whether or not the reader was convinced by them, were certainly not arbitrary. Even Kaufmann admits that we may justify changes in rules of procedure by reference to 'rules of the second order'; "in asking whether a procedural rule of the first order is correct, we presuppose rules of the second order, etc."* But Kaufmann's "etc." involves an infinite regress, in which the justification of an inductive inference is given in terms of a first-order rule of procedure, whose justification is given in terms of a second-order rule of procedure, whose justification is given.... Kaufmann seems to have avoided propounding the 'ultimate justification' so distasteful to him by providing what turns out to be no justification at all.

There is, indeed, an answer to the request for a justification for using the recognized principles of inference which has been provided by some tough-minded philosophers. It is that these principles of inference are *recognized*, i.e. that they are those which are in fact used by reputable scientists to-day. But, apart from the fact that conformity to a majority of customary opinion in intellectual behaviour is no more ultimately justifiable than such a conformity in moral behaviour, this criterion for the validity of an inference —that the inference is made according to a principle used by reputable scientists—is of no avail when different reputable scientists disagree as to the proper principle to use in making an inference from the same empirical data and consequently disagree as to the inductive conclusion which it is proper to draw from these data. This has very frequently happened in the past; and happens at the present time both in the subtleties involved in the selection of principles of 'statistical inference' and in the permanent division of opinion—most obvious in the biological sciences—between those

* Op. cit. p. 233.

who for choice prefer inferences made directly from the observed evidence, in accordance with simple-enumerative and eliminative principles, and those who value more highly inferences made indirectly by the hypothetico-deductive method.

So common consent of reputable scientists cannot be regarded as a satisfactory fundamental criterion for the propriety of using a particular principle of inductive inference. Nevertheless, the fact that there is so much common consent as to the proper principles to be used in establishing scientific hypotheses lies on the route of the quest for justification. For why is there so much common consent? It has surely not arisen merely from scientists imitating one another. There must be some reason behind it.

When the question is put in this form, the answer is obvious. Why scientists use the inductive policies that they do use is the predictive value of these policies—their success in yielding hypotheses from which testable consequences can be deduced which are found to be true. And this is the justification for following a particular inductive policy, Π, of making inferences from observed data according to a particular principle of inference, π—namely, that following this policy yields hypotheses which are in fact confirmed and not refuted by experience. Good inductive policies are those which do what we require of them; they enable us to predict, and thereby partially to control, the future. The justification for an inductive inference from known empirical data, i.e. the criterion for the 'validity' of the inference and for the 'reasonableness' of belief in the hypothesis which is the conclusion of the induction, lies in the principle in accordance with which the inference is made being one the policy of using which has a characteristic which I shall call "predictive reliability" and which C. S. Peirce once called "truth-producing virtue".

THE PREDICTIONIST JUSTIFICATION

This justification of induction was first explicitly proposed by C. S. Peirce in 1877–8;* it has in the last quarter century gained

---

* "Illustrations of the Logic of Science", six articles which appeared first in *Popular Science Monthly*, reprinted in *Chance Love and Logic* (London, 1923) and in *Collected Papers of Charles Sanders Peirce*, vols. 2, 5, 6 (Cambridge, Mass., 1932–5). *The Philosophy of Peirce*, ed. J. Buchler (London, 1940), contains the first, second and fourth articles entire, most of the third and part of the fifth. These works will be denoted by *CLL*, *CP* and *PP* respectively.

many adherents among logicians. In order to discuss it we must express it more precisely than so far has been done.

What is meant by speaking of a policy as being "predictively reliable"? Peirce in 1878 gave a criterion in terms of the proportion among the inferences from true premises covered by the policy of those inferences which lead to true conclusions. Peirce finds the germ of this doctrine in Locke, who having spoken of a man who assents to a mathematical theorem on the authority of a mathematician without taking "the pains to observe the demonstration", went on to say: "In which case the foundation of his assent is the probability of the thing, the proof being such as, for the most part, carries truth with it."* Peirce takes this use of Locke's 'probability' to be the criterion of inductive validity: "in a logical mind an argument is always conceived as a member of a *genus* of arguments all constructed in the same way, and such that, when their premises are real facts, their conclusions are so also. If the argument is demonstrative, then this is always so; if it is only probable, then it is for the most part so. As Locke says, the probable argument is '*such as* for the most part carries truth with it'."† And in the following article in the series he says that, in the case of synthetic inferences (inductions), unlike that of analytic inferences (deductions), "we only know the degree of trustworthiness of our proceeding. As all knowledge comes from synthetic inference, we must equally infer that all human certainty consists merely in our knowing that the processes by which our knowledge has been derived are such as must generally have led to true conclusions."‡

I am sure that this account of Peirce's of the trustworthiness of inductive inference as the criterion of its validity is along the right lines—in that it makes inductive validity depend on some objective fact about the principles in accordance with which the inference is made. But to make this dependence a dependence upon the proportion of inferences covered by the policy which lead to true conclusions is to put it in a form which is, in two ways, badly suited for our purpose.

In the first place, since the conclusion of an induction is a general

* John Locke, *An Essay concerning Human Understanding*, Book IV, Chapter 15, §1.　　† *CLL*, p. 67; *CP*, § 2·649; *PP*, p. 158.
‡ *CLL*, p. 105; *CP*, § 2·693; *PP*, p. 188.

hypothesis, there is no time at which it is conclusively proved. The hypothesis may, of course, be established by the induction, but its establishment at one time will not prevent its refutation at a later time if contrary evidence occurs. It is desirable that our criterion should be such that known evidence will have conclusively proved that the criterion held of some at least of the inductions that have been made in the past.* That the criterion will also hold of some inductions which will be made in the future will, of course, be a proposition that cannot be proved but will be one which is only capable of being established for inductive reasons. But knowledge that the criterion held in the past we wish to be independent of inductive considerations. This can be secured if we substitute for the 'true conclusions' in Peirce's criterion 'conclusions which up to now have been confirmed in experience but never refuted'.

Expressed more exactly, the criterion for the reliability of the inductive policy Π will run: At any time t, more than half of the hypotheses which have been established by the use of Π at a time earlier than t have the joint property (1) of not having been empirically refuted at any time between the time of establishment and t, (2) of having been empirically confirmed at least once at a time between the time of establishment and t.

But this criterion is not yet satisfactory. What we want is a criterion which we can be fairly confident has held of inductions made in the past by the use of some, at least, of the reputable scientific inductive policies. But can we be certain, of any of these policies, that more than half of the hypotheses established by the use of this policy have been empirically confirmed and have not been empirically refuted since their establishment? It would be an extremely rash historian who would venture to maintain such a proposition. To make it at all plausible it would be necessary to diminish the class of reference from being that of all inferences made in the past which were covered by the policy in question to that of all such inferences made by a reputable scientist after the

---

* The past participles in my second quotation from Peirce may show that he had this consideration in mind. Why Peirce did not explicitly treat of this point may be because he intended his criterion to cover inferences with non-general conclusions, and thus, by taking the class of reference to be all possible inferences of the sort in question, to connect the Locke-Peirce 'probability' with a Limiting-Frequency view of probability of events. See *CLL*, p. 68; *CP*, §§2·650f.; *PP*, p. 159.

scientist had tried out a large number of alternative hypotheses which experience had then refuted. For it is one of the best-known facts in the history of science—a fact as notorious as the predictive success of science—that scientific discovery (i.e. the well-establishment of scientific hypotheses) is largely a matter of patience and perseverance in invention, and that there are very few fields in which the scientist expects the first hypothesis which he has thought of to cover the known facts to survive after confrontation with new facts. And even with this qualification we should not have sufficient historical evidence to justify an assertion that most of the hypotheses invented by scientists after many disappointments have been confirmed and not refuted. Moreover, a limitation of the class of reference to include only such hypotheses is far too arbitrary a limitation to be used as a satisfactory criterion for the reliability of an inductive policy.

The escape from this difficulty is to be found, I believe, by the abandonment in Peirce's criterion of the requirement that a *majority* of the hypotheses established by use of the inductive policy should be confirmed and unrefuted. Instead of this requirement all that will be demanded will be that *many* of such hypotheses should be confirmed and unrefuted; however, since this would be satisfied if a bunch of such hypotheses established in the past had this property while newly established hypotheses failed to have it, it is necessary to require that there should be many of these confirmed and unrefuted hypotheses established during each period of time since some fixed date. The criterion thus has to take the somewhat complicated form:

Of every time t later than a fixed time t_0, and of every interval of time of a fixed length of years d lying within the interval $[t_0, t]$, it is true that many of the hypotheses established by the use of policy Π during the interval of d years (unless there are no such hypotheses) have the joint property (1) of not having been empirically refuted at any time between the time of establishment and t, (2) of having been empirically confirmed at least once between the time of establishment and t.

In this criterion there are three arbitrary elements. The first is the fixed time t_0, which can be taken as the date of Babylonian astronomy or of Archimedes or of Galileo according to taste; it is inserted in order that historical evidence may be sufficient to establish the truth of this criterion as applied to the reputable inductive policies. The second is the fixed length d of the intervals, which might be one year or ten years. The third is the meaning of the vague word "many". The second and the third of these arbitraries are related in that, the shorter the interval, the smaller will have to be the least number covered by "many" in order that the criterion, restricted to past times, shall be known to have held of scientifically reputable policies.

There is also an implicit arbitrary element in the class of persons using the policy for whom the truth of the criterion in the past has been established. This class may be taken to be all human beings or all those with a scientific education or some other limited class of persons; and the other arbitrary elements will have to be adjusted to be appropriate to this class.

To save words, let us call a policy satisfying this criterion (suitable values having been assigned to the arbitrary elements) an *effective* policy, and let us call one satisfying the same criterion with "Of every time t not later than the present time and" substituted for "Of every time t" an *effective-in-the-past* policy. It is a historic fact that the inductive policies of good scientific repute are effective-in-the-past policies; it is a general hypothesis that they are effective as well as being effective-in-the-past.

There are two ways in which a policy, Π, may fail to be effective in the future. One way is if enough of the hypotheses established during some period in the past by the use of Π are refuted to contradict the statement that many will not be refuted. The other way is for these old-established hypotheses to continue to be confirmed and unrefuted, but for enough of the new hypotheses established by the use of Π in the future to be refuted to contradict the statement that many of them will not be refuted. An inductive policy, that is, may fail in the future either by its past successes turning out to be failures after all or by its failing to have future successes.

This possibility of refutation in the future dispels the suspicion, which might otherwise be entertained, that the design of the effectiveness criterion in such a way that the inductive policies of good repute are known at present to be effective-in-the-past may have resulted in the effectiveness of a policy being a logical consequence of the policy's effectiveness-in-the-past, so that the reputable inductive policies would as certainly be as effective in the future as they have been in the past. But, although the word "many" is vague, if no old-established hypotheses were unrefuted in the future, or if no newly established hypotheses were unrefuted after their establishment, the inductive policy concerned would be discovered empirically by future facts not to be effective. So the effectiveness of an inductive policy is an empirical proposition which does not logically follow from the policy's effectiveness-in-the-past.

If an inductive policy, Π, is found not to be effective in the future, this does not imply that at some time it will then be unreasonable to believe a hypothesis the only reason for believing which is its establishment by the use of Π. It will not be reasonable to believe the hypothesis in the sense of "reasonable" for which the effectiveness of the policy is the criterion; but it may well be reasonable in some other way. Or rather, the truth is that, if a policy Π which is at present effective-in-the-past turns out in the future not to be effective-in-the-past so that it is not effective, we just shall not know what to say as to the reasonableness or unreasonableness of belief in the hypotheses established by the use of this policy. If the failure in effectiveness of Π were not due to the refutation of hypotheses established by its use in the past, but were due to the refutation of new hypotheses established by its use, there would be an inclination to say that it was still reasonable to believe in the old hypotheses, but that it would be unreasonable to believe new hypotheses established by its means. This state of affairs might be expressed by saying that the old successful inductive policy had done all that it could do in wresting her secrets from Nature, but that it was now played out and new policies must be discovered and used. The state of affairs would appear not so much as a breakdown of a particular policy as an exhaustion of its field of profitable application. On the other hand, there would also be an inclination to say that the failure of the inductive policy to yield new unrefuted hypotheses showed that we had been unjustified in using it in the past, and that its supposed past successes had been just lucky coincidences. We have no satisfactory way of choosing between these two opposing considerations. There would be similar opposed tendencies as to what should be said if the failure in effectiveness of Π were due to the refutation of hypotheses established by its use in the past while it continued to yield confirmed hypotheses in the future.

If the inductive policy broke down in both these ways at once so that neither the old nor the new hypotheses established by its means continued to be unrefuted, there would be a strong inclination to say that it would then be positively unreasonable to believe these hypotheses. But suppose that all the inductive policies broke down simultaneously. It would then, of course, not be reasonable in the sense which used the criterion of effectiveness, or in the modified senses described in the last paragraph, to believe an inductive hypothesis; but it might well be called "reasonable" in some new sense of the word. The sense of "reasonable" as applied to belief in inductive conclusions is different from the sense of "reasonable" as applied to belief in logically necessary propositions; the former use, in connexion with policies of inductive inference, was developed exactly because these policies were found to have been effective-in-the-past. A discovery that these policies were not effective would remove the occasion for applying the epithet "reasonable" to beliefs in hypotheses established by their means; but another use might then be found for the epithet as applied to an inductive hypothesis, and another rationale for that use. It is futile to speculate as to what inductive beliefs we should call "reasonable" and what "unreasonable" if all our present inductive policies proved ineffective; our language is fixed on the assumption of their effectiveness, and if they are all ineffective we have no criterion for the application of the term.

The situation can be inadequately expressed by saying that the criterion we have been expounding for the validity of an inductive inference and for the reasonableness of belief in its conclusion—the criterion of the effectiveness of the inductive policy concerned —is a *sufficient*, but not a *necessary*, criterion. Why this is an inadequate way of expressing the situation is that this way of talking presupposes that there is a necessary criterion for the reasonableness of belief in an inductive hypothesis. And this would be equivalent to being a sufficient criterion for the unreasonableness of a hypothesis—which it would be rash to take as being given by the ineffectiveness of the inductive policy concerned.

The thesis maintained in this chapter is that the effectiveness of the inductive policy concerned is a sufficient condition for the adjective "valid" to be applied to an inductive inference from known evidence and for the adjective "reasonable" to be applied to belief in the conclusion of the inference. But, if this condition—the effectiveness of the inductive policy—fails, we do not necessarily use this failure as a sufficient condition for the adjective "invalid" to be applied to the inference or the adjective "unreasonable" to belief in the conclusion; we may, if this situation ever occurs, make use of some new criterion unthought-of at present. Books on logic have almost always considered only definitions of terms where the definition holds whatever the facts may be. Here we are concerned with a partial definition, in the form of a sufficient criterion, where the applicability of this partial definition depends upon the truth of an empirical matter of fact. If policy Π is effective, then the use of policy Π in inferring an inductive conclusion from empirical data makes the inference a 'valid' one and the belief in the conclusion a 'reasonable' belief. But if policy Π is not effective, then here nothing is said. A justification of induction requires a criterion for the validity of an inductive inference; it does not require one for the invalidity of such an inference.

THE WEAKNESS OF THE EFFECTIVENESS CRITERION

The Locke-Peirce criterion for the trustworthiness of an inference in terms of a majority of inferences of a certain sort being confirmed has been abandoned in favour of a criterion which substitutes "many instances" for the "most instances" of the Locke-Peirce criterion. This new criterion constitutes a very much weaker condition than the Locke-Peirce one. And, even though we have strengthened it into the requirement that there should be many confirmed and unrefuted hypotheses established during every fixed interval of time since a fixed date in the past, the criterion may still be criticized for being too weak. The criticism may be put in the form that, although "valid" and "reasonable" can, if we like, be defined in this way, this definition will make the concepts so weak as to be pragmatically valueless. What a man wants, it will be said, is to use the adjective "reasonable" in such a way that its application to belief is related to its application to action, so that it will be reasonable for him to do an action which it is reasonable for him to believe is a means to a goal at which he is aiming. But if all that can be said in favour of using an inductive policy is that it frequently predicts successfully, is this a justification for basing actions upon beliefs obtained by such a policy? Other policies for obtaining beliefs might be more predictively successful; in which case surely it would be better to use them rather than the inductive policies, or at any rate to prefer them to the inductive policies when there is a conflict of results.

The reply to this criticism is that why it is possible plausibly to propose such a weak condition as effectiveness (as specified in this chapter) for the validity of inductive inference is because no policy for establishing scientific hypotheses other than the scientifically reputable inductive policies is effective in even this very weak sense. It is not as if there were competitors to the inductive policies in the predictive-reliability race so that it would be unreasonable to prefer the inductive policies unless we could depend on their swiftness in the race. The non-inductive policies are not starters. There is no general policy other than an inductive policy which there is good reason to believe has been effective in the past, i.e. has, during every interval of time of a fixed length since some fixed date in the past, established many hypotheses which have been confirmed and not refuted—the fixed lengths of the intervals and the fixed past date being the same as those used in the specification of the effectiveness criterion according to which there is no doubt that the reputable scientific inductive policies have been effective in the past.

Some logicians of the Peircean school (e.g. William Kneale*) say that there is no other way, or at least no other systematic way, of attempting to make true predictions except by pursuing an inductive policy. This, I think, is too sweeping; we can *try* to make true predictions by a policy of consulting a soothsayer selected in some predetermined way or by a policy of deep breathing followed by free-association or by any other systematic non-inductive policy we fancy. But experience has taught us that we shall not succeed by any of these ways, that none of these non-inductive policies are effective-in-the-past, and so none of them are effective. The case for employing the recognized inductive policies is thus not the negative fact that there is no other systematic way of *trying*, but the negative fact that there is no

* *Probability and Induction*, pp. 234, 235, 259.

other way of *succeeding* in making true predictions, combined with the positive fact that pursuing inductive policies frequently does succeed. The justification for the use of an inductive policy in terms of its effectiveness must be read in the context of other predictive policies being known to be ineffective.

It may be objected to this line of thought that it involves supposing that the inductive policies of good repute among scientists at present are the only policies that it ever will be reasonable to think effective. But this is not the case. That policy Π is effective is an inductive hypothesis to be established by induction according to a principle of simple enumeration on the basis of its effectiveness-in-the-past. A policy which has never been tried before may be tested for its ability to yield confirmed and unrefuted hypotheses; if many of the hypotheses which it yields are confirmed and not refuted, then it will satisfy the criterion for effectiveness-in-the-past and its introduction into our inductive repertoire for establishing hypotheses can be justified by the effectiveness of the policy being established in accordance with an effective inductive policy of simple enumeration. For example, suppose that in the first instance I do not accept hypotheses on the strength of their being asserted by Savant M, but, through curiosity perhaps, record the hypotheses he asserts throughout a period of time. If many of these hypotheses are confirmed and unrefuted during this period, the simple-enumerative inductive policy may make it reasonable for me to believe hypotheses asserted by M on the grounds that they have been asserted by M; and I shall thereby have obtained a new predictive procedure which will be good while it lasts. To say that in fact no predictive policy other than the recognized inductive policies is at present known which is effective-in-the-past does not imply that no such policy will be discovered. Indeed, the eliminative inductive policies and the policy of establishing functional laws have been so discovered, the latter only some 350 years ago; the justification for their use is their effectiveness, which was in each case established by a simple-enumerative policy. Similarly, a simple-enumerative policy may establish the effectiveness of new predictive policies in the future; if it does so, these new policies will enter into competition with the present inductive policies of good repute, and we shall be compelled to choose as to which policy to use if they lead to conflicting results. But at present there is no competition.

THE ALLEGED CIRCULARITY IN THE PREDICTIONIST JUSTIFICATION OF INDUCTION

The thesis of this chapter may be expressed not quite precisely by saying that the justification for inductive inference consists in the fact that a policy of passing, in accordance with an inductive principle, from true beliefs to beliefs in general hypotheses frequently enables us to accept hypotheses which are confirmed and not refuted by experience. This thesis is thought by many philosophers to involve a viciously circular way of looking at the matter. Their argument runs as follows: On the predictionist thesis the reason for believing an inductive conclusion consists in two premisses, one being the evidence for the conclusion appropriate to the inductive principle concerned, the other being the proposition that the policy of making inferences in accordance with the inductive principle concerned is an effective one. And reasonableness of belief in the conclusion is due to the reasonableness of the belief in each of these premisses. But, so it is said, this second premiss is itself a general hypothesis, reasonableness of belief in which can only be established by another inductive argument. This second induction will similarly require as premiss the proposition that the policy of making inferences in accordance with the inductive principle used in it is an effective one; and this premiss will again be a general hypothesis which will require justifying by a third inductive argument. Thus, either there will be an infinite regress with an infinite series of inductive policies the establishment of the effectiveness of each policy in the series requiring the establishment of the effectiveness of the succeeding policy in the series, or we shall arrive, in ascending the series, at one inductive policy the establishing of whose effectiveness will require the establishment of its own effectiveness.

Since the reason that could be given for the effectiveness of any inductive policy except those of induction by simple enumeration would be that it had frequently been predictively reliable in the past, i.e. that it had proved to be effective-in-the-past, the establishment of the effectiveness of any other policy would require the establishment of the effectiveness of a simple-enumerative inductive policy. (For the purpose of this argument it is unnecessary to distinguish different simple-enumerative policies; so they will be referred to in the singular as the policy of induction by simple enumeration.) Thus to establish the effectiveness of the policy of induction by simple enumeration would require its own effectiveness to be taken as a premiss; and the circularity horn of the dilemma would be the one upon which we should be impaled. If we care to use the word "presupposition" the argument against the predictionist justification may be expressed by saying that, according to it, the validity of every inductive inference *presupposes* the validity of induction by simple enumeration, and the validity of induction by simple enumeration *presupposes* its own validity; and this, it is alleged, is a viciously circular justification for induction.

Before trying to answer this charge of vicious circularity, the predictionist may be permitted a *tu quoque* retort. The accusation of circularity does not lie solely against the predictionist justification of induction but lies equally against any account of the validity of induction which makes such validity depend upon a premiss which can only be established inductively. Keynes's attempt to justify induction by means of a theory of probability falls into this class. On Keynes's theory inductive confirmation serves only to increase the probability of a hypothesis by multiplying it by another probability, so that if the hypothesis has a 'prior probability' of zero its 'posterior probability' remains zero, however much evidence there may be for it. And the only way to secure that every hypothesis should have a prior probability greater than zero is to assign some probability greater than zero to a proposition which limits the number of possible hypotheses (e.g. Keynes's Principle of Limited Independent Variety).* But such a proposition is itself an empirical hypothesis, which calls for an inductive justification. So Keynes's justification for induction is viciously circular—unless one cuts the circle by the improper expedient of 'postulating' the empirical hypothesis required.

The fact, however, that many other attempted justifications of induction are open to the accusation of vicious circularity does not excuse the predictionist from trying to show that the charge does not lie against his account. This rebuttal must now be attempted.

The first move in the rebuttal is that the proposition 'presupposed' in the predictionist justification of an inductive inference does not function in the inference as an additional premiss. The

---

* J. M. Keynes, *A Treatise on Probability*, Chapter XXII.

inductive inference to the proposition that induction by simple enumeration is an effective policy does not make use of this proposition as a premiss, and so is not circular in the *petitio principii* sense of professing to infer a conclusion from a set of premisses one of which is the conclusion itself.

This point is so important that it is desirable to make it as precisely as possible. What we are concerned with is the sort of circularity, if any, involved in the establishment of the validity of induction by simple enumeration by means of an induction by simple enumeration. A few symbols will abbreviate the discussion.

Let Π be the inductive policy of adding belief in a hypothesis h to belief in a set of propositions which collectively constitute π-evidence for h, i.e. adequate evidence for inferring h in accordance with the principle of inference π, π being the principle of induction by simple enumeration.*

Let e be the proposition that the policy Π is effective.

Now to say that the truth of e is the justification for employing the inductive policy Π is not to say that the principle π of induction by simple enumeration requires that e should be added as an additional premiss to the otherwise adequate evidence; were this to be the case, the inference of e by means of the principle π would require the inclusion of e itself among the premisses believed. The alleged inference would then not be an inference at all, let alone a valid inference, since it would be professing to establish belief in a proposition which was already one of the believed premisses. But since e does not function in the argument as a premiss which has to be believed along with the other premisses, the argument by means of the principle π from the π-evidence for e to e does not commit the fallacy of *petitio principii*, and it is a genuine inference in which belief is acquired in a new proposition which was not believed before.

The circularity in the argument is of a more sophisticated character. It is the circularity involved in the use of a principle of inference being justified by the truth of a proposition which can only be established by the use of the same principle of inference. To express the matter in the symbols we have used:

* The Greek capital letter Π has been used earlier in this chapter to denote any inductive policy. It will henceforth be restricted by this definition to denote only the simple-enumerative inductive policy.

The truth of e justifies the use of policy Π; i.e. for every h the truth of e justifies an addition of belief in h to reasonable belief in π-evidence for h.

From which there follows, by the substitution of e for h: The truth of e justifies an addition of belief in e to a reasonable belief in π-evidence for e. In other words, if e is true, an addition of belief in e to reasonable belief in π-evidence for e is justified.

I do not wish to deny that there is a sort of circularity involved in this statement, but it is a peculiar sort of circularity whose viciousness is by no means obvious. The statement does not commit the fallacy of *petitio principii* as would have been the case had it said that, *if e were reasonably believed*, an addition of belief in e to the believer's body of reasonable belief would be justified. For in the statement the sufficient condition for the addition of belief in e is, not that a premiss is believed, but that an empirical proposition is true. The peculiar circularity consists in the *truth* of the conclusion of an inference being a sufficient condition for the validity of the inference. Let us call this type of circularity "effective circularity", since in the cases in which we are interested it is the effectiveness of the inferential policy which is the sufficient condition. The question before us is whether or not the presence of effective circularity renders an inference invalid, or whether it prevents an inference from being a genuine inference at all?

At this point it is worth remarking that there are deductive inferences in which a sufficient condition for the effectiveness of the principle of inference is exactly that proposition which is the conclusion of the inference itself.* But here the proposition which is the condition for the effectiveness of a deductive principle of inference must be a logically necessary proposition; and though, in the case we are considering, this logically necessary proposition is supposed to be established by being deduced from other logically necessary propositions within a pure deductive system which uses as principle of inference a principle the condition for whose effectiveness is the logically necessary proposition itself, yet this proposition

---

* An example of this would be the deduction within a deductive system of the proposition $(p(p \supset q)) \supset q$ from the two propositions

$$(p(p \supset q)) \equiv pq, \qquad ((p(p \supset q)) \equiv pq) \supset ((p(p \supset q)) \supset q)$$

by the use of an 'implicative' detachment principle the condition for whose effectiveness is the necessary truth of $(p(p \supset q)) \supset q$.

could always be established in ways which do not involve this effectiveness. For there is no necessary order in the deducing of logically necessary propositions from one another. However, in our inductive case, the proposition stating the effectiveness of the induction-by-simple-enumeration policy is a logically contingent proposition which can be established in no other way than by the use of an induction-by-simple-enumeration policy.

Now for a consideration of 'effective circularity'. Is there anything wrong in the effectiveness of a policy of inference being a justification for the inference to this effectiveness as conclusion? Is there anything wrong in obtaining a belief in a proposition by inference according to a principle whose validity as a principle of inference is attested by the proposition which is itself the conclusion of the inference? In order to answer these questions it is necessary to consider what exactly we mean by "inference" and by an inference being "valid".

THE CONDITIONS FOR VALID INFERENCE AND FOR THE REASONABLENESS OF AN INFERRED BELIEF

Inference is the passage of thought from belief, or rational belief, in a set of propositions, collectively called the *premiss* of the inference, to belief, or rational belief, in a proposition called the *conclusion* of the inference, the premiss and the conclusion being related in accordance with some principle of inference. The question of the validity of a process of inference is the same as the justification for adding a belief in the conclusion to the believer's body of reasonable beliefs (his 'rational corpus'), in the case in which the premiss of the inference forms part of this rational corpus, or for associating a belief in the conclusion with the belief in the premiss, in the case where this belief is not part of the believer's rational corpus, in such a way that belief in the conclusion will be inferentially supported by the belief in the premiss. And vice versa the question of the reasonableness of a belief, except in the case in which the reasonableness of the belief consists in the proposition believed being known directly to be true, is that of the validity of an inference by which the belief could be supported.*

* Not 'has been supported', for a belief may be reasonable if the believer would support it by an inference if the reasonableness of his belief were to be disputed, although in fact he had not arrived at it by inference.

Different sorts of criteria can be given for the validity of an inference, or for the reasonableness of an inferentially supportable belief. In order to avoid complications due to differences of nuance in the meaning of the words concerned, these different possible sets of criteria will be considered in the first instance in the form of an abstract classification.

The two relevant propositions are the premiss p of the inference and the proposition r asserting the effectiveness of the inferential policy which uses the principle of inference concerned in passing from the premiss p to the conclusion q. The possible criteria for the validity of an inference from p to q are obtained by considering the possible combinations of a belief in p, or a rational belief in p, with the truth of r, or with a belief in r, or with a rational belief in r. Possible combinations with the truth of p will not be considered, since the truth of the premiss is not relevant to the validity of an inference from it, though it is relevant to the question as to whether or not an inference to a conclusion constitutes a *proof* of the conclusion.*

There are ten possibilities for sufficient criteria for the validity of an inference by a person B of the conclusion q from the premiss p in accordance with a principle of inference the effectiveness of whose use is asserted by the proposition r:†

    I B believes p and believes r;
   II B believes p and reasonably believes r;
  III B believes p, and r is true;
  IV B believes p and believes r, and r is true;
   V B believes p and reasonably believes r, and r is true;

  VI B reasonably believes p and believes r;
 VII B reasonably believes p and reasonably believes r;
VIII B reasonably believes p, and r is true;
  IX B reasonably believes p and believes r, and r is true;
   X B reasonably believes p and reasonably believes r, and r is true.

* My treatment of the validity of inference, which owes much to W. E. Johnson's discussion in his *Logic, Part II* (Cambridge, 1922), Chapter I, disagrees with him here.

† Criteria sufficient, that is, except for the possibility of a vicious circularity to the discussion of which this classification is a preliminary.

Criteria VI–X are possible criteria for the validity of an inference regarded as justifiably adding a belief in q to B's body of reasonable beliefs; criteria I–V are possible criteria for the validity of an inference regarded as justifiably carrying a belief in q along with B's belief in p. Which of the five criteria in each case is chosen as *the* criterion for the validity of the inference is to some extent a matter of taste. The usage of the expression "valid inference" and "reasonable belief" is not sufficiently fixed by common usage for any of the possibilities to be excluded from the outset.

But some of these possibilities will make some inferences circular, and these possibilities will have to be excluded for the case of these inferences; otherwise these inferences will be invalid, not in failing to satisfy one of the criteria, but in involving a vicious circularity. The inferences that will be invalid through circularity will be those satisfying any of the conditions

I–X if B's belief in p includes a belief in q.
I, II, IV, V, VII and X if B's belief in r includes a belief in q.

It is important to notice that B's belief in r may include a belief in q without making inferences satisfying conditions VI and IX circular, for these inferences are allegedly adding q to B's body of reasonable beliefs, and the fact that q is already a proposition which B believes does not invalidate an inference which proposes to move q from B's body of *beliefs* to his body of *reasonable beliefs*.*

Let us now turn to the case of inductive inference. Here p, the evidence for the inductive hypothesis q, is assumed to be reasonably believed; so the criterion for the validity of the inference will be one of the possibilities VI–X. Suppose that the conclusion q of the inference is the proposition stating the effectiveness of a policy of induction by simple enumeration, and that it is according to induction by simple enumeration that the inference is being made. Then the inference will be invalid through circularity only if its condition for validity is taken to be either VII or X. For it is only in these two cases that B's reasonable belief in the conclusion of the inference will be part of his reasonable belief in the effectiveness of the inferential policy. So if any one of the other three possible

* Some accounts of inference would limit it to being a passage of thought leading to belief in a proposition not previously believed; but this would seem to involve an undesirably narrow definition of inference.

criteria for the validity of an inference is taken, i.e. any one of VI, VIII or IX, an inference to a conclusion stating the effectiveness of the inferential policy will be valid without any circularity.

Having reduced the number of possibilities to three, we can conveniently give them specific names. Let us call an inference (leading to reasonable belief in a conclusion) "subjectively valid" if it satisfies condition VI (i.e. if VI is a sufficient condition for its validity), "objectively valid" if it satisfies condition VIII and "both subjectively and objectively valid" if it satisfies condition IX. To justify the use of the policy of induction by simple enumeration by its effectiveness, as we have done, is to use criterion VIII for the objective validity of the inference to the hypothesis of its effectiveness. The inference made by a person B is then objectively valid in that B reasonably believes the evidence for the hypothesis and that the hypothesis justifying the principle of inference is itself true; no question of his belief, reasonable or non-reasonable, in the hypothesis itself is part of the condition for the validity (the 'objective validity') of the induction.

It may, however, be felt that such objective validity is not enough, and that an inference cannot be a properly valid inference unless the inferrer is in some way cognizant of the principle according to which he is making the inference and of the propriety of the principle for this purpose. Whether or not objective validity is thought to be enough depends upon whether the validity of the inference is being considered, as it were, from the outside or from the inside. When considered from the outside the person making the inference is being regarded as a reasoning machine. The machine would first have fed into it a set of propositions which together make up π-evidence for e, and would thus take up a 'position' which would correspond to having a reasonable belief in this π-evidence for e. The machine would then be put in operation according to its principle of working, and would acquire a new position which would correspond to a reasonable belief in the proposition e. Obviously there is nothing objectionable in the machine arriving at a new position which corresponds to having a reasonable belief in a proposition asserting some general property of the method of working of the machine. From the external point of view the machine is making 'valid inferences' if it is working according to its working 'principle of inference' starting with a 'reasonable' position, and it can quite well arrive by such working at a position which corresponds to having 'validly inferred' a proposition which asserts some general property of its working 'principle of inference'.

Similarly, from an external point of view, a man may be considered to be making valid inferences from reasonably believed premisses if his policy of inference is in fact an effective one, quite independently of whether or not he believes or knows that it is effective; and in this case there is no vicious circularity in his arriving by a valid process of inference at the conclusion that the policy of using the principle of inference according to which he is making this inference is an effective one.

An inference will be valid in the sense of being *objectively valid* (in the sense explained) if the inference proceeds from reasonably believed premisses to a conclusion according to a policy of inference which is in fact effective, whether or not the inferrer knows or believes that it is effective or indeed whether he considers the question of its effectiveness at all. The inferrer, that is, is acting like a machine which works according to certain principles without being cognizant of these principles. He is not assumed to be wholly a machine, since he is supposed to start with a reasonable belief in the premisses and to end with a reasonable belief in the conclusion, but the process by which he passes from his original reasonable belief to his final reasonable belief is supposed to be one which does not require his cognitive participation. This process may be regarded as analogous to the free-association used in adding up a column of figures, where the adder has consciously thought of the number written at the head of the page and again consciously thinks of the number to be written at the bottom of the page but does not consciously think of the arithmetical relationships which justify his calculating from the one to the other. The result of a calculation obtained in this automatic way may well be a statement which asserts the effectiveness of the method of calculation.

But if the machine becomes self-conscious and critical of its mode of working, it will not be satisfied with a criterion for the validity of inference which depends upon the effectiveness in fact of the method by which it is working, but will demand a condition, either as an alternative or as an addition, which states its belief in this effectiveness. From the inside, that is, criteria VI or IX will seem more appropriate criteria for the validity of an inference than the criterion VIII giving 'objective validity'. Since criterion IX combines the conditions of criteria VI and VIII, it is criterion VI that needs to be considered. It substitutes the requirement that the thinker should believe that the policy of induction by simple enumeration is effective for the fact that it is effective; in the terms we have used it ascribes 'subjective validity' rather than 'objective validity' to the inference leading to the conclusion that the policy is effective. Though belief in the effectiveness of this policy is one of the conditions for the subjective validity of an inference yielding a reasonable belief in such effectiveness, this fact does not (as we have seen) make the inference *ipso facto* circular, for the thinker is passing from a mere belief in this effectiveness to a reasonable belief in it. He is, as it were, moving it within his body of beliefs into the privileged position of being one of his body of reasonable beliefs.

So a critic who finds the objective validity criterion inadequate, in that it gives no place for the thinker's consciousness of the principle according to which he is making the inference, can be offered instead, without fear of vicious circularity, either the subjective validity criterion or the subjective-and-objective validity criterion for the validity of his inference. In none of these three cases is there any vicious circularity.

At this point, however, the question may well be raised as to whether either of the criteria VI or VIII are at all appropriate criteria for the validity, in any proper sense, of the inductive inference. The objector may perfectly well say that for an inference to yield a new reasonable belief the inferrer must not merely *believe* that the policy represented by the principle of inference is effective, but must *reasonably believe* this proposition. And he can point out that we have required reasonable belief, and not only belief, in the premiss in order that an inference may yield reasonable belief in the conclusion.

On the face of it this objection seems a very cogent one. To obtain a reasonable belief by inference, it says, we must have reasonable belief all along the line, reasonable belief in the effectiveness of the policy of inference no less than reasonable belief in the premiss. But the rejoinder can be made that such a requirement would invalidate the majority of inferences, deductive as well as inductive, that are actually made in the course of reasoning. For the requirement would admit only deductions in which the proposition authenticating the principle of inference used was either seen directly to be true or was seen directly to be a logical consequence (in a chain of proof sufficiently short to be taken in at one glance) of a proposition seen directly to be true. Any other way of attaining belief in the effectiveness of the policy of deduction, e.g. by citing authority, or by remembering that one had satisfied oneself of its truth in the past, would involve inductive steps, and would thus not permit the belief to be 'reasonable', since the inference by which it had been obtained, or upon which it could be based, would not satisfy this stringent condition for validity. So to insist that an inference is only valid, and a belief in the conclusion of the inference only reasonable, if the inferrer's belief in the effectiveness of a policy of inference is already a reasonable one would exclude a great number of the inferences, and a great many of the beliefs, which would normally be considered to be valid or to be reasonable.

The objector may, of course, say at this point that he is not concerned with the application of the terms "valid" and "reasonable" in our ordinary slovenly everyday language; what he is concerned with is a purified use of these terms after the user of them has been purged by a course of treatment with methodological doubt. In a purified sense the objector may decline to admit valid inferences in which the effectiveness of the policy of inference is not reasonably believed—even if this were to exclude most so-called 'valid' inferences. But then it is difficult to see what the argument is about. It started presumably because the objector wished to dispute the justification of induction by simple enumeration in terms of the effectiveness of this policy, the ground of his objection being that such an attempted justification was circular. We then replied that this justification would not be open to this criticism if the criterion of valid inference did not require reasonable belief in the effectiveness of the policy of inference, but only required alternatively either the fact of this effectiveness or belief in this effectiveness. And we maintained that a reasonable belief in this effectiveness was not an essential part of the criterion for validity of the sorts of passages of thought which are normally thought of as valid inferences. If the objector has decided beforehand to decline to admit as a valid inference any inference in which the

COMPARISON WITH 'RIGHTNESS'

effectiveness of the policy is not reasonably believed, his valid inductive inferences will be bound to suffer from circularity (unless, indeed, he takes the desperate course of regarding the effectiveness of the policy as a logically necessary truth). But the vicious circularity will be of the objector's own making.

A COMPARISON WITH DIFFERENT SENSES OF "RIGHTNESS"

It is illuminating to compare the senses in which we have called inferences "objectively valid", "subjectively valid" and "both objectively and subjectively valid" with various senses in which an action may be said to be "right".

Whether the rightness of an action is held to consist in its fittingness to a certain situation or in its having a certain characteristic or producing effects having that characteristic or in some blend of these two, it will in all cases be possible to distinguish two senses of rightness—one an objective sense, when the action in fact is fitting to the situation or in fact has the characteristic or produces effects having the characteristic, whether or not an action is right in this sense being entirely independent of whether or not the agent believes it to be right; the other a subjective sense in which what determines the rightness of the action is whether or not the agent believes the action to be fitting to the situation or to have the characteristic or to produce effects having the characteristic, i.e. whether or not the agent believes the action to be right in the first, objective sense of the word. And a third sense of rightness can then be given in which an action is right if it is both objectively right and subjectively right.*

The objective sense of rightness of an action may be compared with the objective sense of validity of an inference, that in which the policy of inference is in fact effective; the subjective sense of rightness of an action may be compared with the subjective sense of validity of an inference, that in which the policy of inference is believed to be effective. The third, composite sense of rightness may then be compared with the sense of validity of an inference

* Further distinctions can be made for the cases in which an action is subjectively right without being objectively right according as the agent's erroneous belief as to the objective rightness of his action is due to his being in error on a matter of fact (as to the nature of the situation or as to what effects the action will in fact have) or in his moral evaluation. Such further distinctions are not relevant for the purpose of our comparison.

which is both objective and subjective. And there would similarly be comparable senses of reasonable belief in a proposition which had either been derived by inference or would be defended against criticism by citing an inference by which it might have been derived. An objectively reasonable belief, associated with an objectively valid inference, would be compared with an objectively right action; a subjectively reasonable belief, associated with a subjectively valid inference, would be compared with a subjectively right action; and there would be a similar comparison in the case of a belief which was both objectively and subjectively reasonable.

The enlightenment produced by making these comparisons seems to me to be as follows. Whatever be the sense of rightness appropriate to describing other moral situations, it is almost undisputed that the sense of rightness which is appropriate to the imputation of moral praise or blame is the subjective sense of rightness. A man, that is, is not considered blameworthy for doing an action which is objectively wrong provided that, at the time of doing it, he believed it to be objectively right, nor is he considered praiseworthy for doing an action which is objectively right if he believed it to be objectively wrong.

Now to say of a man that he is reasonable in holding a belief q, or that his belief in q is reasonable, is in many contexts to make a judgment which is either a moral judgment or closely resembles one. It is a moral judgment if reasonably holding beliefs is regarded as one of the modes of moral goodness of a man; it is closely related to a moral judgment if reasonably holding beliefs is regarded, not as itself a manifestation of moral goodness, but as a positive symptom for moral goodness in a man. In either case to say of a man that he is unreasonable in holding a belief q is to make, or to imply, a hypothetical moral criticism of him. For it is to imply that the man would be morally better were he not unreasonably to hold the belief q. It does not imply that he ought not unreasonably to hold the belief, for it may not be in his power either to hold it reasonably or to abandon the belief. But, whether these possibilities are in his power or not, the fact that he unreasonably holds the belief makes him worse than would be the case were he not unreasonably to hold it.

In any context in which reasonableness is ascribed with this moral implication, to the extent that the reasonableness of a man's belief is derived by a valid process of inference, this reasonableness must depend not upon the actual effectiveness of the policy of inference but upon the man's belief in this effectiveness. For otherwise a man would be regarded as blameworthy for holding a belief in a scientific hypothesis which he had inferred from reasonably believed evidence by following an inductive policy which, although the man believed it to be effective, was not in fact an effective one. And he would be regarded as praiseworthy for holding an inductive belief which he had inferred by a policy which was in fact effective, even though he did not believe that it was effective. Such judgments would be contrary to our moral sense as displayed in our use of moral language, which makes "unfortunate" and "fortunate" more appropriate epithets than "blameworthy" and "praiseworthy" to ascribe to these two cases. So, in the contexts in which reasonableness is associated with praiseworthiness and unreasonableness with blameworthiness (and these are the most frequent contexts), what we have called the subjective sense of reasonableness will have to be taken—just as it is the subjective sense of rightness that has to be associated with praiseworthiness.

This comparison with subjective rightness further supports the view that the sufficient condition for the subjective reasonableness of an inductively supported belief, so far as the effectiveness of the inductive policy is concerned, is simply the belief that this policy is effective, without the qualification that this belief should be a reasonable one. For we think that a man is acting rightly if he does what he believes to be the objectively right action, irrespective of whether or not this belief of his is a reasonable one. If we think that his belief is an unreasonable one, and that he might, for example, by a previous more diligent study of the facts of the situation, have prevented himself from having this unreasonable belief, and instead have acquired a different and reasonable belief, we may blame him for his past sin of omission in not having taken the steps he might have taken to acquire a more reasonable belief. But we do not blame him for acting on his present belief, whether this be reasonable or unreasonable. Similarly, we should consider a man reasonable in following an inductive policy which he believed to be effective, independently of the question as to whether or not his belief was a reasonable one. Thus, in the sense of reasonableness

THE JUSTIFICATION OF INDUCTION

which is comparable with subjective rightness, it is belief in the effectiveness of the inductive policy, whether or not this belief is well grounded, that is a condition for the validity (in the subjective sense) of the inductive inference and for the reasonableness (in the subjective sense) of belief in the inductive conclusion.

But may not this comparison with subjective rightness be pushed further so that the reasonableness of belief in the premiss is as irrelevant to the reasonableness of belief in the inductive conclusion as is the reasonableness of belief in the effectiveness of the inductive policy? If so, the criterion for the reasonableness of an inductive belief would be criterion I, namely, that the thinker believes both the premiss of the inductive inference and that the inductive policy is effective. In defence of selecting this criterion it might be argued that a man might defend himself against criticism for holding an inductive belief q by saying that he believed the evidence from which he derived his belief in q by following an inductive policy in whose effectiveness he believed, without thinking it necessary to maintain that his belief in the evidence was a reasonable one. But I do not think that this would be a good defence. If the reasonableness of a belief is to be defended by the belief's having been inferred or being able to be inferred from other beliefs, these other beliefs must themselves be reasonably held. The process of obtaining new reasonable beliefs by inference is one of adding new beliefs to the thinker's body of reasonable beliefs (his rational corpus) on the basis of some of the beliefs already in this corpus. If the beliefs to which the new belief is added by the inference fall within the body of the thinker's beliefs but not within his rational corpus, the inference may well be valid in the sense of justifiably supporting his belief in q, but the inference does not justify him in including q in his rational corpus.

The situation in the criterion for reasonable belief in an inductive conclusion is thus different in respect of the roles played by belief in the inductive premiss and by belief in the effectiveness of an inductive policy pursued. The former belief has to be a reasonable one, in order that the inference should build upon a stable foundation. But what is required of the latter proposition—that the inductive policy pursued is effective—is, for the subjective sense of reasonable, that this proposition should be believed, for the objective sense of reasonable, that this proposition should be true.

In neither the subjective sense nor the objective sense nor the combined sense is a reasonable belief in the policy's effectiveness a requirement for a belief which has been obtained by the use of the inductive policy from a reasonably believed premiss to be a reasonable belief.

The result of this discussion is, I hope, to uphold the thesis that there are three proper criteria yielding three proper senses for the 'validity' of an inductive inference made by a man B, and also three corresponding senses for the 'reasonableness' of B's belief in a conclusion arrived at, or that would be based upon, an inductive inference. All three criteria agree in requiring that the premiss of the inference—the inductive evidence—should be reasonably believed by B; they agree in requiring something which is concerned with the effectiveness of the inductive policy pursued. They differ as to what is this something required. The objective criterion requires that the proposition asserting the effectiveness of the inductive policy should be true, the subjective criterion requires that this proposition should be believed by B. The both-objective-and-subjective criterion requires both that the inductive policy should in fact be effective and that its effectiveness should be believed by B. Let us now re-examine the accusation of vicious circularity in the light of this triple distinction.

CIRCULARITY RE-EXAMINED

The conditions for it to be subjectively valid for a man to infer e, the proposition that the policy Π of induction by the principle π of simple enumeration is effective, from π-evidence for e, and correlatively that this belief obtained or supportable by this inference is a subjectively reasonable belief are, first, that he should reasonably believe the evidence for e, and secondly, that he should believe e. Since neither of these conditions includes the requirement that his belief in e should be a reasonable one, there is no explicit circularity in his reasoning. Nor is there any implicit circularity, since he can quite well reasonably believe both that he is reasonable in believing the π-evidence for e and that he is believing e without reasonably believing, or indeed believing, that he is reasonable in thus believing e. The critic will thus be compelled to withdraw his charge of circularity. But, of course, he will fall back upon saying that this criterion for the validity of an inductive inference and for the reasonableness of an inductive belief is too weak a one.

In which case the critic can be offered the stronger criterion for the man's inference to be both subjectively and objectively valid and for his inductive belief to be both subjectively and objectively reasonable. This criterion adds a third condition to the two conditions for subjective validity, namely, that e should in fact be true. Since this third condition no more than the other two includes the requirement that the man should reasonably believe e, there is again no explicit circularity. But the critic will then insist that here there is an implicit circularity in that to have a reasonable belief in this third condition for the validity of the inference requires an inference of exactly the same sort to establish it.

There is one consideration which is worth mentioning at this point, since it may perhaps serve to mitigate this implicit circularity. Let us consider the new inference whose premiss is the conjunction of the three conditions for the both subjective and objective validity of the inductive inference whose conclusion is e, i.e. whose premiss is the conjunctive proposition

(B reasonably believes the π-evidence for e) and (B believes e) and (e is true),

and whose conclusion is the proposition that it is both subjectively and objectively reasonable for B to believe e. Since with this sense of "reasonable" it is a logically necessary proposition that the premiss in this new inference is a sufficient condition for the conclusion, this new inference is a deduction. Now think of this new inference, not as adding a belief in the conclusion to B's body of reasonable beliefs but as carrying with it B's belief in the conclusion by deducing the conclusion from a premiss which B believes. Criteria I–V (p. 279) then become the relevant criteria for validity; and, if B believes the conjunctive premiss, belief in which is equivalent to a conjunction of three beliefs held by B— a belief that he reasonably believes the π-evidence for e, a belief that he believes e, and a belief in e—this triple belief justifiably carries with it a belief that it is both subjectively and objectively reasonable for B to believe e, provided that, if criteria I, II, IV or V are used, B believes that the conjunctive premiss is logically a sufficient condition for the conclusion, and, if criteria II or V are used, that this belief of B's is a reasonable one. So B's second-order belief that he is both subjectively and objectively reasonable in believing e is a belief justifiably carried along with a first-order belief in e, together with a second-order belief that he has this belief, a second-order belief in the reasonableness of a belief in the π-evidence for e and—in the case of some of the criteria—a belief or a reasonable belief in a logically necessary proposition. If we take the second-order belief that he has a belief in e as going along automatically with a belief in e, and if we take for granted the second-order belief in the reasonableness of a belief in the π-evidence for e and the belief or reasonable belief in the logically necessary proposition, what we have said can be reduced to the statement that, if a man believes e, this belief justifies his holding along with it the second-order belief that he is both subjectively and objectively reasonable in believing e. Thus a belief in e is self-rationalizing—not, of course, in the sense that believing e makes this belief itself reasonable, but in the sense that believing e carries along with it a belief that this belief in e is reasonable.

The critic may object that all this farrago is like taking in one another's washing, and goes no way to producing any argument which is not circular for a belief in e to be reasonable in a sense of "reasonable" which is not merely subjective. I can say no more than that the account of objective validity of an inference which has been given is in terms of the right working of an inference-machine, and that the implicit circularity only arises from the inference-machine becoming self-conscious about the way in which it operates. The predictionist can offer to the circularity-mongering critic two alternatives—a weak subjective sense of reasonableness with no suspicion of circularity, or a stronger sense of reasonableness, objective as well as subjective, with no explicit circularity but with an implicit circularity which depends essentially upon the inferrer being regarded as an inference-machine and the validity of his inference depending upon his operating, *qua* inference-machine, according to an efficient mode of operation (with the rider that, if the inferrer believes that he, *qua* inference-machine, is operating efficiently, that belief is self-rationalizing in the way explained in the last paragraph). If neither of these alternatives, nor the third alternative of the purely objective sense of reasonableness, satisfies the critic, and if he is not prepared to be satisfied by one sense of reasonableness in some contexts and another in other contexts but still demands a method of establishing the effectiveness of an inductive policy which is not to be obtained by following an inductive policy, he must be told outright that what he is demanding is that the effectiveness of an inductive policy should not be an empirical proposition. But if so, induction would be deduction, there would be no inductive problem to puzzle our heads over, and this chapter tediously worrying at the problem would have been altogether unnecessary.

CHAPTER IX

LAWS OF NATURE AND CAUSALITY

Scientific hypotheses have been taken to be general empirical propositions whose generality is not restricted to limited regions of space or of time. For the reasons explained in the first chapter scientific laws, corresponding to true scientific hypotheses, have been taken to assert no more than constant conjunctions of properties, so that the scientific law that everything which is A is B asserts no more than that all the things which are A as a matter of fact are also B. It is now time to examine this assumption in order to see whether the Humean analysis of scientific laws is adequate, or whether it is necessary to suppose that the sort of necessity of scientific law ('nomic' necessity) requires some extra element of 'necessary connexion' over and above a merely factual uniformity. To express the matter in another way, everyone will agree that everything which is, was or will be A is, was or will be B is a logical consequence of the scientific law expressed apodeictically as "Every A nomically must be B"; the question is whether or not it is justifiable to regard the former proposition, not as a consequence, but as an analysis of the meaning of "Every A nomically must be B".

I have used W. E. Johnson's adjective "nomic" rather than the more usual "causal" to express, without prejudging the analysis, the characteristic sort of necessary connexion with which we are here concerned, because the notion of causality might well be held to involve considerations of temporal precedence and of spatio-temporal continuity which are irrelevant to the present issue. For here we are concerned with the nature of the difference, if any, between 'nomic laws' and 'mere generalizations'; in Johnson's language, between 'universals of law' and 'universals of fact'.*

David Hume maintained that objectively there is no difference, but that a psychological fact about the way in which our minds work causes us to ascribe necessity to scientific laws, the 'idea of necessary connexion' being derived from our experience of the

* W. E. Johnson, *Logic, Part III* (Cambridge, 1924), Chapter I.

constant conjunction of properties and not from anything in nature over and above constant conjunction. In common with most of the scientists who have written on the philosophy of science from Ernst Mach and Karl Pearson to Harold Jeffreys, I agree with the principal part of Hume's thesis—the part asserting that universals of law are objectively just universals of fact, and that in nature there is no extra element of necessary connexion. The time has now come to defend this thesis against philosophers who disagree with it.

The discussion has been postponed until this point because the principal argument used by philosophers against a Humean analysis of nomic necessity into constant conjunction is that such an analysis of scientific laws makes justification of induction impossible. But such a criticism thinks along the lines of assimilating induction to deduction. If we do not attempt such an assimilation and instead propound a Peircean account of induction, as was done in the last chapter, the argument hangs fire. For the Peircean account bases itself upon the inductive policies which are used in making the inferences, and to pursue these policies certainly does not presuppose that the conclusions of the inferences made by pursuing them are anything more than universals of fact. So the positive justification of induction given in the last chapter is my answer to the argument that a Humean view of scientific laws makes their establishment invalid.

If this book were concerned with attacking rival views, cogent criticisms could be made of attempts to justify induction which build on a nomic necessity distinct from constant conjunction. Those philosophers, for example, who wish to identify nomic necessity with logical necessity lay themselves wide open to the charge that, since all the premisses in a valid inference to a logically necessary conclusion must be logically necessary propositions, to treat scientific laws as being logically necessary propositions removes all possibility of basing them upon empirical data. And those philosophers who wish to make nomic necessity a third ultimate category distinct both from logical necessity and from constant conjunction lay themselves open to the charge that their nomic connexion is, as 'substance' was for Locke, "something I know not what"; and that a philosopher is shirking his duty who uses Butler's maxim "Every thing is what it is, and not another thing" to avoid having to consider whether the difference between universals of law and universals of fact may not lie in the different roles which they play in our thinking rather than in any difference in their objective contents.

For it cannot be disputed that we do make a distinction of some sort between those empirical general propositions which we dignify with the name of "laws of nature" or "natural laws" and those which we call, sometimes derogatorily, "mere generalizations". A Humean philosopher may well deny that this distinction is one of objective fact; but if he denies that there is any distinction whatever, he runs counter to ordinary usages of language.

SUBJUNCTIVE CONDITIONALS

One of the relevant usages of language to which philosophers have recently given a great deal of attention has been the use of conditional sentences of the form "If a thing is A, it is B" under circumstances in which nothing is A and the use of hypothetical sentences "If p then q" under circumstances in which p is false. These sentences have been called conditionals or hypotheticals which were "contrary to fact", "counterfactual", or, since the subjunctive mood is one way of expressing them in English, "subjunctive". We shall use the term "subjunctive conditional" for an assertion of the form: Although there are no A's, if there were to be any A's, all of them would be B's, e.g. If there were to be a gas whose molecules had zero extension and did not attract one another (although in fact there are no such gases) its pressure and volume would be related by Boyle's law. We shall use the term "subjunctive hypothetical" for an assertion of the form: Although p is false, if p were to be true, q would be true, e.g. If the picture-wire had broken (although it didn't), the picture would have fallen to the ground.*

* Many contemporary logicians use the terms "conditional" and "hypothetical" synonymously, and call what I am calling a "conditional" a "general conditional" or "general hypothetical". My distinction between the use of "conditional" and "hypothetical" was suggested by that made by J. N. Keynes (*Studies and Exercises in Formal Logic*, fourth edition (London, 1906), pp. 249 f.). In the considerable amount of recent literature dealing with subjunctive conditionals and the related question of natural laws solutions resembling the account to be given here have been published by Hans Reichenbach (*Elements of Symbolic Logic* (New York, 1947), Chapter VIII) and by J. R. Weinberg (*Journal of Philosophy*, vol. 48 (1951), pp. 17 ff.). F. P. Ramsey (*The Foundations of Mathematics and other logical essays*, pp. 237 ff.) and David Pears (*Analysis*, vol. 10 (1950), pp. 49 ff.) have also approached the questions along lines similar to mine.

At present we are concerned with subjunctive conditionals. The problem which they present to a Humean is the following dilemma. The constant-conjunction analysis leaves two choices open for the analysis of «If a thing is A, it is B». One choice is «Every A is B» taken, as traditional logic would say, 'existentially', i.e. understood in such a way as to assert the existence of at least one thing which is A. The other alternative is «Every A is B» taken non-existentially, i.e. understood as not to assert the existence of an A. On the first interpretation, «Every A is B» is equivalent to the conjunction of «Nothing is both A and non-B» with «Something is A». A subjunctive conditional would combine this conjunctive assertion with the assertion that there is nothing which is A, and would thus be self-contradictory. On the second interpretation «Every A is B» is equivalent to the single proposition «Nothing is both A and non-B». A subjunctive conditional would conjoin this assertion with the assertion that there is nothing which is A. But since, if nothing is A, *a fortiori* nothing is both A and non-B, the conjunction of these two propositions is logically equivalent to the former one alone. Thus a Humean analysis, it is alleged, makes the assertion of a subjunctive conditional either self-contradictory or one which adds nothing to the assertion that nothing is A, which is expressed by the subjunctive mood being used. Each horn of the dilemma is equally uncomfortable, since neither horn will account for the fact that we make subjunctive conditional assertions freely and without consciousness of paradox. The opponents of the constant conjunction view conclude that "If a thing is A, it is B", used to express a nomic connexion, must mean more than «Nothing is both A and non-B» (with or without the conjunction of «Something is A») in order to account for the function played in our thinking by subjunctive conditionals.

This criticism can be met without requiring that the proposition expressed by the sentence "If a thing is A, it is B", used nomically, should be distinguished from the proposition that nothing is both A and non-B. What is required is that what is involved in *asserting* the subjunctive conditional expressed by such a sentence as "Although nothing is A, yet, if a thing were to be A, it would be B" is distinct from what is involved in *asserting* a conjunction of «Nothing is A» with «Nothing is both A and non-B». We can make this distinction by taking the assertion of «Nothing is both

A and non-B» involved in asserting the subjunctive conditional as being not simply the assertion of «Nothing is both A and non-B» as being a true proposition, but also the assertion of «Nothing is both A and non-B» as being deduced from a higher-level hypothesis in a true and established scientific deductive system. To put the matter metaphorically, the generalization «Nothing is both A and non-B» enters into an assertion of a subjunctive conditional accompanied by a certificate of origin. Though the generalization itself conjoined with the proposition «Nothing is A» is logically equivalent to this latter proposition alone, a belief in the truth of the generalization which is accompanied by a belief about its origin, conjoined with a belief that nothing is A, is by no means equivalent to this latter belief alone.

Perhaps it is easiest to think of the matter in terms of the temporal order in which beliefs are acquired. Suppose that a person who has never considered whether or not there are any A's has come to accept a scientific deductive system in which the proposition that nothing is both A and non-B is deducible from higher-level hypotheses in the system which have been established by induction from evidence which does not include any instances of the generalization «Every A is B». If the person then makes this deduction in the scientific system, he will have confirmed the proposition «Nothing is both A and non-B» indirectly; if he regards the higher-level hypotheses as established, he will also regard it as established that nothing is both A and non-B, and will add this proposition to his body of rational belief. Now suppose that he subsequently discovers that in fact there are no A's. Had he acquired reasonable belief that there are no A's before he had acquired his reasonable belief that nothing is both A and non-B, he could have deduced this latter proposition from the former, and would not have required to establish it by deducing it from higher-level hypotheses in the scientific deductive system. But he did not do this; he arrived at his reasonable belief in the generalization «Nothing is both A and non-B» quite independently of his subsequently acquired belief that this generalization was 'vacuously' satisfied. The assertion of a subjunctive conditional may be regarded as a summary statement of this whole situation.

Take, for example, the statement "Although there are no gases whose molecules have zero extension and do not attract one another, yet if there were to be such gases, all of them would obey Boyle's law, $PV = $a constant". The assertion of this statement envisages a situation in which, before it was known that there were no such gases, a functional law had been established relating the pressure and volume of a gas by examining gases with extended molecules which did attract one another, e.g. van der Waals's equation $(P + a/V^2)(V - b) = $a constant. From this functional law the special law for gases whose molecules have zero extension and do not attract one another can be deduced by putting $a = b = o$; i.e. from van der Waals's equation deducing Boyle's law. This special law will then have been established quite independently of any knowledge as to whether or not there are any gases whose molecules have zero extension and do not attract one another. To assert the subjunctive conditional is then to refer to the fact that the proposition that no gases whose molecules have zero extension and do not attract one another fail to obey Boyle's law has been established independently of the fact, which the subjunctive conditional also asserts, that there are no such gases.

Since «Nothing is A» is logically equivalent to the conjunction of «Nothing is both A and non-B» with «Nothing is both A and B», to establish «Nothing is A» after «Nothing is both A and non-B» has been established is to establish in addition only «Nothing is both A and B». The evidence for «Nothing is both A and non-B» provided by the evidence, direct or indirect, for «Nothing is A» will, of course, be additional to the evidence for «Nothing is both A and non-B» provided by the evidence for higher-level hypotheses from which this generalization logically follows; but, since this generalization is supposed to have been already established by a deduction from the established higher-level hypotheses, the additional evidence will not serve to establish it. In Freudian language, its establishment is 'over-determined': there are two sets of evidence each sufficient to establish it, and the set which in fact establishes it is the one which gets in first.

Let us now remove the condition that the generalization «Nothing is both A and non-B» has been established by deducing it from established higher-level hypotheses before it has been considered whether or not it is vacuously true. We then have the situation that there are two ways of establishing the generalization: I can choose which of the two I regard as having got in first and as being the genuine establishment. One way is to deduce the generalization from the proposition, supposed to have been established, that nothing is A—call this the "vacuous" establishment; the other way is to deduce the generalization from the supposedly established higher-level hypotheses—call this the "hypothetico-deductive" establishment. The assertion of the subjunctive conditional «Although there are no A's, yet if there were to be any A's all of them would be B's» asserts that there are no A's, that nothing is both A and non-B, and that the latter of these propositions is establishable hypothetico-deductively without reference to the establishment of the former. The peculiarity of the subjunctive conditional is that to assert it is not only to assert two propositions, one of which is a logical consequence of the other, but is also to assert that this former proposition, though vacuously establishable by deduction from the latter, is also hypothetico-deductively establishable in an independent way. The assertion of the subjunctive conditional makes a remark about the relation of two of the propositions asserted in regard to the way that they can be established. This analysis, it seems to me, satisfactorily explains the peculiarity of subjunctive conditionals without our having to suppose that the sentence "Every A is B", used nomically, need mean any more than that nothing is both A and non-B.

To consider without asserting a subjunctive conditional is to perform a highly sophisticated activity. It is essentially to consider two propositions which are logically related but to consider them separately within two scientific deductive systems. The proposition «Nothing is both A and non-B» has to be considered as a deduction from higher-level hypotheses which do not include the proposition «Nothing is A», and not as a deduction from this proposition itself. The proposition «Nothing is both A and non-B» appears as a consequence in two deductive systems; but the fact that it appears in one of these systems has to be attended to, that it appears in the other unattended to. Thus the meaning of a subjunctive conditional sentence cannot be given simply by stating the proposition which is entertained whenever the sentence is used; the function of the sentence in our language is primarily to be used assertively, and in this case, as has been explained, it refers to the origin of the asserter's belief in one of the propositions asserted. Logicians have naturally concentrated their interest upon the meaning of sentences in which consideration of the meaning of the sentence—entertainment of the proposition—can be separated from what is, in addition, involved in using the sentence assertively; subjunctive conditional sentences cannot be treated in this way, since they have a very definite and important use when used assertively, which cannot be broken up into a set of unrelated assertions.

It is important to notice that we have no normal use for subjunctive conditionals which would deny the existence of things specified by a theoretical concept, in the sense of theoretical concept of Chapter III. This is because it is self-contradictory to conjoin a statement about a theoretical concept with a statement asserting that there are in fact no instances of the concept, since the truth of a statement about a theoretical concept is sufficient to endow the concept with existence. The only use of subjunctive conditionals in such cases is as a means for insisting that the statement about the theoretical concept, though it appears to be a contingent proposition, is in fact being used as a sterile formula (in the sense of Chapter IV) defining the meaning of the symbol for the concept. Consider, for example, the subjunctive conditional statement "Although there are in fact no hydrogen atoms, yet, if there were, all of them would be systems composed of one proton and one electron each". On the supposition that the term "hydrogen atom" in this sentence is to be used to stand for a theoretical concept, I can think of no use for this sentence except as used by someone who was using it as a sterile formula to give a definition of a new theoretical term "hydrogen atom" as a logical construction out of the theoretical concepts proton and electron already in use in his mode of thinking about physics.

LAWS OF NATURE

Our solution of the problem of how our use of subjunctive conditionals is consistent with a constant conjunction analysis of nomic generalizations will enable us to solve a related problem which is posed to Humeans, namely, that of distinguishing between what are laws of nature (or natural laws) and what anti-Humeans contemptuously call "mere generalizations". Surely, they say, this distinction must be admitted: for us it consists in the laws of nature asserting principles of nomically necessary connexion; since you decline to admit such principles how can you make the distinction? The problem is sometimes put in the form that we all distinguish between uniformities due to natural law and those which are merely accidentally true, "historical accidents on the cosmic scale";* if natural laws are just uniformities, how can this distinction be made?

It seems to me foolish to deny (as some Humeans do) that such a distinction is made in common speech; but it also seems perfectly sensible to try to give a rationale for this distinction within the ambit of a constant conjunction view. The distinction will then have to depend upon knowledge or belief in the general proposition rather than in anything intrinsic to the general proposition itself; but this is exactly how we have solved the related problem of subjunctive conditionals. Let us try to use this solution to pick out some among true contingent general propositions to be given the honorific title of "natural law".

Let us tentatively try the following criterion: A true contingent general proposition «Every A is B» whose generality is not limited to any particular regions of space or of time will be called by a person C a *law of nature* or *natural law* if either the corresponding subjunctive conditional «Although there are no A's, yet if there were any A's they would all be B's» is reasonably believed by C, or this subjunctive conditional would be reasonably believed by C if he were reasonably to believe that there are no A's. In terms of the notion of C's rational corpus of knowledge and reasonable belief, the criterion requires that the corresponding subjunctive conditional should form part of his rational corpus if this includes a belief that there are no A's, or, if it does not include this belief, would form part of his rational corpus if it did include this belief. In terms of the notion of assertion used in the discussion of subjunctive conditionals, an assertion of a natural law together with an assertion that there are no A's would come to the same thing as an assertion of the corresponding subjunctive conditional.

In addition all true hypotheses containing theoretical concepts will be given the title of natural laws.

The condition for an established hypothesis h being *lawlike* (i.e. being, if true, a natural law) will then be that the hypothesis either occurs in an established scientific deductive system as a higher-level

* William Kneale, in *Analysis*, vol. 10 (1950), p. 123.

hypothesis containing theoretical concepts or that it occurs in an established scientific deductive system as a deduction from higher-level hypotheses which are supported by empirical evidence which is not direct evidence for h itself. This condition will exclude a hypothesis for which the only evidence is evidence of instances of it, but it will not exclude a hypothesis which is supported partly directly by evidence of its instances and partly indirectly by evidence of instances of same-level hypotheses which, along with it, are subsumed under a higher-level hypothesis. This account of natural law makes the application of the notion dependent upon the way in which the hypothesis is regarded by a particular person at a particular time as having been established: "lawlike" may be thought of as a honorific epithet which is employed as a mark of origin. If the hypothesis that all men are mortal is regarded as supported solely by the direct evidence that men have died, then it will not be regarded as a law of nature; but if it is regarded as also being supported by being deduced from the higher-level hypothesis that all animals are mortal, the evidence for this being also that horses have died, dogs have died, etc., then it will be accorded the honorific title of "law of nature" which will then indicate that there are other reasons for believing it than evidence of its instances alone.

This criterion for lawlikeness has the paradoxical consequence that the hypothesis that all men are mortal will be regarded as a natural law if it occurs in an established scientific deductive system at a lower level than a hypothesis (e.g. All animals are mortal) which has other lower-level hypotheses under it which are directly confirmed by experience; whereas the higher-level hypothesis that all animals are mortal, if it appears as the highest-level hypothesis in the established deductive system, will not be regarded as a natural law, since the ground for its establishment is solely the evidence of its instances. However, it seems to me that this corresponds to the way in which, generally speaking, we use the notion of natural law. A hypothesis to be regarded as a natural law must be a general proposition which can be thought to *e plain* its instances; if the reason for believing the general proposition is solely direct knowledge of the truth of its instances, it will be felt to be a poor sort of explanation of these instances.* If, however, there

* But see below, p. 322.

is evidence for it which is independent of its instances, such as the indirect evidence provided by instances of a same-level general proposition subsumed along with it under the same higher-level hypothesis, then the general proposition will *explain* its instances in the sense that it will provide grounds for believing in their truth independently of any direct knowledge of such truth. And this connexion with a notion of explanation fits in well with the honorific title of natural law being ascribable to every hypothesis containing theoretical concepts, whether or not such a hypothesis stands at the highest level in the established scientific system. For even if the hypothesis with theoretical concepts is not deducible from an established higher-level hypothesis, yet it will not have been established simply by induction by simple enumeration; it will have been obtained by the hypothetico-deductive method of proposing it as a hypothesis and deducing its testable consequences. The case for accepting any particular higher-level hypothesis containing theoretical concepts is exactly that it serves as an explanation of the lower-level generalizations deducible from it, whereas the case for accepting a particular generalization not containing theoretical concepts and not deducible from any higher-level hypothesis is the fact that it covers its known instances rather than that it explains them.

I do not wish to emphasize unduly this relation between explanation and natural law: the marginal uses of both of these concepts are indefinite, and the boundaries of their uses will certainly not agree. Generally speaking, however, a true scientific hypothesis will be regarded as a law of nature if it has an explanatory function with regard to lower-level hypotheses or its instances; vice versa, to the extent that a scientific hypothesis provides an explanation, to that extent will there be an inclination to endow it with the honourable status of natural law.

To consider whether or not a scientific hypothesis would, if true, be a law of nature is to consider the way in which it could enter into an established scientific deductive system. As with the case of subjunctive conditionals, to consider this question is a sophisticated activity; the question cannot be answered by a straight yes or no without reference to how the hypothesis is related to other hypotheses which are used in our scientific thinking.

THE CASE OF THE BLACK RAVENS

This way of looking at the matter will, it seems to me, dispose satisfactorily of the cases brought forward recently by philosophers to show that there must be a distinction between laws of nature and "matters of fact with accidental universality".* Kneale says truly that there is no incompatibility between all ravens in fact being black and "the suggestion that if ravens had been tempted to live in a very snowy region they would have produced descendants that were white although still recognizably ravens"; and he uses this as an argument to prove that there must be a third possibility besides the two envisaged by Humeans—"either (i) it is a law of nature that all ravens are black, or (ii) there has been or will be somewhere at some time a raven that was not or is not black"—namely, that it is a "historical accident on the cosmic scale" that all ravens are black.† But the blackness of all ravens is surely "accidental" if no reason can be given for such blackness; and this is equivalent to saying that there is no established scientific system in which the generalization appears as a consequence. If a reason can be given for the blackness of all ravens by exhibiting such a scientific system, this generalization will be regarded as lawlike. Kneale is right in thinking that there is a trichotomy rather than a dichotomy; but the trichotomy arises from a second dichotomy following the first objective dichotomy of general propositions into those which are true and those which are false (i.e. into Kneale's cases (i) and (ii)). The second dichotomy divides those general propositions which are true into those ('natural laws') which occur as lower-level hypotheses in an established scientific deductive system or as highest-level hypotheses containing theoretical concepts and those which do not so occur ('matters of fact with accidental universality'). This second dichotomy is relative to the body of knowledge and reasonable belief (the rational corpus) of a particular person at a particular time; but since the rational corpora of many people with the same sort of scientific education

* William Kneale, *Probability and Induction*, p. 195.
† *Analysis*, vol. 10 (1950), p. 123. The fact that Kneale uses the word "accident" seems to show that he is half-way to the doctrine of this chapter. For what is a 'historical accident on the cosmic scale' or a 'matter of fact with accidental universality' except something not known to be covered by a more general law, i.e. not deducible in an established deductive system?

and culture agree in the scientific deductive systems which these corpora contain at a given time, these people will all agree in making statements that such and such scientific hypotheses, if true, are laws of nature and that other scientific hypotheses are 'mere generalizations', though each of the people will mean something different by his statement.

THE CASE OF THE BRAKELESS TRAIN

An argument used by Johnson against the identification of 'universals of law' with 'universals of fact' is that it is our belief in the universal of law that every brakeless train is dangerous which makes us insist that all trains should have brakes, and consequently to ensure that there should be no brakeless trains, and hence that the proposition «Nothing is both a brakeless train and undangerous» should be true. "This illustration suggests a wide class of cases which indicate human foresight or prudence; and in all such cases the distinction between the nomically necessary and the factual universal is quite apparent."* But the distinction we have made between generalizations which are, and those which are not, conclusions in established deductive systems will resolve any paradox involved in the making of a generalization true by the fact of believing it. For what is believed is not merely the generalization by itself, but is the generalization in the context of an established deductive system in which it appears as a deduction from higher-level hypotheses which have been established independently of the generalization. It is belief in these higher-level hypotheses (that momentum increases with mass and with velocity, that bodies with large momenta moving against only small retarding forces are dangerous, etc.) which is responsible for our insistence that all trains should have brakes. The method by which this complex belief secures the truth of the proposition that all trains have brakes is by providing a premiss from which someone who did not already know it could deduce the proposition that all brakeless trains are dangerous. But he would not be able to deduce the higher-level hypotheses about momentum, etc.; and it is these propositions belief in which has caused to be vacuously true the generalization that all brakeless trains are dangerous.

* W. E. Johnson, *Logic, Part III*, p. 12.

THE CASE OF THE TWO FACTORY HOOTERS

Suppose that there are two factories a and b, a being situated in Manchester and b in London, and that each factory has one and only one hooter. Consider the following five general propositions which are supposed to be true and to have been reliably established:

p Every sounding of the hooter of factory a at about noon is followed by workers leaving factory a;

q Every sounding of the hooter of factory b at about noon is followed by workers leaving factory b;

r Every sounding of the hooter of factory a at about noon is followed by workers leaving factory b;

s Every sounding of the hooter of factory b at about noon is followed by workers leaving factory a;

t Every sounding of the hooter of factory a at about noon is simultaneous with a sounding of the hooter of factory b, and every sounding of the hooter of factory b at about noon is simultaneous with a sounding of the hooter of factory a.

The logical relationships of these five propositions are that r is a logical consequence of the conjunction of q with t, and that s is a logical consequence of the conjunction of p with t.

This example was used by Bertrand Russell to discredit the notion of an event's having a unique cause* and by C. D. Broad to insist upon the distinction between a regular sequence and a causal connexion.† We are not at present concerned with causality: the paradox for us in this example lies in the fact that, while p and q are properly regarded as natural laws, r and s are not so regarded, whereas they are lower-level hypotheses in the deductive systems in which q and t, p and t respectively are established highest-level hypotheses, and thus r and s appear to satisfy the criterion for a natural law of being establishable hypothetico-deductively.

The solution of the paradox lies in the fact that, although each of the general propositions q and t may be perfectly well established, they will not be established in the same scientific deductive system. The established system in which q is established is one in which this generalization, though it may be supported by direct evidence, is also supported hypothetico-deductively by being deduced from

---

* *The Analysis of Mind* (London, 1921), p. 97.
† *The Mind and its Place in Nature* (London, 1925), pp. 454 ff.

the higher-level hypotheses that workers will leave work when they hear the pre-arranged signal for leaving work, and that the hooter is near enough for them to hear it. The generalization p equally follows from these two higher-level hypotheses. The established deductive system in which p and q occur is a highly complicated edifice in which physico-psychical hypotheses about the association of sensory stimuli with sensations, psychological hypotheses about recognition and psycho-physical hypotheses about voluntary action are highest-level hypotheses; the generalizations p and q take a very insignificant place in this deductive system of psychological theory.* But the general proposition t—that the hooters of the two factories always sound at about noon simultaneously— though it may be well established by induction by simple enumeration, has no place in the deductive system of psychological theory. Thus, although r is deducible from the conjunction of q and t, both of which propositions may be separately well established, they are not established within the same deductive system: consequently, since r is not deducible either from q or from t alone, r does not qualify for endowment with the status of natural law.

The epistemologically relative character of the ascription of the title of natural law shows itself if we consider the circumstances under which we would promote the proposition r to that status. If our social organization were as closely knit and as involuntary as that of ants, the proposition t might well appear in an established sociological deductive system as a conclusion from higher-level hypotheses which were supported independently, and so would itself have the status of natural law. If now the sociological deductive system and the psychological deductive system containing the propositions p and q were such they could be conflated into a unified system (which might well happen through the highest-level hypotheses of the sociological system being subsumable under the highest-level hypotheses of the psychological system), then the proposition r would be a lower-level hypothesis in this unified and established deductive system, and would thus be supported indirectly in the way required to make it into a law of nature.

* The aesthetic defect of psychological theory as a scientific deductive system is that it has a vast number of separate highest-level hypotheses; up to the present it has succeeded hardly at all in unifying itself by the use of colligating concepts, as physical theory did in the seventeenth century.

This case of the two hooters was used by Russell and by Broad in a discussion of causality and causal laws, whereas we have used it to discuss the simpler notion of natural law. But the two discussions are intimately connected. Why we refuse to say that the sounding of the hooter in Manchester is the cause, or part of the cause, of workers leaving the London factory is the absence of a spatio-temporally continuous chain of events between the event in Manchester and the event in London. Now it is the presence of such an intermediate chain of events between the sounding of the hooter in Manchester and the workers leaving the Manchester factory—an intermediate chain of events some of which are outside the body, some of which are in the ear, some of which are in the auditory nerve, some of which are in the brain—which enables us to incorporate proposition p—that every sounding of the Manchester hooter at about noon is followed by workers leaving the Manchester factory—into the psychological (including physico-psychical and psycho-physical) deductive system.

CAUSAL LAWS

The notion of natural law is indefinite enough for it to be a thankless task to try precisely to distinguish natural laws within the class of true constant conjunctions. The case is even worse if one attempts to pick out from within the class of natural laws those that should, in some special sense, be called "causal laws". I shall not attempt this thankless task, but shall merely mention a few criteria which could be used by anyone who wishes to make such a distinction and which do not assume a special unanalysable causal relation, in order that the possibility of making this distinction should not be used as an argument against the Humean view of general hypotheses.

Many of the plausible criteria have reference to the way in which time enters into the natural law; so it is necessary at this point to abandon the simplification which has up to now been possible in our discussion, the simplification of treating all generalizations alike as asserting a concomitance of properties in the same thing or event, i.e. as considering all generalizations as being of the form «Everything which is A is also B» or «Nothing is both A and non-B». It is now necessary to distinguish among these generalizations those which genuinely assert *regular concomitances* and those which on a further analysis will be seen to assert constant associations of properties in different events, and which are of the form «Every event which is A is accompanied, later, simultaneously, or earlier, by an event which is B». These statements will be said to assert *regular sequences, regular simultaneities* and *regular precedences* respectively.

Let us leave aside for the moment the regular concomitances, and consider the cases in which the natural law asserts a constant conjunction of properties in two distinct events. There are several plausible criteria with claims to be used for selecting, out of the class of natural laws of this two-event sort, a subclass of causal laws. Here are some of the most plausible ones:

I. The weakest criterion is perhaps to exclude only natural laws asserting regular precedences and to admit those asserting regular simultaneities as well as regular sequences.

II. A stronger criterion would admit only natural laws asserting regular sequences, but would admit all these whatever the interval of time between the event having the property A (the 'cause-event') and the event having the property B (the 'effect-event').

III. A yet stronger criterion would admit only those natural laws asserting regular sequences where these regular sequences were deducible in an established deductive system from regular sequences in which there was no time-interval between the cause-event and the effect-event, i.e. in which the cause-event and the effect-event were temporally continuous.

IV. An even stronger criterion would admit only these temporally continuous regular sequences.

The difference between criteria III and IV is that criterion III is satisfied if there is a continuous chain of events between the cause-event and the effect-event the pairs of successive members of which chain obey a law of regular sequence satisfying criterion IV. Criterion III, that is, allows to be causal those regular sequences in which there is 'action at an interval of time' provided that such action at an interval can be explained by filling in the interval with a continuous chain of events between successive members of which there is no action at an interval. Criterion IV admits only regular sequences in which there is no action at an interval at all.

Criteria III and IV can be strengthened into III′ and IV′ by adding the requirement, in IV′ that the cause-event and the effect-event should also be spatially continuous (i.e. 'action at a distance' should also be excluded);* in III', that the explanatory continuous chain of events should be spatially as well as temporally continuous. It is because the regular sequence formed by the soundings of the Manchester hooter followed by workers leaving the London factory does not satisfy this criterion III' that Broad is so certain that it is not an instance of a causal law.

A condition proposed by some philosophers in addition to the criteria already suggested would require that the two events which are constantly conjoined should be events 'in the same substance', or 'in the same kind of substance'—whatever may be meant by these phrases. This criterion has been used to deny the name of causal law to laws of psycho-physical and physico-psychical association, on the ground that mind and matter are such different kinds of substance that they cannot 'interact', so that the accepted regular sequences and simultaneities holding between events in the mind and events in the body will have to be regarded as examples of 'parallelism' and not of 'causal interaction'.

The brief recital of these criteria, all of which have (I believe) been actually used to exclude some of the constant conjunctions which, according to the account of this chapter, would be natural laws from being causal laws or 'laws of necessary connexion', will, it may be hoped, suffice to make it clear that the distinctions made in ordinary speech between causal laws and other natural laws do not require the postulation of any specific 'causal' relationship not analysable into some combination of constant conjunction and temporal (or spatio-temporal) relations. This is recognized by most of the modern critics of the constant-conjunction view whose arguments against the view have been based upon its alleged inadequacy to provide a justification for induction or to account for the use of subjunctive conditionals rather than on its inadequacy to allow for the features peculiar to causal as contrasted with non-causal natural laws.†

---

* This would correspond to Hume's first definition of cause: *A Treatise of Human Nature*, Book I, Part III, §14.

† Regular sequences, as contrasted with regular simultaneities or concomitances, present a serious additional difficulty to the view which identifies nomic necessity with logical necessity. For an event's having a property A can only entail a later event's having a property B if an event's having the property A entails that a later event exists, i.e. if it entails that the world does not come to an end immediately after the first event. But that this should happen is not logically impossible.

All these features that may be taken to be peculiar to causal laws can be related to the place of a causal natural law in an accepted scientific deductive system. The criteria which have to do with temporal or spatio-temporal continuity (III, IV, III', IV') owe their appeal to the fact that the procedure of constructing a deductive system by asserting the existence of intermediate events forming a continuous chain, and by explaining a hypothesis by deducing it within such a system from hypotheses about successive events in such chains, has proved so successful, especially in the physical sciences, that we feel much happier about a natural law if it can be embedded in such a system. The criteria which have to do with events being the same (or the same kind of) substance owe their plausibility to the many deductive systems which have been established dealing with the same kind of thing as compared with the few that have been established dealing with things of different sorts.

CAUSE AND EFFECT

The principal use of causal notions, however, lies less in the notion of causal law as distinguished from that of natural law than in the distinction, in the case of an instance of a natural law, between one event as cause and the other event as effect. A philosopher might agree with Bertrand Russell in holding that "in a sufficiently advanced science the word 'cause' will not occur in any statement of invariable laws",* while yet using causal language (as Russell agrees it is convenient to do) in the application of the natural laws of the science to common-sense matters of fact. The importance of the language of cause and effect in everyday life is that it is the most convenient way to express the facts of intentional action directed towards non-immediate ends—when we do one thing in order that another thing may follow. In the applications of natural laws what is called "cause" and what is called "effect" is determined by the possibility of using the first as a means for the production of the second—the theoretical possibility, since a means to an end is taken to be a cause with the end as effect even if the means-cause cannot be produced by human agency.

This fact comes out very clearly in our use of causal language in reference to laws of regular concomitance and of regular simultaneity. In the case of regular concomitance the thing or event

* *Our Knowledge of the External World* (London, 1914), p. 220.

which has the property A is the same as the thing or event in question which has the property B; so the law does not hold of pairs of events, and the events concerned cannot be distinguished into cause and effect. But a causal distinction can be applied to the holding of the two properties. Suppose «Every A is B» is established as a natural law but «Every B is A» is not so established. Then if we were to make a thing A, doing this would ensure that it should also be B, whereas the reverse would not be the case. So in the application of the law «Every A is B» to a particular thing c, c being A may be distinguished as the cause from c being B as the effect. Similarly in an established case of regular simultaneity, where every event having the property A is accompanied by a simultaneous event having the property B, then if the converse law is not also established, the former event will be considered to be the cause and the latter event the effect. In both cases, however, if the converse laws are also established, we shall not be able to make a distinction between cause and effect, and shall either decline to use the cause-and-effect language at all or will use it symmetrically by calling both facts or events "causes" and both "effects".

In the case of a law of regular sequence, however, we call the earlier event the cause and the later event the effect; and we do this even if the law is reciprocal so that the later event nomically determines the earlier event quite as much as the earlier determines the later. Conversely, in the case of a law of regular precedence, we decline to call the later event the cause and the earlier event the effect, even if the law is not reciprocal so that, although every event which is A is preceded by an event which is B, it is false that every event which is B is followed by an event which is A. Here the spirit of our language forbids us, in a particular instance of this law, to call the A-like event the cause and the B-like event the effect, just because to say this would be to make a cause later in time than its effect. If we want to describe the later event in relation to the earlier event, we shall use such language as, for example, that it is a 'nomically sufficient condition' for, or that it 'nomically determines', the earlier event; but we shall not say that it is its 'cause'.

A natural explanation for the refusal of our cause-and-effect language to allow a cause to succeed its effect, whether in regard to particular cases of causation or in regard to the criteria for distinguishing causal laws from other laws of nature, is to suppose that it has arisen from our predominant interest in voluntary action. If an earlier event's occurring is a nomically sufficient condition for a later event to occur, we can (in suitable cases) ensure that the later event should occur by taking steps to see that the earlier event does occur. For this purpose it is irrelevant whether or not the later event's occurring is a nomically sufficient condition for the earlier event to occur. It may be the case that the later event can be produced in other ways, that it has a 'plurality of causes' (to use Mill's language); this possibility does not prevent us from indirectly producing the later event by producing the earlier event. But, if a later event's occurring is a nomically sufficient condition for the earlier event to occur, we cannot indirectly produce the earlier event by producing the later event, since by the time that we should be producing the later event the earlier event would irrevocably either have occurred or not have occurred. This difference between the case of regular sequence and that of regular precedence is, I think, the reason why we are prepared to call a nomically sufficient condition for an event a cause of that event if it precedes the event but are not prepared to call it a cause if it succeeds the event. For by our actions we can frequently affect the future, but we cannot affect the past.

This explanation of the elements peculiar to causality, as distinguished from nomic legality, in terms of what is possible and what is impossible in voluntary action does not require us to adhere to a 'primitive animism' which would endow nature with human desires and volitions and would explain causation in the physical world on the analogy of volitional action. Indeed it enables an explanation to be suggested of the way in which the 'idea of necessary connexion' may have originated in the phenomena of volitional action without supposing a projection of such action on to the physical world. For the process of indirect volitional action, where the end intended can only be attained by first producing a means to that end, will naturally be regarded as a chain of two processes—the volitional action resulting in the production of the means, followed by the means causing the production of the intended end.* For example, a primitive man's volitional action in driving off an approaching bear by throwing a stone at it will

* The means causing the end is a case of R. G. Collingwood's second sense of the word "cause" (*An Essay on Metaphysics* (Oxford, 1940), pp. 285 ff.).

naturally be regarded by him as falling into two processes, the first being his throwing the stone, the second being the stone's hitting the bear. Since the necessary connexion will, in a pre-critical stage of thought, be attributed to the relationship between the volition (or the intention) and the end which can only be indirectly attained no less than to the relationship between the volition (or the intention) and its immediate effect of producing the means, consistency will require that necessary connexion should also be attributed to the relationship between the means and the end produced by it. If the primitive man believes that there is a necessary connexion between his intention to frighten off the bear and the stone thrown by him hitting the bear as well as a necessary connexion between his intention to throw the stone at the bear and the stone's being thrown towards the bear, he will be forced to believe (if he wishes his beliefs to hang together) that there is also a necessary connexion between the stone's moving towards the bear and its hitting the bear. Once the notion of necessary connexion has become attached to regular sequences, it will easily get transferred to regular simultaneities and regular concomitances, since these also serve the function of enabling an intended end to be attained indirectly by first producing the means to that end, the means being such that it as cause can produce, without human intervention, the humanly intended end as an effect.

PARTICULAR CAUSAL PROPOSITIONS, INDICATIVE AND SUBJUNCTIVE HYPOTHETICALS

It remains to say a few words upon particular causal propositions of the form «q because p», where p and q are particular contingent propositions and the statement "q because p" is used to assert that p is a cause, or part of a cause, of q, and is not used either to state that a relation of logical consequence holds between q and p or merely to state something about belief in q, e.g. that knowledge of p makes it reasonable to believe q. With particular causal propositions of the form «q because p» (e.g. This picture fell to the floor at noon yesterday because its wire broke then) we may also consider indicative hypotheticals of the form « If p then q» (e.g. If the wire of this picture breaks at noon tomorrow, the picture will then fall to the floor) and also subjunctive hypotheticals of the form «Although p is false, yet if p were to be true, q would be true» (e.g. Although the wire of this picture did not break at noon yesterday, yet, had it done so, the picture would have then fallen to the floor), p and q in all these cases being particular contingent propositions and the hypotheticals being used nomically.

Let us start by considering indicative hypotheticals of the form «If p then q», the analysis of which has been much discussed recently by logicians. It seems to be certain that, as used in normal empirical contexts, the *assertion* of «If p then q» involves more than the mere *assertion* of «Not both p and not-q», whether the meaning of the statement "If p then q" is the same as or comprises more than that of "Not both p and not-q". The assertion of «If p then q» asserts, besides the proposition «Not both p and not-q», the proposition that this proposition has been established, or could be established, by deducing it from hypotheses in a scientific deductive system which is both true and established, together, perhaps, with certain other propositions p_1, p_2, etc., which are implicitly assumed to be common property to the asserter and his hearer. To assert, for example, that if the picture wire breaks at noon tomorrow, the picture will then fall to the floor, asserts not only that it is false both that the wire will break tomorrow and that the picture will not fall to the floor, but also that this proposition is deducible from an established scientific law (that unsupported bodies fall) together with certain propositions taken to be common knowledge to both the asserter and the hearer (e.g. that there will be no solid object between the position of the picture just before the wire breaks and the floor, that the picture will not be supported by its standing on a ledge as well as by the picture wire, etc.). The difficulty in saying exactly what are the scientific hypotheses and the additional propositions involved in any particular assertion of a proposition of the form «If p then q» should not lead us to think that there are not definite hypotheses and definite extra particular propositions involved in the assertion of the "If p then q" statement. These hypotheses and propositions may differ in each case of the same indicative hypothetical. But unless the asserter is prepared to specify, more or less precisely, what are the hypotheses and extra particular propositions involved, he can with doubtful propriety be said to be asserting the indicative hypothetical statement.*

---

* The objection to an analysis of the meaning of the statement "If p then q" into «There are a set of true hypotheses and a set of true propositions which are

The assertion of a subjunctive hypothetical is similar to that of an indicative hypothetical except that the assertion includes an assertion that p is false (indicated by the use of the subjunctive mood or by some other device) and that the assertion that the asserted proposition «Not both p and not-q» is deducible in an established deductive system must be qualified by adding that such a deduction must be independent of the falsity of p.

The assertion of a particular causal proposition of the form «q because p» (e.g. This picture fell to the floor at noon yesterday, because its wire broke then) is also similar to that of an indicative hypothetical except that the assertion includes assertions both that p is true and that q is true. (p and q will have to be subject to certain temporal, or spatio-temporal, restrictions in relation to one another in order that p may be the cause-proposition and q the effect-proposition; but this restriction, whatever exactly it may be, is extraneous to the present argument.) Though the conjunction of p with «Not both p and not-q» is logically equivalent to the conjunction of p with q, yet the assertion of «q because p» involves more than the joint assertion of p and of q. The extra element, as in the case of indicative and subjunctive hypotheticals, is the assertion that the proposition «Not both p and not-q» is deducible within a true and established deductive system from hypotheses of that system together, usually, with propositions about cause factors which have not been explicitly mentioned. My example of the assertion that this picture fell to the floor at noon yesterday because its wire then broke omits explicit reference to other cause factors in the absence of which the picture would not have fallen or, if it had fallen, would not have fallen to the floor, the hypotheses in the deductive system concerned remaining unchanged. Many of these cause factors are the 'permanent conditions' of Mill. Why it is usually unnecessary to mention them is because they are, as it were, fixtures; these fixtures may be negative, as in our example (there was no rail upon which the picture rested, there was no sofa under the picture, etc.). Nevertheless for a full account of the logic

such that «Not both p and not-q» is a logical consequence of the conjunction of these hypotheses and these propositions» is that, on any occasion of the assertion of «If p then q», something more definite is implied about the hypotheses and the extra propositions than merely that there are such true hypotheses and propositions.

of the assertion they will have to be mentioned explicitly. It is because of the fact that the relevant causal conditions are hardly ever all explicitly mentioned that it is difficult to hold that a particular causal proposition explicitly mentions the general hypothesis of which it is an instance. But what is certain, I think, is that the *assertion* of a particular causal proposition «*q* because *p*» involves the assertion of a general hypothesis in the sense that the asserter must always reply to a demand for the citing of the hypothesis upon which he has based his particular causal proposition—on pain of being accused of doing no more than asserting the conjunction of *q* with *p* if he is unable to meet this demand.

CONCLUSION

The argument of this chapter cannot be regarded as a knock-out blow to those who wish to maintain that nomic necessity is, quite apart from epistemological considerations, objectively something over and above constant conjunction. The thesis which has been maintained is that the genuine differences between assertions of constant conjunction and assertions of natural law arise out of the way in which the propositions concerned in the assertions are related to other propositions in the deductive systems used by the asserter. This thesis makes the notion of natural law an epistemological one and makes the 'naturalness' of each natural law relative to the rational corpus of the thinker.* Thus it is impossible to say in general that to assert that «Every *A* is *B*» is a law of nature, or that if a thing were to be *A*, it would also be *B*, is to assert exactly so-and-so or such-and-such; and the anti-Humean who demands a precise answer to the question as to what exactly is the difference between a specific law of nature and its corresponding 'mere generalization' cannot properly be given one. The imprecise answer which has been given in this chapter, though far superior in subtlety to the solution I propounded twenty-four years ago (according to which the distinction lay almost entirely in the fact

---

* C. D. Broad says that it seems to him "fairly certain on inspection" that he does not mean by "causal laws" propositions of the form «*A* is always accompanied by *B*» "limited by conditions about spatio-temporal and qualitative continuity and decked out with psychological frillings" (*Aristotelian Society Supplementary Volume* 14 (1935), p. 93). The 'frillings' with which the constant-conjunction propositions are decked out in this chapter in order to elevate them to higher rank are epistemological, not psychological.

that natural laws were generally believed whereas mere generalizations were not),* is, I suspect, still not subtle enough to do justice to the complexities of the situation. But the complexities are those of the nuances in our use of language to describe, explicitly or implicitly, the scientific deductive systems with which we think about the empirical world: there is no need to suppose that they spring from anything transempirical in the world itself.

* *Mind*, n.s., vol. 36 (1927), pp. 467 ff.; vol. 37 (1928), pp. 62 ff.

CHAPTER X

CAUSAL AND TELEOLOGICAL EXPLANATION

Any proper answer to a 'Why?' question may be said to be an explanation of a sort. So the different kinds of explanation can best be appreciated by considering the different sorts of answers that are appropriate to the same or to different 'Why?' questions.

What is demanded in a 'Why?' question is intellectual satisfaction of one kind or another, and this can be provided, partially or completely, in different ways. Frequently the questioner does not know beforehand what sort of answer will satisfy him. And what gives partial or complete intellectual satisfaction to one person may give none whatever to a person at a different stage of intellectual development. A small child, for example, is frequently satisfied by a confident reassertion of the fact about which he has asked the 'Why?' question. This is not foolishness on his part. The child is prepared to accept the fact without question on authority; what he is doubtful about is whether the authority is good enough, and a confident reassertion by the person to whom he has asked the 'Why?' question may serve to strengthen the authority sufficiently for him to feel complete satisfaction in accepting it.

When an adult wishes for satisfaction of this purely confirmatory sort, he phrases his question in the form "Is it really the case that...?", reserving his 'Why?' questions for cases in which he requires for satisfaction something more than a repetition of the 'Why?' sentence with the omission of the "Why". What he requires is explanation in the proper sense of the proffering of an explicans-proposition as an explanation of the explicandum-fact about which he has asked the 'Why?' question, the explicans being required to be different from the explicandum.

Different sorts of explicanda call for different sorts of explanation—or, at least, for different sorts of first-stage explanations. The primary explicanda for science are particular empirical facts; and first-stage explanations of these are of two types. When an adult asks "Why f?" of a particular matter of fact f, he is usually wanting either a *causal explanation* expressed by the sentence "Because of *g*" or a *teleological explanation* expressed by the sentence "In order that *g*". Each of these types of explanation will involve an explicit or implicit reference to scientific laws; a 'Why?' question asked of a scientific law (and this may well be a second-stage 'Why?' question asked of a particular matter of fact) will be a request, not for a cause or for a teleological goal, but for a reason for the scientific law being what it is. This chapter will be devoted to the first-stage explanation of particular empirical facts, a discussion of the explanation of scientific laws themselves being postponed until the next chapter.

CAUSAL EXPLANATION

Our discussion in the last chapter of the meaning of the sentence "*q* because *p*" covers most that need be said of causal explanation. When a person asks for a cause of a particular event (e.g. the fall of this picture to the floor at noon yesterday), what he is requesting is the specification of a preceding or simultaneous event which, in conjunction with certain unspecified cause factors of the nature of permanent conditions, is nomically sufficient to determine the occurrence of the event to be explained (the explicandum-event) in accordance with a causal law, in one of the customary senses of "causal law".* The 'Why?' question is not expected to be answered by detailing all the events which together make up a total cause, i.e. a set of events which collectively determine the explicandum-event; all that is usually expected is the part-cause which is of most interest to the questioner—which presumably is that of which he is ignorant. One sense of giving a *complete* explanation would be that of specifying a total cause; in this sense, as indeed in most senses of complete explanation, a 'complete explanation' will not be unique, since (in almost all senses of "cause") the same event can perfectly well have many different total causes.

---

* By saying that an event having the property B is *nomically determined* by an event having the property A, in conjunction with events having property A_1, having property A_2, etc., no more is meant than that the generalization «Every conjunction of an event having A with events having A_1, A_2, etc., is associated with an event having B» is a true generalization. Use of the language of nomic determination does not, therefore, presuppose any non-Humean analysis either of nomic or of causal statements.

There are various complications about causal explanations considered as answers to 'Why?' questions which need not detain us long. The formal explanation just given is that in which the questioner is taken to be asking for a *sufficient condition* for the explicandum-event, or for part of a sufficient condition, the other part being supposed to be already known. The explicans in such an explanation is an event the occurrence of which possessing a certain property, in conjunction with other events with suitable properties, nomically determines the occurrence of the explicandum-event with a certain property. So the existence of the explicans event ensures the existence of the explicandum-event. But the 'Why?' question is sometimes a request for a *necessary condition* for the explicandum-event; it then asks for the specification of an event which is such that, had it not occurred, the explicandum-event would also not have occurred. In this case it is the explicans-event which is nomically determined by the explicandum-event instead of the other way round. And frequently the 'Why?' question requires as answer the specification of an event which is both one of a set of events which together form a sufficient condition and one which in the presence of the rest of the set of events is a necessary condition for the occurrence of the explicandum-event.

There is one type of causal explanation which, rightly or wrongly, gives great intellectual satisfaction to those who have been educated in the contemporary natural sciences—namely, causal explanations making use of causal laws which are causal according to criterion III' (p. 310). Here an event to be acceptable as explicans must be the first member of a *causal chain* of events ending with the explicandum, a spatio-temporally continuous chain of events being said to form a causal chain if every event in the chain nomically determines its neighbours in the chain in such a way that the causal law relating the explicans-event with the explicandum-event is a consequence within a true deductive system of higher-level laws which relate only spatio-temporally continuous events. The intellectual satisfaction provided by an explanation which cites a 'causal ancestor' is due partly to the great success of such explanations in the physical sciences, but partly also to the fact that, if the deductive system whose highest-level hypotheses relate spatio-temporally continuous events is unrefuted by the evidence, there will be a great deal of evidence which supports it. For an unlimited number of lower-level hypotheses about causal ancestries will be deducible from the highest-level hypotheses, and the falsity of some, at least, of these would be expected to leap to the eye if the highest-level hypotheses were false.

At the other extreme there is a type of explanation based upon generalizations which have been established by direct induction without any indirect hypothetico-deductive support. These generalizations, therefore, do not satisfy the conditions for being entitled "natural laws", and consequently cannot be classed as causal laws by even the weakest criterion of the last chapter (criterion I). So these explanations cannot be called causal explanations. Nevertheless they can give some intellectual satisfaction, for they give information on one point about which the questioner may be ignorant. Molière was right in laughing at the doctors who offered the *virtus dormitiva* of opium as the answer to the question as to why opium produced sleep.* But it would not be foolish to answer the question as to why a particular specimen of powder produced sleep by replying that it was because that powder was opium, and that opium had the property of producing sleep. For this would inform a questioner who did not know it already that the powder produced sleep, not by virtue of its colour, degree of powderedness, etc., but by virtue of its chemical composition. Similarly, when a child asks "Why is this bird white?", the reply "Because it is a swan, and all English swans are white" tells him that the whiteness is not a peculiarity of this particular bird, and thus shows the particular case to be an instance of a general proposition.

TELEOLOGICAL EXPLANATION

We must now turn to a type of explanation which has so far not been discussed—a type which has given rise to a great deal of discussion among philosophers and philosophically minded biologists, because it has been thought to raise peculiar scientific and philosophical difficulties. This type of explanation is that in which the 'Why?' question about a particular event or activity is answered by specifying a goal or end towards the attainment of which the

* In the third ballet scene of *Le Malade Imaginaire*.

event or activity is a means. Such explanations will be called "teleological explanations".* If I am asked why I am staying in Cambridge all through August, I should reply "In order to finish writing my book"; to reply thus would be to give a teleological explanation. If I am asked why my cat paws at the door on a particular occasion, I might well reply "In order that I should open the door for it"—another teleological explanation. If an ornithologist is asked why a cuckoo lays its egg in the nest of another bird, and replies "So that the other bird may hatch out and nurture its young", or if a physiologist is asked why the heart beats, and replies "To circulate the blood round the body" or (in more detail) "To convey oxygen from the lungs to the tissues and carbon dioxide from the tissues to the lungs" or (in terms of an ultimate biological end) "In order that the body may continue to live", he will be giving in each case a teleological explanation of the action in terms of the goal or end of the action. The explanation consists in stating a goal to be attained: it describes the action as one directed towards a certain goal—as a 'goal-directed activity' (to use E. S. Russell's convenient phrase†), the word "directed" being used (as it will here be used) to imply a direction but not to imply a director.

If we take an explanation (as we are doing) to be any answer to a 'Why?' question which in any way answers the question, and thereby gives some degree of intellectual satisfaction to the questioner, there can be no doubt that teleological answers of the sort of which I have given examples are genuine explanations. The fact that they all may give rise to further questions does not imply that they are not perfectly proper answers to the questions asked. My answer as to why I am staying in Cambridge all through August would almost certainly not lead to a further question, unless my friend wished to start a philosophical discussion as to the correct analysis of the motives of rational action. My answer as to why my cat paws the door might lead to the further question as to why the cat (to use common-sense language) 'wants to be let out', to which another teleological answer would be appropriate, or to the

---

* The remainder of this chapter follows, with some alterations and additions, the text of my 1946 Presidential Address to the Aristotelian Society (*Proceedings of the Aristotelian Society*, n.s., vol. 47 (1946–7), pp. i ff.).

† *The Directiveness of Organic Activities* (Cambridge, 1945).

question as to how the cat has learnt to paw the door to show that he wants to be let out, which would lead to a description, which might or might not be in teleological terms, of the processes of learning in cats. But all these would be regarded as further and different questions; the first simple teleological answer would be taken as what the questioner was asking for, and if it did not give him adequate intellectual satisfaction, he would expect not to repeat the question but to ask another.

But, having insisted that teleological explanations are perfectly good first-stage explanations, we have to admit that they have one feature which distinguish them from causal explanations, and that this feature has proved very puzzling to philosophers, whether concerned with philosophical psychology or with the philosophy of biology. In a causal explanation the explicandum is explained in terms of a cause which either precedes or is simultaneous with it: in a teleological explanation the explicandum is explained as being causally related either to a particular goal in the future or to a biological end which is as much future as present or past. It is the reference in teleological explanations to states of affairs in the future, and often in the comparatively distant future, which has been a philosophical problem ever since Aristotle introduced the notion of 'final cause'; the controversy as to the legitimacy of explanations in terms of final causes rages continually among philosophers of biology and, to a less extent, among working biologists.

Now there is one type of teleological explanation in which the reference to the future presents no difficulty, namely, explanations of an intentional human action in terms of a goal to the attainment of which the action is a means. For my teleological answer to the question as to why I am staying in Cambridge all through August— that I am doing so in order to finish writing my book—would be regarded by my questioner as equivalent to an answer that I am doing so because I intend to finish writing my book, my staying in Cambridge being a means to fulfil that intention; and this answer would have been an explanation of the causal sort with my intention as cause preceding my stay in Cambridge as effect.* Teleological

---

* Jonathan Cohen has pointed out (*Proceedings of the Aristotelian Society*, n.s., vol. 51 (1950–1), pp. 262 ff.) that such an explanation would differ from an ordinary causal explanation in that it is more difficult to specify the total cause of which the intention is only a part than it is to specify the total cause in an ordinary causal explanation. But this difference is only one of degree (as Cohen

explanations of intentional goal-directed activities are always under-stood as reducible to causal explanations with intentions as causes, to use the Aristotelian terms, the idea of the 'final cause' functions as 'efficient cause'; the goal-directed behaviour is explained as goal-intended behaviour.

This is not to say that there is no philosophical difficulty about intentional action; there is the problem—fundamental for philo-sophical psychology—as to the correct analysis of the intention to act in a certain way. But this is different from our problem as to how a future reference can occur in an explanation, unless indeed an extreme behaviouristic analysis is adopted, according to which there is no conscious element in an intention, and goal-intended behaviour is simply what we call goal-directed behaviour in the higher animals. But for this extreme behaviourism psychology reduces to biology, and intentional action falls under biological goal-directed activity and the type of teleological explanation we meet in the sciences concerned with life in general and not especially with mind.

The difficulty about the future reference occurs then in all teleological explanations which are not reducible to explanations in terms of a conscious intention to attain the goal. Here one cannot obviously reduce the teleological answer, which explains a present event by means of a future event, to a non-teleological answer in terms of a present or past cause. It is teleological explanations which cannot obviously be so reduced which present the philosophical problem; and the rest of this chapter will be devoted to this type of teleological explanations and to the pro-blems raised by them.

CURRENT ATTEMPTS TO ELIMINATE FINAL CAUSES

There are two ways of solving this problem which are fashionable to-day. Both are attempts to reduce all teleological explanations to causal explanations, and thus to eliminate the special puzzle pre-sented by future reference; but the attempts are made in opposite directions.

The first way is to emphasize the similarity between teleological

---

seems prepared to admit; loc. cit. p. 268 n.); and, however partial the intention may be as a factor in a total cause, it will not be later in time than the action which it is put forward to explain.

explanations of the type with which we are now concerned and the teleological explanations of intentional actions in which the future reference can be explained away, and to argue by analogy that in all cases the teleological explanation is reducible to one in which an intention, or something analogous to an intention, in the agent is the 'efficient cause', so that goal-directed activity is always a sort of goal-intended activity. My cat's behaviour in pawing at the closed door, it may be said, is sufficiently similar to a man's behaviour in knocking at a locked door for it to be reasonable to infer that the cat, like the man, is acting as it does because of a conscious intention, or at least a conscious desire, to be let through the door. Similarly a neurotic's goal-directed behaviour may be explained by his having an unconscious intention or desire; a bird's nest-building by its having an instinct to do so. When the goal-directed activity to be explained is that of a part of a whole organism, as in my example of the heart's beating, the analogue to the intention—the drive or conatus or nisus or urge—is usually posited not in the separate organ but in the organism as a whole—an urge towards self-preservation, for example. Sometimes the analogy is pressed so far that a purposiveness similar to that of voluntary action is assumed in all teleological behaviour. William McDougall, for instance, after explaining that by 'purposiveness' in human movements he means not only that "they are made for the sake of attaining their natural end" (i.e. that they are teleological in my sense), but that "this end is more or less clearly anticipated or foreseen", goes on to speak of a "scale of degrees of purposiveness", at the lower end of which there is a "vague anticipation of the goal" which may also be ascribed to an animal's goal-directed behaviour.*

Other writers (e.g. E. S. Russell) would reject as unduly anthropomorphic the attribution of purposiveness to such activities, and would describe the efficient cause as a conatus or drive. But all writers who deal with the problem of teleological explanation in the first way agree in postulating something in the organism which is present whenever goal-directed behaviour is taking place and which is to explain it in the ordinary causal way, and agree in supposing that this something cannot be analysed purely in physico-chemical terms.

* W. McDougall, *An Outline of Psychology* (London, 1923), pp. 47 f.

The biological orthodoxy of to-day, however, would say that the postulation of this 'something', not explicable in physico-chemical terms, to account for teleological behaviour was an assumption which was either methodologically vicious (if the 'something' was supposed to have no properties other than that of being the cause of the goal-directed behaviour) or metaphysical and non-empirical (if it was supposed to have additional properties such as McDougall's purposiveness). And orthodox biologists would go on to say that satisfactory explanations had been given of many goal-directed activities in physico-chemical terms, and that as the new sciences of biochemistry and biophysics advance, there is less and less reason to suppose that there will be any teleological action (or at any rate any teleological action in which consciousness is not involved) that will not be explicable by means of the concepts and laws of chemistry and physics alone.

This attitude is equivalent to an attempt to solve the problem of teleological explanations in a second way, by reducing them to physico-chemical explanations of the ordinary causal sort. It is admitted that biochemistry and biophysics at the moment cannot effect this reduction in the great majority of cases, but it is expected that some day they will be able to. Teleological explanations must be accepted as irreducible to causal explanations at present, but not as in principle irreducible. Thus the philosophical problem presented by the reference to the future in such explanations is a temporary problem only, to be solved by the progress of science. A teleological explanation is to be regarded as a very poor sort of explanation indeed, to be discarded as soon as the real, physico-chemical causes have been discovered.

It seems to me that the orthodox biologists are right in rejecting the postulation of a conatus or drive which is non-physical and *sui generis* in order to explain the goal-directed behaviour which they meet in their biological studies, but wrong in minimizing the intellectual satisfaction to be derived from teleological explanations. I believe that we can go on the orthodox assumption that every biological event is physico-chemically determined, and yet find an important place in biology for such explanations. So what I propose to do is to try to give an account of the nature of teleological explanations which will resolve the philosophical difficulty about the apparent determination of the present by the future without either contravening the usual determination principles of science or reducing all biological laws to those of chemistry and physics.*

TELEOLOGICAL CAUSAL CHAINS

If we make the ordinary determination assumptions of physical science, the apparent determination of the explicandum by a future event in a teleological explanation is not direct, but works by means of a causal chain of events lying between the explicandum and the goal. Even in intentional action the intention does not directly produce the goal: it starts a chain of action whose final stage is attainment of the goal. In non-intentional goal-directed action the goal-directedness consists simply in the fact that the causal chain in the organism goes in the direction of the goal, unless one wishes to suppose that there is always an extra 'something'—conatus or drive—involved in goal-directedness, an assumption which I do not wish to make. Thus the notion of causal chain is fundamental. Of course this notion is equally fundamental in the non-teleological explanations provided by the physical sciences, where the explaining cause is frequently given not as a preceding event continuous with the explicandum but as a preceding event connected with the explicandum by a causal chain. Let us approach our problem, therefore, by asking what (if any) is the peculiarity of the causal chains which are involved in teleological explanations.

Bertrand Russell, in his behaviouristic account of desire, approached our problem in the same way as I am doing by asserting that the peculiarity of teleological causal chains of actions is that they form 'behaviour-cycles'. But the only criterion he gave to enable us to pick out the behaviour-cycles from other repeated series of events in the life of an animal was that the final stage in a behaviour-cycle is "normally a condition of temporary quiescence".† He illustrated this by an animal falling asleep after

---

* Analyses of teleological explanations which have more or less resemblance to my analysis have been given by E. Rignano, *Mind*, n.s., vol. 40 (1931), p. 337; A. Rosenblueth, N. Wiener and J. Bigelow, *Philosophy of Science*, vol. 10 (1943), p. 24 ("Teleological behavior becomes synonymous with behavior controlled by negative feed-back"); L. von Bertalanffy, *British Journal for the Philosophy of Science*, vol. 1 (1950), p. 157. The field of study called by Wiener "cybernetics" is largely concerned with 'teleological mechanisms'.

† *The Analysis of Mind*, p. 65.

it has eaten. But temporary quiescence is quite inadequate to serve as the *differentia* for which we are seeking. After a bomb has exploded, or a volcano ceased to erupt, a state of temporary quiescence is attained. Here no teleology is concerned in the causal chains. It seems impossible to find any characteristic of the final state by itself of a teleological causal chain which is general enough to cover all the goals of goal-directed actions and yet specific enough to differentiate such actions from other repeated cycles of behaviour. It is necessary, I think, to look at the whole causal chain and not merely at its final state.

It seems to me that a distinguishing criterion can be found in one of the characteristics which biologists have emphasized in their descriptions of goal-directed behaviour, namely persistence towards the goal under varying conditions. To quote E. S. Russell: "Coming to a definite end or terminus is not *per se* distinctive of directive activity, for inorganic processes also move towards a natural terminus. ...What *is* distinctive is the active persistence of directive activity towards its goal, the use of alternative means towards the same end, the achievement of results in the face of difficulties".* Examples of the 'plasticity' of goal-directed behaviour will spring to every mind. To give one example only, Lashley's rats who had learnt to obtain their food by running his maze were still able to traverse the maze without false turns in order to obtain food after their powers of motor coordination had been seriously reduced by cerebellar operations, so that they could no longer run but could only crawl or lunge.† Plasticity is not in general a property of one teleological causal chain alone: it is a property of the organism with respect to a certain goal, namely that the organism can attain the same goal under different circumstances by alternative forms of activity making use frequently of different causal chains. Let us try to elucidate the logical and epistemological significance of this plasticity, in order to see whether it will serve our purpose of preserving the importance of teleological explanations without introducing extra-physical causation.

Consider a chain of events in a system *b*. The system may be a physical system of more or less complexity (a pilotless plane or an electron) or it may be an organic system (a complete organism

---

* *The Directiveness of Organic Activities*, p. 144.
† K. S. Lashley, *Brain Mechanisms and Intelligence* (Chicago, 1929).

or a relatively isolable part of a complete organism, e.g. the kidneys). Make the ordinary determination assumption that every event in the system is nomically determined by the whole previous state of the system together with the causally relevant factors in the system's environment or field (which will be called the "field-conditions"). Then the causal chain c of events in b throughout a period of time is nomically determined by the initial state e of the system together with the totality of field-conditions which affect the system with respect to the events in question during the period. Call this set of field-conditions f. Then, for a given system b with initial state e, c is a one-valued function of f; i.e. for given b and e, the causal chain c is uniquely determined by f—the set of field-conditions.

Now consider the property which a causal chain in a system may possess of ending in an event of type Γ without containing any other event of this type. Call this property the Γ-goal-attaining property, and the class of all causal chains having this property the Γ-goal-attaining class γ. Every causal chain which is a member of γ contains one and only one event of type Γ, and contains this as its final event.

Define the variancy ϕ with respect to a given system b with given initial state e, and to a given type of goal Γ, as the class of those sets of field-conditions which are such that every causal chain in b starting with e and determined by one of these sets is Γ-goal-attaining. To express this more shortly with the symbols already used, the variancy ϕ is defined as the class of those f's which uniquely determine those c's which are members of γ. According to this definition, to say that a causal chain c in a system b starting from a state e ends (without having previously passed through) a state of type Γ is logically equivalent to saying that the set of field-conditions is a member of ϕ. The variancy is thus (to repeat the definition in a looser form) the range of circumstances under which the system attains the goal.

The variancy ϕ defined in relation to b, e and Γ may have no members, in which case there is no nomically possible chain in b starting from e and attaining a goal of type Γ. Or ϕ may have one member, in which case there is exactly one such chain which is nomically possible. Here the system starting from e has no plasticity; there is only one set of field-conditions which, together with e, is nomically sufficient for the attainment of a goal of type Γ.

The case in which we are interested, in which the system has plasticity, occurs when the variancy ϕ has more than one member, so that the occurrence of any one of alternative sets of field-conditions is, together with e, sufficient for the attainment of a goal of type Γ. It is important to notice that the variancy may have many members and yet there be only one nomically possible chain: it is because the size of the variancy may be greater than the number of possible causal chains that the notion of the variancy has been introduced. For it may be the case that there are various sets of field-conditions each of which, together with e, determines exactly the same causal chain. This might happen if the ultimate causal laws concerned are such that each of the events in the chain might be determined by two or more alternative field-conditions. But it more frequently happens when the events in the chain are taken as being events which attribute properties to the system as a whole, and when, although alternative field-conditions determine different part-events in the system or in parts of the system, these part-events are causally so connected that the whole event determined by them remains unchanged. For example, if the causal chain of events with which we are concerned is the chain of body temperatures throughout a period of time of one of the higher animals, a change in the relevant environmental conditions (e.g. external temperature and available sources of food) will produce changes in the activities of the animal (both changes in its total behaviour, e.g. its feeding and migration habits, and changes in its parts, e.g. its sweat glands), yet these changes will be such as to compensate for the changed environmental conditions so that the animal's body temperature does not vary. Another example would be the path of a pilotless plane, in which the machine is fitted with 'feed-back' devices so designed that the plane will maintain a straight course at the correct height to the desired goal irrespective of the weather conditions it may encounter.

But usually when the variancy has more than one member, there is more than one nomically possible chain in the system in question which attains the required goal. An animal can move to get its food in many ways, a great variety of physiological processes can be called into play to repair damaged tissue, a bird can adapt its nest-building to the kind of material available. Nevertheless the essential feature, as I see it, about plasticity of behaviour is that the goal can be attained under a variety of circumstances, not that it can be attained by a variety of means. So it is the size of the variancy rather than the number of possible causal chains that is significant in analysing teleological explanation.

Let us now take the standpoint of epistemological or inductive logic and consider what are the types of situation in which we reasonably infer that there will be a goal-attaining chain of events in a system. To predict that, starting from an initial state e of a system b, there will be a causal chain which will attain a goal of type Γ is, by the definition of variancy, equivalent to predicting that the set of field-conditions which will occur will be a member of the variancy ϕ. So the reasonableness of the prediction depends upon the reasonableness of believing that ϕ is large enough to contain every set of field-conditions that is at all likely to occur.* Call the class of these sets of field-conditions ψ. For simplicity's sake I shall for the moment assume that we know that any set of field-conditions that will occur will be contained in ψ; that is, that the system will not in fact encounter a very unlikely environment (e.g. the next Ice Age starting suddenly to-morrow). Then the reasonableness of the prediction that the system will attain the goal depends upon the reasonableness of believing that ψ is included in ϕ.

Now there are two ways in which we may have derived our knowledge of the variancy ϕ. We may have deduced ϕ from knowledge of the relevant causal laws, or we may have inferred it inductively from knowledge of the sets of field-conditions under which similar causal chains had attained their goals in the past. In the first case, that in which the members of ϕ have been obtained by deduction, there are two interesting subcases in which we take positive steps to secure that ψ—the class of the sets of field-conditions at all likely to occur—is included in the variancy ϕ. The first subcase is that in which ϕ is small, but we deliberately arrange that ψ shall be smaller still. This happens when scientific demonstrations are performed for students in a laboratory, when elaborate precautions are taken (e.g. the experiment is done in a vacuum or distilled water is used) in order to eliminate unwanted relevant causal factors (air-currents or chemical impurities) and thus to secure that every set of conditions that may occur will fall

---

* The phrase "at all likely to occur" can be interpreted in different ways; but their difference does not affect the argument.

within the known variancy ϕ so that the demonstration will be a success. The second subcase is that in which ψ is large, but we deliberately arrange that ϕ shall be larger still. This happens when a machine is deliberately designed to work under a large variety of conditions. This object may be achieved by using suitable materials: a motor-car is built to stand up to a lot of rough and careless treatment. Or it may be achieved by incorporating in the machine special self-regulating devices to ensure that the machine adjusts its method of working according to the conditions it encounters, as in the pilotless plane.

When our knowledge of the relevant variancy has been obtained by deduction from previous knowledge of the causal laws concerned, a teleological explanation of an event in terms of its ' goal-directedness is felt to be almost valueless.* For in this case, that the causal chain which will occur will lead to the goal—the 'teleology' of the system—has been calculated from its 'mechanism'. To give a teleological answer to the 'Why?' question would require forming (and suppressing) an ordinary causal answer, which would (if expressed) have given intellectual satisfaction to the questioner, in order to deduce from it a teleological answer. This would be an unprofitable, and indeed disingenuous, way of answering his question.

The situation is entirely different when our knowledge or reasonable belief about the variancy ϕ has not been derived from knowledge of the causal laws concerned. In this case our knowledge as to what sets of conditions make up the variancy has been obtained either directly by induction from previous experience of goal-attaining behaviour that was similar to the behaviour with which we are concerned, or indirectly by deduction from general teleological propositions which have themselves been established by induction from past experience. Neither of these ways makes use of laws about the mechanisms of the causal chains. The variancy ϕ is inferred—inductively inferred—from knowledge of classes similar to ψ; that is, from past observation of the conditions

---

* In the case of a machine or of a laboratory demonstration a teleological explanation can of course be given of the action of a man in starting and controlling the machine or demonstration. Derivatively we can apply such a teleological explanation (as a 'transferred epithet') to the working of the machine itself. But these explanations are all in terms of intentions as efficient causes, and so do not raise the special problem with which we are here concerned.

under which similar teleological behaviour has taken place. For example, my knowledge of the conditions under which a swallow will migrate is derived from knowledge about past migrations of swallows and of other migrants, fortified perhaps by general teleological propositions which I accept about the external conditions for self-preservation or the survival of the species, themselves derived inductively from past experience.

It is when our knowledge of the relevant variancy has been obtained independently of any knowledge of the causal laws concerned that a teleological explanation is valuable. For in this case we are unable, through ignorance of the causal laws, to infer the future behaviour of the system from our knowledge of the causal laws; but we are able to make such an inference from knowledge of how similar systems have behaved in the past.

It should be noted that in all cases of teleological explanation of a present event by a future event, whether reducible or irreducible, inductive inferences occur at two stages of the argument. One stage is in the inference of the variancy, whether this itself is obtained inductively, or whether it is obtained deductively from causal laws or teleological generalizations which have themselves been established inductively. The other inductive stage in the argument is the inference that the set of relevant conditions that will in fact occur in the future will fall within the variancy. Every teleological answer, however reasonable, may be mistaken in each of these two ways.

But in general irreducible teleological explanations are no less worthy of credence than ordinary causal explanations. A teleological explanation of a particular event is intellectually valuable if it cannot be deduced from known causal laws: other things being equal, it is the more valuable the wider the variancy of the conditions, and hence the greater the plasticity of the behaviour concerned. It is because we are acquainted with systems—organisms and parts of organisms—which exhibit great plasticity that we make use of teleological explanations. Such an explanation may be regarded as merely another way of stating the fact of the plasticity of the goal-directed behaviour. But to state this fact is to bring the explicandum under a general category; moreover it enables us to make reliable predictions as to how the system will behave in the future. It seems ridiculous to deny the title of explanation to a statement which performs both of the functions characteristic of scientific explanations—of enabling us to appreciate connexions and to predict the future.

The analysis which has here been given of teleological explanation of non-intentional goal-directed activities supposes that the goal to which the activity is directed is later in time than the action (this indeed creates the philosophical problem) and makes great use of the notion of causal chain. It has been objected that this analysis will not cover the case of explanations of biological facts which are given in terms, not of a future goal, but of a biological end which is as much present as future; but that this case, as well as that in which the explanation is in terms of a future goal, will be covered by the more general notion of *functional explanation* in which the explanation is in terms of another part of a whole of which the explicandum is a part.* But the questions which seem to call for a more general functional explanation rather than for a causal-chain teleological explanation turn out on examination to be ambiguous questions. If a physiologist is asked why the heart beats, he may take this question as a request for an explanation of a particular fact, the beating of a particular heart on a particular occasion, in which case the explanation "In order to circulate the blood round the body" will also refer to the movement of the blood in a particular body on a particular occasion. But, on this interpretation, the particular movement of blood outside the heart due to a particular beating of the heart is an event whose beginning is later in time than the beginning of the event which is the heart's beating: the latter event is connected with the former event by a causal chain of events, and is a teleological explanation of it in terms of a future goal. But the physiologist may more naturally take the question, not to be a question about one particular heart on one particular occasion, but to be a question about all beatings

* Jonathan Cohen, *Proceedings of the Aristotelian Society*, n.s., vol. 51 (1950–1), pp. 270, 292. Cohen holds that "a functional explanation asserts the explanandum to be [a] necessary condition (logically, causally or in any other generally recogr ised way) of the explanans and thereby also of the persistence under varying circumstances of a whole of which both explanans and the explanandum are parts" (loc. cit. p. 292). But the beating of my heart is not a *necessary* condition for the circulation of my blood: it is only because my anatomy includes a heart but no other mechanism for circulating my blood that it is causally necessary that the heart should *beat* in order that the blood should circulate.

of all hearts (or of all human hearts, or of all mammalian hearts, or etc.), in which case the question is a request for a teleological generalization of which the particular teleological explanation of the beating of one particular heart on one particular occasion would be an instance. In both cases, however, the explanation would be in terms of goal-directed activities with future goals. The peculiarity of a biological end is that it is a permanent goal; at all times during the life of the organism there are activities of the organism to be explained in terms of the biological end. My heart's beating at one moment is responsible for the circulation of my blood a short time afterwards; and my heart will have to *continue* beating for my blood to *continue* circulating. The teleological generalization of which the particular teleological explanation of the beating of my heart on a particular occasion is an instance will have instances at every moment of my life, unlike teleological laws concerning goals upon the attainment of which the animal sinks into a 'temporary quiescence'.*

TELEOLOGICAL LAWS

We have referred in several places to teleological generalizations. Just as particular causal explanations are instances of causal propositions, of more or less generality, so particular teleological explanations are instances of teleological generalizations of more or less generality; that is, they (if true) are instances of laws according to which an event of a certain sort in a system of a certain sort is nomically determined by a later event of a certain sort in the same system. Such a teleological law will be valuable as an explanation if it has not been deduced from non-teleological laws, and it will be the more valuable both intellectually and predictively the wider the range of the variancy associated with it.

The special philosophical difficulty about teleological, as contrasted with causal, explanations of particular events—namely, that in them the present appears to be determined by the future—does not arise in the case of teleological as contrasted with non-teleological laws, considered as laws of nature without regard to their

---

* The term "functional explanation" may sometimes also be used to cover a mere description of the *modus operandi* of an organ like the heart. This would correspond to J. H. Woodger's third sense of 'function' (*Biological Principles* (London, 1929), p. 327).

applications to yield particular explanations. For many non-teleological laws of nature, e.g. Newton's laws of mechanics, are symmetrical with respect to the earlier and later times occurring in the laws: they state that the present is determined by the future just as much as it is determined by the past. Nor do teleological laws in general differ from non-teleological ones in having a time-interval between the two related events: many non-teleological laws are about what happens during a period of time taken as a whole, e.g. the Law of Least Action. The difference between the two types of law seems to consist simply in the way in which the related variancy is discovered.

Here a comparison may be made with another type of law of a somewhat peculiar nature which occurs in psychology and in biology, and which shares with the teleological type the two characteristics that there is an interval of time between the determining and the determined event, and that the law holds under a wide variety of conditions which have been discovered inductively and not deductively. I refer to the laws governing what Bertrand Russell called "mnemic phenomena", laws which he called "mnemic laws".* The simplest example of such a law is that of memory recall, in which a present memory-image is determined (partially determined) by the occurrence in the rememberer of an experience of which the present memory-image is an image; but there are plenty of non-psychological examples in biology. The Mendelian laws of heredity state that sometimes some of the present characteristics of an organism are determined very precisely by the characteristics of its parent or parents at the time when the reproduction process commenced. More frequently the Mendelian determination is only statistical. In all the mnemic laws an earlier event is said to determine a later event without the intervening causal chain being specified or indeed known. We may postulate, if we wish, persistent genes to explain the facts of heredity, and traces in the brain or unconscious ideas in the mind to explain memory; but these are extra explanatory hypotheses going beyond what the mnemic law itself states. Or we may follow Russell's suggestion of supposing that there is a type of causation (which he called "ultimate mnemic causation") in which a past event directly determines a future event without there being any

* *The Analysis of Mind*, pp. 77 ff.

intermediate causal chain.* To suppose this would be almost as alien to our usual ways of thinking as to suppose that the future goal directly determines the present goal-directed action; and I agree with Russell in being unprepared to accept it if any way of escape is possible. As it is, physiologists are in the process of discovering strong independent evidence for the existence of the genes which the geneticists postulate; and the neurologists or the experimental and clinical psychologists may in time discover satisfactory independent evidence for cerebral traces or for a persistent Unconscious. But in the meantime we have our mnemic laws; and the best account of them seems to me to be given by treating them in exactly the same way as teleological laws and making the inductively inferred variancy the distinguishing feature. This variancy is frequently large in the case of mnemic laws as it is in teleological: Lashley's rats retained their acquired skill in running his maze after large, and different, portions of their brains had been removed.

Both teleological and mnemic laws, then, assert that there is a causal chain connecting the determining and the determined events which holds under a wide range of conditions, i.e. that the system in question has a large variancy; and this variancy or plasticity has not been deduced from non-teleological or non-mnemic laws but has been established inductively by observation. The difference between them, that in a teleological law the determining event succeeds the determined event whereas in a mnemic law it precedes it, seems unimportant in comparison with their similarity; and I shall therefore class both types together under what I will for the lack of a better name call "biotic laws". I choose this name because the biotic laws which have struck our attention are those which apply to living systems, but my definition of "biotic law" includes no reference to life. Sometimes both a mnemic and a teleological law can be subsumed under one more general biotic law which is itself both mnemic and teleological: Mendel's laws of heredity state that characteristics of a set of organisms both are statistically determined by those of its set of parents and also statistically determine those of its set of offspring.†

---

* Ultimate mnemic causation would satisfy criterion II of the last chapter (p. 309), but no stronger condition.

† Rignano's inclusion of the 'finalist manifestations of life' in a 'mnemonic property' goes far beyond my simple comparison.

This general notion of biotic law, of which teleological law is a species, is therefore offered as an attempt to settle the dispute between the biological 'mechanists' and the biological 'teleologists'. But I fear that it will satisfy neither party. The teleologist will say that the whole account of teleological law in terms of causal chains and variancy of conditions presupposes the mechanist assumption that every event is physico-chemically determined, and that to admit teleological explanations *faute de mieux* is to ignore the essentially irreducible character of teleological law for which he is contending. The mechanist will declare that he has no use for teleological laws unless they are ultimate and irreducible; and that it is methodologically vicious to introduce new types of law just because we do not know all the laws of nature of the ordinary type. And both parties will join forces in criticizing my treatment as being unduly epistemological: the controversy, they will both say, is not as to how we derive our knowledge of general propositions about goal-directed activity, but is about the content of these general propositions; it is a question of the ultimate elements in the biological facts, not of the organization of our present biological knowledge.

All these criticisms, and the joint one particularly, are based upon what must be regarded as a naïve attitude to the function of a scientific law. For this function is just exactly that of organizing our empirical knowledge so as to give both intellectual satisfaction and power to predict the unknown. The nature of scientific laws cannot be treated independently of their function within a deductive system. The world is not made up of empirical facts with the addition of the laws of nature: what we call the laws of nature are conceptual devices by which we organize our empirical knowledge and predict the future. From this point of view any general hypothesis whose consequences are confirmed by experience is a valuable intellectual device; and the profitable use of such a hypothesis does not presuppose that it will not at some future time be subsumed under some more general hypothesis in a more widely applicable deductive system, nor that the facts which it explains will not some time be explicable by a quite different hypothesis in another deductive system.

Biotic hypotheses behave exactly like other scientific hypotheses in that they can frequently be treated as lower-level hypotheses in a new deductive system in which they are deducible from a set of higher-level hypotheses. For example, the special teleological law about a particular food (e.g. grass) as goal and a particular species of animal (e.g. horses) is deducible from a less special teleological hypothesis about food-seeking in general together with biochemical hypotheses about the conditions for the digestibility of grass. Frequently one goal-directed activity in an animal (e.g. the building of a nest) is followed by another type (e.g. sitting on the eggs laid in the nest); and the succession of these two types of teleological activity falls under some general teleological law (e.g. the mode of propagation of the species). Discussions as to the proper classification of instincts are largely discussions as to which is the best general deductive system containing higher-level biotic hypotheses for explaining the special instinctive modes of behaviour. When E. S. Russell puts forward the generalization that "the goal of a directive action or series of actions is normally related to one or other of the main biological ends of maintenance, development and reproduction",* he is suggesting that a deductive system whose highest-level hypotheses include teleological hypotheses about these three ends will be able to absorb all the systems which are in terms of particular goals. Of course none of these more elaborate deductive systems have been worked out in detail; but the possibility of constructing them makes teleological explanatory hypotheses like the non-teleological ones in another respect also, namely, that we can hope to provide further explanations, and a deeper intellectual satisfaction, by incorporating special laws in a unified system.

I have given as part of the definition of a biotic law that it is not incorporated in a physico-chemical deductive system.† But if such a law, already incorporated in a biotic deductive system (i.e. one with biotic laws among its highest-level hypotheses), were to be found capable of physico-chemical explanation, we should not by that mere fact be estopped from continuing to make use of its place in the biotic system whenever we found it profitable to think in this way. The chemical deductive system with Dalton's laws of atomic combination as highest-level hypotheses has in recent years

* *The Directiveness of Organic Activities*, p. 80.
† By this phrase is meant a deductive system whose highest-level hypotheses are physical or chemical.

been included more and more within the more general deductive system of physics; but chemists find it far more convenient to treat most of their problems in terms of atoms and molecules than in terms of electrons or wave-functions. And there is one feature of both teleological and mnemic explanations which will almost certainly make them continue to be useful (whatever they would then be called) even if they could be superseded by physico-chemical ones. This feature is that usually teleological explanations make no reference to the exact length of time taken in attaining the goal, and mnemic explanations no reference to the exact length of time since the determining event. Indeed the unimportance of the time taken in reaching the goal is implicit in the persistency feature of goal-directed activity emphasized by biologists.* Teleo-logical explanations do not specify the length of the causal chain, only that it attains the goal. So even if the biological mechanists can provide us with a complete explanation of life in physico-chemical terms, we shall probably continue to give teleological (or what would previously have been called teleological) explanations when-ever the exact time taken in reaching the goal does not interest us.

I will conclude this chapter by summarizing the biological part of the argument. An account has been given of the distinguishing feature of teleological explanations which does not assume that such explanations are ultimately irreducible to chemistry and physics and which does not require any novel concept of causal law. To do this I have followed biologists in emphasizing the plasticity of goal-directed behaviour, and have analysed the peculi-arity of a teleological explanation in terms of the related notions of the multiplicity of the causal chains by which the goal may be attained and of the variety of conditions under which the goal-directed activity may occur. These notions have been found to be also involved in mnemic explanation. What has been drawn is only an outline sketch which will need much working upon to make a convincing picture. But I have done enough to convince myself that what I have been trying to do is possible; and that the realm of biology will not have to sacrifice the autonomy proper to it if physics should succeed in establishing the claim, made by many biologists on its behalf, to be the Emperor of all the Natural Sciences.

* For example, E. S. Russell, *The Directiveness of Organic Activities*, p. 110: "If the goal is not reached, action usually persists."

CHAPTER XI

EXPLANATION OF SCIENTIFIC LAWS

The last chapter has described how scientific explanations are given, either in terms of causes or in terms of teleological goals; such an explanation always involves a reference, explicit or implicit, to a scientific law. We must now discuss the type of explanation demanded in asking "Why?" of a scientific law. This may well be the second-stage explanation requested by someone who has already been given the cause, or the goal, of a particular event, but asks the reason why it is that the cause causes the event or that the goal is the goal for the particular event. In this case he is asking for an explanation of the law, causal or teleological, reference to which, either explicit or implicit, has been made in the first-stage explanation he has already been given.

A 'Why?' question asked of a scientific law may be a request for several different sorts of answers. It may be a request for the reasons for believing the law, in which case either a recital of the empirical evidence or a justification for the inductive policy used in establishing the law on the basis of that evidence may be a satisfying answer. The question might even be a request for the psychological or sociological cause for the general acceptance of the law, and thus not a request for a reason at all; in this case the question would be better expressed in the form "What makes people believe the scientific law?" All these requests are covered by what has already been said. The question which raises a new problem is the request, neither for a cause nor for a reason for the acceptance of the law, but for a reason for the law itself. That this is what is required comes out most clearly if the question is put in the expanded form "Why is the scientific law in question what it is?" Why, for example, is the law of gravitation an inverse square law rather than an inverse cube law?

The way in which I shall answer this question will be no surprise to a reader who has got so far in this book. The question will be answered by citing an established deductive system in which the law in question is a lower-level hypothesis deducible within that system from a set of higher-level laws. To explain a law is to exhibit an established set of hypotheses from which the law follows. It is not necessary for these higher-level hypotheses to be established independently of the law which they explain; all that is required for them to provide an explanation is that they should be regarded as established and that the law should logically follow from them. It is scarcely too much to say that this is the whole truth about the explanation of scientific laws; the rest of this chapter will be mostly of the nature of a commentary on this truth.

ALTERNATIVE EXPLANATIONS

Let us first remark that the fact that a scientific generalization can be explained as a consequence of one set of higher-level laws does not exclude its also being explicable as a consequence of another set of higher-level laws. The two explanations would only be incompatible if the two sets of higher-level laws were mutually inconsistent. But if one of these sets contains theoretical concepts, or if both sets contain theoretical concepts but used within different deductive systems, the two sets of higher-level laws will not be incompatible with one another. As was shown in the last chapter, a teleological explanation of a particular fact is in no way inconsistent with a causal explanation of the same fact. Similarly, the explanation of an established generalization by subsuming it under a teleological law does not exclude another explanation of it by subsuming it under a non-teleological law. To accept two first-stage explanations of the same lower-level law at the same time is to regard the law as being incorporated in each of two established scientific deductive systems which work with different sorts of concept.

An example of a preparedness to use double explanations is to be found in the work of those psychologists and psychiatrists who, following Freud, would explain established generalizations about some kinds of human behaviour by hypotheses which make use of theoretical concepts such as 'unconscious wishes'. The best of these (including Freud himself) insist that to give an explanation in terms of processes in an 'Unconscious' does not preclude the possibility of also giving an explanation in terms of physico-chemical processes in the brain (or, more generally, in the body). What they claim is that, although they believe, with the physiologists, that such a physico-chemical explanation may well be discovered in the future, at the present time, with our ignorance of the detailed working of the brain, an explanation in terms of an Unconscious is more profitable in enabling predictions to be made. Some rash psychologists, in their devotion to such a psychological explanation, have thought it necessary to deny in advance the possibility of a physico chemical explanation. On the other hand, many neurologists have rejected out of hand any explanation in terms of unconscious states of mind on the ground that these are unobservable and are therefore mythical entities, without appearing to realize that, if such entities occur as theoretical concepts in the highest-level hypotheses of a scientific deductive system conclusions from which are empirically confirmed, they have exactly the same epistemological status as the electrical and chemical concepts in terms of which the neurologists would wish to give their physico-chemical explanation. Both these negative attitudes are unjustified. To give one type of explanation does not preclude also giving another type—provided that the two explanatory hypotheses would not be inconsistent were they to occur in the same deductive system. Two alternative explanations are only *exclusively* alternative if they are incompatible.*

When two alternative explanations of the same lowest-level law, or set of laws, are incompatible, no hard and fast principles can be laid down for preferring one to the other. Generally speaking, of two such explanations that explanation will be preferred which, besides covering the laws it was put forward to explain, also covers the greater or more important range of other established lowest-level laws. If two explanatory hypotheses each cover the same range of established laws, or cover equally important ranges of

---

* C. D. Broad, in illustration of a "*prima facie* axiom about causation", writes: "Surely it is self-evidently absurd to say that this mental change is a total cause of this movement, *and* that this bodily change is *also* a *total* cause of this same bodily movement"; and he uses the allegedly self-evident uniqueness of a total cause as an argument against the Humean view of causality (*Examination of McTaggart's Philosophy*, vol. 1 (Cambridge, 1933), pp. 232 ff.). Broad's 'total cause' is, of course, taken as not including a causal ancestor. But it seems to me a mere prejudice to rule out the possibility of two compatibly alternative causal explanations of a particular piece of behaviour of mine, such as a slip of the tongue—one in terms of events in my brain, the other in terms of processes in my 'Unconscious'.

established laws, but one covers ranges of generalization not yet empirically tested which the other does not cover, while it would be foolish to regard the wider hypothesis as the better until its further consequences have been tested and confirmed, yet it can be regarded as being more fertile in that it covers more generalizations which might be established in the future. And if these new generalizations had not been thought of until they had been seen to be consequences of the explanatory hypothesis (as was the case with me in my construction of the three-factor theory of Chapter III), the value of considering the explanatory hypothesis would not be entirely wiped out if these consequences were found to be refuted by experience, though the explanation will of course have to be rejected as giving rise to conclusions known to be false.

If two explanations both cover exactly the same range of lowest-level generalizations, whether established or not, the deductive systems in which they occur will be *empirically equivalent* in the sense of Chapter IV (p. 97). In this case there will be nothing to choose between the two explanations except in so far as one of the explanatory sets of hypotheses, but not the other, admits of a possibility of extension by its incorporation into a set of hypotheses forming a theory with a wider range, in the manner illustrated by the example given on p. 97. Moreover, two empirically equivalent explanatory hypotheses will only be incompatible in that they use their theoretical terms in different and incompatible ways; with adequate self-consciousness about what is happening, it will be possible to accept both hypotheses, i.e. to think by means of both deductive systems, at the same time.

EXPLANATIONS IN AN ASCENDING HIERARCHY

These comments on the relative merits of two alternative explanatory hypotheses have omitted one highly relevant consideration, namely, as to whether or not the explanatory hypothesis, which stands as a highest-level hypothesis in the deductive system whose lowest-level hypotheses are the established generalizations to be explained, can itself be explained as being a consequence in a more comprehensive deductive system of more general laws. Other things being equal, in order that we should regard an explanation which can be thus incorporated in a more comprehensive deductive system as better than one which cannot, it is not necessary that the better explanatory hypothesis should receive independent indirect confirmation through being deducible from a more general law which independently is directly confirmed; it is enough that there should be some explanation of it in the form of a more general law, or set of laws, from which it can be deduced —even if the evidence for believing the more general law is no more than that for believing the hypothesis that it explains. Just as to subsume a generalization under a higher-level law raises the intellectual status of the generalization by providing some explanation for it (so that, as was maintained in Chapter IX, it may be proper to call it a "natural law"), so the subsumption of this higher-level law under a still-higher-level law raises its intellectual status, even if the evidence supporting all three propositions is exactly the same. In order that the propositions should form an ascending hierarchy they must, of course, cover a wider and wider range of possible experience, so that a higher-level hypothesis could be refuted by observations which would not refute a hypothesis standing at a lower level. But, provided that there is no evidence refuting the highest-level hypothesis, the evidence establishing it need be no more than that establishing a lower-level hypothesis for it to be regarded as providing an explanation for the latter and as thereby raising the latter's intellectual status in relation to comparable hypotheses for which we can provide no such explanation.

To one who was studying physics when Einstein's General Theory of Relativity of 1915, and its confirmation by the eclipse expeditions of 1919, burst upon the learned world, the paradigm of explanation of a law will always be the explanation of gravitation provided by this theory. Newton explained the moon's revolution round the earth by subsuming it (together with other phenomena) under his law of gravitation; Einstein explained Newton's law of gravitation by showing that it was a consequence (an approximate consequence) of a law concerning the structure of space and time. The only way in which Einstein's theory can be said to have superseded Newton's was that in which, where their consequences differed (as about the motion of Mercury's perihelion), the evidence showed that Newton's law was only true as a very near approximation. In every other way Einstein's theory gave powerful support to the inverse-square law of gravitation. What had previously appeared as a 'historical accident on the cosmic scale' disconnected from other aspects of physics was now seen to be intimately connected with the framework of space and time within which all physical phenomena take place.

To explain a law, as we have seen, is to incorporate it in an established deductive system in which it is deducible from higher-level laws. To explain these higher-level laws is similarly to incorporate them, and the deductive system in which they serve as premisses, in an established deductive system which is more comprehensive and in which these laws appear as conclusions. To explain the still-higher-level laws serving as premisses in this more comprehensive deductive system will require their deduction from laws at a still higher level in a still more comprehensive system. At each stage of explanation a 'Why?' question can significantly be asked of the explanatory hypotheses; there is no ultimate end to the hierarchy of scientific explanation, and thus no completely final explanation.

IS SCIENTIFIC EXPLANATION DEFECTIVE?

Some philosophers have taken the fact that, whenever a higher-level law is presented as an explanation of a lower-level law, an explanation can be demanded of this higher-level law itself, as indicating that there is something inherently defective in the process of scientific explanation. The demand for explanation, they would say, is the demand for an ultimate reason; all that subsumption of a lower-level law under a higher-level law can do is to prove that, if a reason could be found for the higher-level law, this reason would also provide an explanation of the lower-level law. But until a genuine reason can be produced for the higher-level law, there has been no explanation in a sense which should satisfy a philosopher. Einstein's General Theory of Relativity, the critics would say, throws a great deal of light upon *how* gravitation works, but it tells us nothing of *why* it works. For it has only converted the question as to why the law of gravitation is very approximately an inverse-square law into the question as to why the structure of physical space-time possesses a Riemannian geometry. And it will be no use to explain this fact, as may well be done in the future, by subsuming it under some still more general fact, if this still more general fact again calls for an explana-tion. The only way in which the law of gravitation could properly be explained would be if an ultimate reason for it could be found which required no further explanation; until this can be done, gravitation will remain inexplicable and 'irrational'.

This criticism of scientific explanation is related to what has been called the "descriptive" view of science. It is sometimes said, by those who wish to compliment science for sticking close to facts as well as by those who wish to depreciate it, that science provides descriptions of facts but never explanations of facts, and that the questions which science answers are essentially 'How?' questions and only superficially 'Why?' questions. Now there is no possible objection to calling scientific laws descriptions of the course of experience, or to saying that they describe how Nature behaves—provided that such a way of speaking is not taken as implying that the laws have no other function. For the most important fact about our acceptance of a scientific law is that of enabling us to make reliable predictions, and this predictive function of a scientific law would be ignored if the function of the law were taken as being purely descriptive. In its origin the descriptive view of science was closely associated with the views of philosophers of science like Mach who wished to emphasize the intellectual function of a scientific law in enabling us to effect an 'economy of thought' by subsuming a number of particular instances under a general law. A scientific law, however, is of unlimited generality; it not only covers instances that have been observed, but an unlimited number of possible instances that have not been observed. To give an account of the function in our thinking of believing a scientific law as being that of enabling us to economize thought is grossly to under-estimate the importance for us of such a belief. Similarly, to describe this function as that of describing the course of experience is dangerously deceptive unless it is clearly realized that the description is that of the future as well as of the past, of the unobserved as well as of the observed, and that to call it description does not preclude its also being properly called explanation.*

For an explanation, as I understand the use of the word, is an answer to a 'Why?' question which gives some intellectual satis-

---

* It would be particularly inappropriate to call a law containing theoretical concepts merely a description, since theoretical concepts cannot, by the argument of Chapter III, be regarded as 'logical constructions' out of observable entities.

faction. And it is absurd to suppose that the scientific explanation of a lower-level law by deducing it from a higher-level law does not give some intellectual satisfaction even if no explanation is known of the higher-level law itself. I count my appreciation of Einstein's explanation of the law of gravitation as one of the keenest intellectual pleasures of my life. Any incorporation of a fact—be it a particular instance of a law or the law itself—into a deductive system in which it appears as a conclusion from other known laws is, by virtue of that incorporation, an explanation of that fact or law. Whether we take the incorporation as being an answer to a 'How?' or to a 'Why?' question is immaterial; what matters is that we know more than we did before of the connectedness of the fact or law with more fundamental laws covering a wider range. We have not only attained more knowledge of the inter-connectedness of Nature, but we have also acquired the possibility of a power of making predictions that was not open to us before. To deny that this achievement should give us intellectual satisfaction is to forbid us to glory in one of the excellences of the human intellect.

Our critic may retort that, since the scientific explanation is never an ultimate one, the intellectual satisfaction is incomplete and imperfect as contrasted with the perfect satisfaction to be provided by a complete and final explanation. But what sort of complete explanation could there be of a scientific law? What sort of answer to a 'Why?' question asked of a scientific law could there be which did not permit of a similar 'Why?' question being asked of the answer itself?

It seems to me that philosophers who have desired that a final explanation should be given of a scientific law have been thinking of scientific laws either as being analogous to the laws of a legal system or as being analogous to mathematical theorems. The laws of a legal system are explained in terms of the fact that they have been enacted by a lawgiver, either an individual or a society. If the legal system is one in which lower-level laws are deducible from certain general principles, as in the Code Napoléon, the lower-level laws will be explained in terms of these general principles, which themselves will be explained in terms of their enactment. That the particular laws, or the general principles from which they follow, have been enacted is a complete explanation in the sense that no further question can be asked of the same sort. We can ask what was the cause of the enactment, or of the acceptance, of a code of laws; but this is a question of a different sort—a demand for a scientific explanation, in fact; and the possibility of asking it does not prevent the explanation in terms of the enactment of the code of laws from being a complete explanation of the laws in the code being what they are.

But to consider the reasons for scientific laws being what they are as analogous to the reasons for legal laws being what they are is an entirely false analogy. Scientific laws have neither been enacted by a society nor are they consequences of general principles which have been enacted by a society. One form of the theistic hypothesis would posit a divine lawgiver who had enacted the laws of Nature; though acceptance of this hypothesis would provide a complete explanation of scientific laws, it would itself raise the questions as to why the divine lawgiver had enacted just these laws and as to why Nature obeyed the laws thus enacted, questions which would themselves require answers which would be scientific explanations and hence essentially incomplete.

MATHEMATICAL AND SCIENTIFIC DEDUCTIVE SYSTEMS

However, the consideration that has been most responsible in the history of modern thought for the demand for an ultimate explanation of the laws of Nature has been the alleged analogy, not with the laws promulgated by a lawgiver, but with the logically necessary propositions of pure mathematics. Here there is a genuine and important resemblance. Just as scientific laws of increasing generality can frequently be arranged in a deductive system in which less general laws are consequences of sets of more general laws, so the theorems of a branch of mathematics can be arranged in a deductive system in which less fundamental theorems follow from a set of more fundamental theorems. As we saw in Chapter II, a scientific deductive system and a mathematical deductive system can be represented by the same calculus, the formulae of which, interpreted in one way, stand for the empirical laws of a science, while, interpreted in another way, they stand for the theorems of a branch of pure mathematics. Moreover, the two deductive systems, scientific and mathematical, represented both by the same calculus, will use exactly the same principles of deduction to deduce the later propositions from the earlier ones. Thus there may be a complete structural similarity between a scientific and a pure deductive system; a person opening the pages of a systematic treatise in the 'Mathematics and Physics' section of a bookshop will not be able to tell, from a glance at the sequence of formulae, whether the treatise is about physics or is about pure mathematics.

A mathematical deduction starting from first principles provides complete intellectual satisfaction. The complete explanation of a theorem in the Theory of Numbers, for example, is that the theorem is deducible from the fundamental laws of arithmetic. This complete satisfaction is produced because these laws are truisms; they are seen to be logically necessary and there is no question of testing them by experience. In common with many contemporary philosophers I believe that the truth of such logically necessary propositions arises, not from any fact about the world, but from the way in which we use language and other techniques for thinking about the world, and, in particular, for deducing contingent propositions from other contingent propositions. But this controversial thesis is not the subject of this book; what is relevant for our purpose is that the propositions of pure mathematics should be recognized, as they have been by almost all philosophers except extreme empiricists like Mill, to have a different epistemological status from that of empirical propositions so that our grounds for believing them are essentially different. My ground for believing in the first principles of arithmetic which stand at the head of the deductive system of the Theory of Numbers does not depend upon my ground for believing the surprising theorems (e.g. Lagrange's theorem that every non-negative integer is the sum of the squares of four integers) which appear a long way lower down as deduced theorems in that system. On the contrary, my ground for believing these surprising theorems is solely that they are logical consequences of the elementary laws of arithmetic, so that it would be inconsistent for me to believe the latter without also believing the former. But in the case of a scientific deductive system whose initial propositions are all contingent ones (an 'impure deductive system' in the sense of Chapter II), although the later propositions are logical consequences of the earlier ones, yet the grounds for believing the later propositions are not that they are deducible from the earlier; on the contrary, the grounds for believing the earlier more general propositions are that lowest-level propositions about observable events are deducible from them. Of course, the ground for believing any one scientific hypothesis may well be only the indirect support given to it by its deducibility from a higher-level hypothesis which is directly supported in an independent way. But, taking all the highest-level hypotheses of a scientific system together, the grounds for believing them are no more and no less than the fact that the lowest-level hypotheses deduced from them are confirmed by experience. Thus there is an essential difference in the way in which we think by means of a mathematical and by means of a scientific deductive system. In the former we start from the beginning and go on to the end, both logically and epistemologically; in the latter we start from the beginning and go on to the end only logically, the epistemological order being from the end to the beginning. To use again the metaphor of the zip fastener, the truth-value of truth (i.e. of formal truth) for mathematical propositions is assigned first at the top and then by working downwards, in a scientific system the truth-value of truth (i.e. of conformity with experience) is assigned at the bottom first and then by working upwards.*

The fact that the notion of a deductive system was introduced into European thought in a mathematical context—Euclid's *Elements*, in which geometrical theorems are proved by being deduced from a set of first principles ('definitions', 'postulates' and 'axioms' or 'common notions')—has had the effect that the first explicitly deductive scientific systems, including the greatest of them, Newton's *Principia*, were modelled on the Euclidean analogy and professed to prove their later propositions—those which were confirmed by confrontation with experience—by deducing

---

* For simplicity's sake I have considered mathematical deductive systems like that of Higher Arithmetic (the Theory of Numbers) in which the theorems are not all intuitively obvious, whereas the premises are the intuitively obvious first principles of arithmetic. Mathematical logicians during the last century have been concerned to develop this arithmetical deductive system backwards by finding more 'primitive' propositions—which, in Whitehead and Russell's *Principia Mathematica*, are logical propositions which cannot be called arithmetical propositions at all—from which the first principles of arithmetic themselves logically follow. Such propositions are, of course, not more intuitively obvious than the arithmetical first principles deducible from them; the purpose of the construction of a deductive system in which the arithmetical principles appear as consequences is not to strengthen our belief in them but to determine their 'logical geography' by showing their relationships with formally true propositions of other kinds.

them from original first principles.* And it took a long time for scientists to realize that the hypothetico-deductive inductive method of science was epistemologically different from the prima facie similar deductive method of mathematics; and that, in properly imitating the deductive form of Euclid's system, they were not *ipso facto* taking over his deductive method of proof. The enormous influence of Euclid has been so good in inducing scientists to construct deductive systems as more than to counterbalance his bad influence in causing them to misunderstand what they were doing in constructing such systems; the good genius of mathematics and of unself-conscious science, Euclid has been the evil genius of philosophy of science—and indeed of metaphysics.

The irreducible difference between the propositions of logic and mathematics and those of a natural science are that the former are logically necessary and the latter logically contingent. The demand for a final explanation of a scientific law, to the extent that it is based upon the fact that a pure and an applied deductive system can both be represented by the same calculus, ignores the essential difference in the interpretations of the formulae in the calculus. We can ask for a reason for a logically necessary proposition being true, in the sense of asking why that proposition rather than some incompatibly alternative proposition is the formally true one; and such a reason, when it can be provided, gives complete intellectual satisfaction.† But it does not make sense to ask in the same way for a reason for one logically contingent proposition (such as a scientific hypothesis) rather than an incompatible alternative being the true one, for it is experience which decides. The grounds for believing in the hypothesis rather than in some alternative are, not any logical reason or any fact about the use of language, but the

* Newton starts with eight *definitiones* and three *axiomata, sive leges motus.*
† The simplest case is that of what Mill called "verbal propositions", e.g. «Every brother is male», where complete intellectual satisfaction is given by the explanation that the application of the word "brother" is, by the conventions of the English language, limited so that the word is applicable only to members of a subclass of the class of male persons. The next simplest case is that of what Wittgenstein called "tautologies", e.g. «Either it is the case that I now have a pencil in my hand or it is not the case that I now have a pencil in my hand», where this statement places no restriction upon the possible truth-values (truth, falsity) of the proposition «I now have a pencil in my hand». The exact way in which arithmetical propositions are logically necessary is a matter of high controversy among mathematical logicians.

logically contingent fact that it, or a deductive system in which it is a conclusion, is firmly based upon experience.

To use another metaphor, a pure deductive system, like that of arithmetic, hangs from its summit and can be indefinitely extended downwards; an impure deductive system, like that of a natural science, is supported on its empirical basis and can be indefinitely extended upwards. To ask for an ultimate explanation for a scientific theory is to confuse the function of science—that of organizing our empirical knowledge in a way which will enable us to make reliable predictions—with that of mathematics, whose function, apart from that of giving intellectual satisfaction in its own right, is to serve as the cement between the various storeys of scientific systems. The peaks of science may appear to be floating in the clouds, but their foundations are in the hard facts of experience.

There remains to allude to two of the matters relevant to the subject of this book upon which more requires to be said, and to indicate the directions which might profitably be followed in their investigation.

IS REASONABLENESS OF INDUCTIVE BELIEF MEASURABLE?

The account of degrees of probability expounded in Chapter VI of this book was an account solely of the probability concept as used within the body of a science, all the statements assigning definite numerical values to probabilities being empirical propositions of the nature of scientific hypotheses. This empirical sense of probability, it was emphasized, was quite different from that in which a scientific hypothesis H may *itself* be said to be probable. Such a statement has been treated as being equivalent to the statement that it is reasonable to believe H, and the meaning and justification for such a statement has been discussed at length in Chapter VIII. But no meaning has been given to the notion of assigning numerical values to degrees of reasonableness of belief; indeed, no meaning has been given to saying that, of two beliefs, both of which are reasonable, one is more reasonable than the other. A criterion was advocated in Chapter VII for preferring one statistical hypothesis H_1 to another statistical hypothesis H_2; this criterion could be used to define a sense of "more reasonable than" in which the criterion for preferring H_1 to H_2 was used also as the criterion for making belief in H_1 more reasonable than belief in H_2. But even if this were to be done (and I should wish not to make this translation, since it would confuse what we do in selecting among statistical hypotheses), such a comparison of the reasonableness of scientific beliefs would only apply to comparisons of beliefs in statistical hypotheses which assigned different values to the same probability-parameter.

Can any satisfactory general criterion be given for arranging beliefs in scientific hypotheses in an order so that one can be said to be more reasonable than another? To ask for this is less than to ask for a criterion for assigning numerical values to degrees of reasonableness of belief, since it may be perfectly possible to compare things by arranging them in a serial order without being able to measure them. Nevertheless, I do not believe that any criterion can be given for even a non-numerical comparison which is at all satisfactory.

The occasion upon which there is most inclination to speak of one belief as being more reasonable than another is one in which we are comparing two beliefs in the same hypothesis H, when both beliefs have been obtained solely by simple-enumerative induction and when the instances which form the evidence for H in the case of the first belief include as a subclass all the instances which form the evidence for H in the case of the second belief. Here the hypothesis believed is the same in the two cases, but the evidence for it is, in a straightforward sense, stronger in the first case than in the second—in that the evidence in the first case is all that there is in the second case and more. Now if this is the whole of what is meant by saying on such an occasion that the first belief is more reasonable than the second, well and good. But such language may be dangerously misleading, since it may conceal the fact that the reasonable acceptance of a hypothesis is an all-or-none reaction.* Whether it is reasonable for a man to add a belief in the hypothesis H to his body of knowledge and reasonable belief (his rational corpus) depends upon whether his rational corpus contains sufficient evidence for H for it to be valid for him to infer H from the evidence when he is following a particular simple-enumerative inductive policy. If his corpus contains evidence which is sufficient

* The all-or-none character of reasonable acceptance has been emphasized by Felix Kaufmann, *Methodology of the Social Sciences*, Chapters III, IV.

according to the policy, the man is reasonable in adding H to his rational corpus if the simple-enumerative policy concerned is effective (or is believed by him to be effective, or both); if the evidence is insufficient and the man nevertheless adds H to his rational corpus, his action is not covered by that particular simple-enumerative policy. It is not a question of being more or less reasonable in adding H to his rational corpus; if the inductive policy covering his inference is effective (or believed by him to be effective, or both), the addition is a reasonable one and the belief a reasonable belief. Thus two people will both be reasonable in acquiring beliefs in H by following the same simple-enumerative policy even if the evidence which one has to support his belief includes all, and more than all, of the evidence which the other has to support his belief, provided that the inductive policy which they are both following is effective (or believed to be effective, or both).

But, it may be objected, an alternative line of argument will show that it is more reasonable to believe a hypothesis if it is supported by a greater amount of evidence of the same kind. Even if the criterion of reasonableness depending on whether or not the inference is covered by an effective inductive policy is one that does not admit of degree, surely the inductive policies themselves can be arranged in order in such a way that inferences made according to some are more stringent than those made according to others? For example, a policy of simple-enumerative induction which requires one thousand confirming instances (and no refuting instances) to establish a generalization can surely be said to be more stringent than a simple-enumerative policy requiring only one hundred confirming instances. But does it follow that it is more reasonable to believe an inductive conclusion obtained by following the more stringent policy? The same considerations are relevant here as those which settle what value of k is used in a k-rule-of-rejection for statistical hypotheses. If we assume that the hundred-instance simple-enumerative policy and the thousand-instance policy are both effective, a rational decision as to which policy to use should depend upon the practical importance of accepting the hypothesis in question rather than upon whether one policy is more stringent than the other. If the acceptance of the hypothesis will have the most desirable practical consequences, it would surely be foolish to delay its acceptance after enough instances have been examined for it to be established by following one effective policy until more instances had been collected so that it could also be established by following a more stringent policy. And although we could say, if we wished, that it would nevertheless have been more reasonable, independently of all practical considerations, to delay acceptance until it could be established by a more stringent policy, this would be to divorce the notion of reasonableness as applied to belief in the hypothesis from that of the reasonableness of acting upon the belief.

The view taken in this book—that the question of the reasonableness of an inductive belief is bound up with that of the validity of the inductive inferences by which it has been or could be derived—thus leaves no room for the comparison of the reasonableness of different inductive beliefs. If the inference is valid, incorporation of belief in its conclusion into a man's rational corpus is justified, and it is reasonable, having the old reasonable belief, to acquire the new one. There is no question of its being more or less reasonable to do so.

It follows from this view that the content of a person's rational corpus changes discontinuously by jumps. Beliefs that are included in it are found to be false and are discarded, while other beliefs are all the time being added to it. But a belief is discarded at one blow; it does not fade out of the corpus by becoming less and less 'probable' until it falls below the 'probability threshold'. Similarly a belief is added at one blow; it does not fade in by becoming more and more 'probable' until it rises above the threshold. The notion of a scale of probabilities of a hypothesis with a corresponding scale of degrees of reasonableness of belief in the hypothesis is, I believe, a philosopher's myth. The myth has arisen partly from the desire to subsume the 'probability' of hypotheses under a unified theory of probability which will also include the numerically measurable probabilities of events, but partly also from a confusion between the notion of degrees of *reasonableness* of belief and the quite different notion of degrees of *belief*.

Degrees of belief in particular propositions whose truth is now unknown but will be known in the future within some definite period of time can be regarded as being determined by the limit of the odds at which the believer would be prepared to back the truth of the proposition. The stakes in the betting will have to be things which are good in themselves and whose goodnesses are comparable, rather than things which are means-to-a-good like money, where the complication of diminishing marginal utility enters. And an allowance will have to be made for a disinclination to bet at all. But a scheme of hypothetical generalized betting can be devised which will enable degrees of belief to be measured.* If the man's belief is in the occurrence of an event which has an objective probability in the empirical sense of Chapter VI (e.g. that this penny, when next thrown, will fall head upmost), then the odds at which he is prepared to bet, i.e. the degree of his belief, must correspond to this objective probability on pain of the man's standing to lose, by and large, if he bets at these odds on the event's happening, and also bets at these odds on the event's not happening, in a large number of similar cases. If the man's belief is in the occurrence of an event which cannot be regarded as one of a class of similar events (e.g. that a particular horse will win a particular race), there is no objective empirical probability with which the degree of his belief ought to correspond; on the contrary, an average of the degrees of belief of well informed and intelligent betting men who, by and large, do not lose by their betting, as measured by the limit of the odds at which they would be prepared to back the horse, provides the only possibility, so far as I can see, for a definition of the 'probability' of a particular horse winning a particular race.†

However, a man cannot bet on a possibility which takes an infinite future to be realized; so his belief in the truth of a scientific hypothesis cannot be measured by a generalized betting procedure.‡ But what he can bet on is the proposition that the scientific hypothesis which is now believed (by him or by other people) will still be believed in ten years' time, or that a hypothesis

---

* F. P. Ramsey, *The Foundations of Mathematics and other logical essays*, pp. 166ff. See also Rudolf Carnap, *Logical Foundations of Probability*, pp. 165, 235.

† The odds offered by a bookmaker depend less upon his own degree of belief than upon the distribution of degrees of belief among his clients, since he has to protect himself from the risk of ruinous loss by 'making a book'.

‡ Indeed, it would seem difficult to give any meaning whatever to the notion of *partial* belief in a general proposition if we wish to keep any connexion between belief and action, since belief in a general proposition « Every *A* is *B* » goes along with a disposition to act on each occurrence of an *A* as if a *B* had also occurred.

not now believed will come to be believed within the next ten years. Indeed it is possible to bet, not only on the content of a man's corpus of belief, but on the content of his rational corpus of knowledge and reasonable belief. For the discovery of a contrary instance will expel an established universal hypothesis from this rational corpus, and further statistical evidence may require the substitution of an alternative statistical hypothesis for one already included in his corpus. Since betting odds can be attached to the possibility of changes in my rational corpus these will serve to define a subjective sense of "probability" (if we choose to give it that name) according to which it can be said that the probability of Einstein's theory of gravitation being refuted within the next ten years is $\frac{1}{100}$, meaning by this that if I were to bet on its being refuted (or on its not being refuted), I should regard 99 to 1 as fair odds.

The fact that we can bet on changes in a rational corpus within a limited period explains the fact that, although I am unable to recognize differences in the reasonableness of the beliefs which are in my rational corpus, I am able to recognize differences in the *tenacity* with which I hold them. For this tenacity is a symptom of the extent to which I should be prepared to back my belief that the belief which I am holding tenaciously will still be part of my rational corpus in ten years' time.

Since a change in the lowest-level hypotheses in a man's rational corpus will involve a change in some of the highest-level hypotheses, while the highest-level hypotheses can all change with the lowest-level hypotheses remaining the same, the totality of the highest-level hypotheses in his rational corpus is held with less tenacity than are the lowest-level hypotheses. Of course any particular set of highest-level hypotheses may be held very tenaciously, being treated as 'functionally *a priori*' propositions,* but this will only result in the others being held less tenaciously. Generally speaking, a man will be much more prepared to bet at given odds on the lowest-level hypotheses in his rational corpus being the same after ten years than that the highest-level hypotheses will be the same. For the function in his thinking of these highest-level hypotheses is to explain the lowest-level hypotheses which he accepts; after ten years another explanation of the same lowest-level hypotheses may

* Arthur Pap's expression; see above, p. 112.

seem preferable, and he would therefore then substitute these new explanatory hypotheses for the old ones.

Thus, although a negative answer must be given to the original question as to whether the reasonableness of beliefs in general empirical propositions can be measured, or whether such reasonablenesses can even be compared, on the ground that addition to or expulsion from a man's rational corpus is an all-or-none operation, yet a sense can be found for the variations in tenacity with which beliefs are held within the rational corpus, these variations being related to differences in the odds at which the man would be prepared to back his opinion that the beliefs would still be found in his rational corpus after some definite period of years. But I am well aware that there is a great deal more thinking that requires to be done on this subject.

THE CLOSENESS OF FIT OF A DEDUCTIVE SYSTEM

Another way in which the account given in this book of scientific explanation in terms of deductive systems may well be thought to be too cut-and-dried is that the account appears to make no allowance for the looseness of fit to experience of many accepted hypotheses. It is not the case, it may be said, that every universal hypothesis is regarded as having to be rejected by the discovery of one contrary instance; the propounder of the hypothesis may well say that he did not expect it to hold under all circumstances, and that the fact that it does not hold universally does not show that it is not a good approximation to the truth.

The account given in this book will allow for two cases in which a hypothesis will not be regarded as definitively refuted by contrary instances. The first case is that in which the hypothesis is a statistical hypothesis when the rejection of the hypothesis on the evidence of a set of observations is always a provisional rejection which may have to be cancelled on the basis of further evidence. The second case is that in which the hypothesis, though not explicitly of the form of a statistical hypothesis, will be treated as a statistical hypothesis in that it is to be rejected (and only provisionally rejected at that) only if the contrary instances show deviations from the value asserted in the hypothesis which exceed a certain amount. If the deviations are less than this amount, the instances will be taken as being affected by 'errors of observation' rather than as refuting the hypothesis. For example, the experiments performed by Coulomb and others directly to determine the force between two electrically charged bodies, and hence to determine the value of n in the law of electrostatics stating that the force varies with the inverse nth power of the distance between the two bodies, gave values of n most of which were not exactly 2, but which were distributed closely round 2 in a way which would be expected if the observations were subject to random 'errors of observation'. In this case what is happening is that the hypothesis that the electrostatical law is an inverse-square law is taken to be conjoined with the statistical hypothesis that the measurements in the experiments are subject to small random variations round the 'true value'; and it is the conjunction of these two hypotheses, which conjunction is itself a statistical hypothesis, that is subjected to experimental test.

It is worthy of note in this connexion that, if the value that would be given to the inverse power in the law if we were selecting the value from among other possible values by the maximum-likelihood method or some similar method were some number not exactly 2 (e.g. 2·001), this would not imply that it would be most reasonable to accept the hypothesis having 2·001 as the inverse power. For the rationale of the maximum-likelihood method (or of the similar method employed) would be to justify the preference of the value (2·001) which gave maximum likelihood to the experimental results over values (e.g. 2) which gave a slightly less likelihood to these results, other things being equal. But here other things are not equal; for the inverse-square hypothesis, in contrast to the inverse-2·001th-power hypothesis, can take place, along with other highest-level hypotheses, in a more comprehensive deductive system in which other sorts of consequences can be deduced and observed.

TENDENCY STATEMENTS

There is, however, one type of hypothesis which it is more difficult to fit into the view expounded in this book of scientific deductive systems, the meaning of the statements in which is determined by the empirical conditions for their rejection. This is the type of general proposition found in sciences which have not yet reached the present development of the physical sciences and in which the conditions under which a universal law holds are not known. Here all that it is possible to state is that so-and-so has a general tendency to act in such-and-such a way. Scientific hypotheses of this nature will be called "tendency statements". Examples are most of the known laws of psychology,* e.g. that the strength of an association tends to increase with frequency. This statement does not assert that under all circumstances the strength of an association in a person's mind increases with the frequency with which he has experienced it. Nor does it assert that the strength of most associations increases with frequency, which would be equivalent to asserting the statistical hypothesis that the probability is greater than $\frac{1}{2}$ that the strength of an association in a person's mind increases with the frequency with which he has experienced it. What presumably the tendency statement does assert is that the strength of an association *satisfying certain unspecified conditions* increases with its frequency. The peculiarity of a tendency statement is that it states that, if a thing is C, then if it is also A it is also B, where « If a thing is A, it is also B» (i.e. Every A is B) is an ordinary scientific hypothesis, and C is an unspecified property.

The proposition « If a thing is C, then if it is also A it is also B» is equivalent to the proposition « If a thing is both C and A, then it is also B». It is the fact that a tendency statement is a conditional statement with an unspecified antecedent condition which creates the difficulty. We are well accustomed to conditionals in which part only of the antecedent is explicitly mentioned, as in many nomic conditionals, whether indicative or subjunctive (e.g. If a billiard ball is released in mid-air, it will fall to the ground). But whatever may be the case when such a conditional is merely being considered, when such a conditional is asserted the asserter must be prepared explicitly to specify the whole of the antecedent, if asked to do so.† The asserter of a tendency statement, however, is quite unable to mention the part of the condition represented by C, since he has no notion what it is.

This makes it impossible to treat the assertion of a tendency statement as the assertion of a general proposition which would

---

* See the contributions of B. A. Farrell and Margaret Masterman to the Symposium on "Causal Laws in Psychology", *Aristotelian Society Supplementary Volume* 23 (1949).

† Compare the asserting of hypotheticals, discussed on p. 315.

be refuted by the discovery of a contrary instance. For the observation of an A which is not a B will not necessarily be regarded as compelling us to reject the tendency statement. It will always, be possible to say that in the prima facie contrary instance of an A which is not a B, the thing which is A does not satisfy the unknown condition C, so that the existence of an A which is not a B is perfectly compatible with the truth of the general proposition that every A which is a C is a B. The way in which we use tendency statements, therefore, appears to be different from that in which we use ordinary scientific hypotheses, and it may be questioned whether the notion of functioning in our thinking within a scientific deductive system is applicable to them.

It seems to me that the distinction which has been emphasized in this book between a calculus concerned with words and other symbols and a deductive system which is the interpretation of such a calculus will enable tendency statements to find a place within the way of scientific thinking expounded in this book without any necessity for giving a radically different account of their use. Using the calculi developed in Chapter II the general proposition «Everything which is both C and A is also B» would naturally be expressed by the formula

$$(\gamma\alpha) \leftrightarrow (\beta(\gamma\alpha)),$$

where α and β are taken as standing for the class of A's and of B's respectively and γ as standing for the class of C's, which in a tendency statement is unspecified. There is, however, an alternative method for expressing this proposition by means of a different formula occurring in the same calculi, namely the formula

$$\alpha \leftrightarrow (\beta\alpha),$$

provided that the calculi are given a different interpretation so that the Greek letters α and β are taken as standing, not for the class of A's and of B's respectively, but for the class of things which are both C and A, and for the class of things which are both C and B, respectively.* The elements in the calculi are then given an interpretation as representing classes, not any classes whatever but

---

* The new formula will also express the general proposition if β is interpreted as standing, as originally, for the class of B's. But it is simpler to discuss the matter by giving an interpretation which restricts the classes designated both by α and by β to being subclasses of the class of C's.

only classes which are subclasses of the class of C's. In old-fashioned logical language the 'universe of discourse' to which the symbols refer has been restricted to the universe of things which have the unspecified property.*

Now while it is not possible to interpret $(\gamma\alpha) \leftrightarrow (\beta(\gamma\alpha))$ without knowing what is the class of C's, for γ has to be interpreted as standing for it, it is possible to give a sort of interpretation of $\alpha \leftrightarrow (\beta\alpha)$ without knowing what the class of C's may be. We can, as it were, 'interpret' the elements of the calculi as standing, neither for any classes whatever nor for any subclasses of a known class, but for any subclasses of an unspecified class. The 'deductive system' thus obtained will refer only to this unspecified class. Such a system will not consist of propositions, properly speaking: the formula $\alpha \leftrightarrow (\beta\alpha)$ will not be interpreted as the *proposition* that everything which is both C and A is also B, since no meaning will have been attached to the property-symbol "C". But for some purposes we can pretend that such quasi-propositions are propositions. For their logical relationships within the quasi-deductive system will be similar to the logical relationships holding in a deductive system in which $\alpha \leftrightarrow (\beta\alpha)$ is interpreted as the proposition that every A is B.

If $\alpha \leftrightarrow (\beta\alpha)$ is interpreted within the unknown restricted universe of discourse as standing for the general quasi-proposition that everything which both has some unspecified property and has the property A also has the property B, this quasi-proposition will not be refuted by discovering a thing which is A without being B. For the A which is not a B may well lack the unspecified property and so will fail to be a contrary intance of the general quasi-proposition represented by $\alpha \leftrightarrow (\beta\alpha)$. Consequently the quasi-deductive system consisting of such quasi-propositions will not be testable by experience in the way in which a scientific deductive system is testable. Nevertheless, experience will be relevant to the way in which the system is used. For if an A is discovered which is not a B, either the quasi-proposition represented by $\alpha \leftrightarrow (\beta\alpha)$ must be rejected or the observed case of an A which is not a B must be taken as lacking the unspecified property. In the latter

* The 'universe-class' (the complement of the null-class) will be the class of C's, and (if the calculi are extended to include the Huntington Calculus of Chapter III) will be designated by o'.

case something will then be known about this unknown property, namely that it does not apply to the observed case. A large number of such observed cases, combined with a determination not to reject the quasi-proposition on the evidence of these cases, will result in our acquiring the knowledge about the unspecified condition that it does not hold in any of these observed cases. Frequently, after observation of a fair number of these cases, it is possible to make a guess as to what the unspecified condition may be; if a guess can be made an ordinary deductive system can be substituted for the tendency quasi-deductive system, and the tendency statement converted into an ordinary scientific hypothesis which can be tested in the usual way.

Such a change of a tendency statement into an ordinary hypothesis depends upon a determination to retain in the rational corpus the tendency statement that everything which both satisfies an unspecified condition and has the property A also has the property B even though cases of things which are A without being B are observed. There would be little purpose in continuing to retain the tendency statement in the face of such prima facie contrary instances unless the tendency statement did not stand alone as the sole 'hypothesis' of a quasi-deductive system. But if there are a number of tendency statements all of which can be regarded as belonging to the same quasi-deductive system with the same restricted universe of discourse, their logical relationships can be examined independently of a knowledge of the restricting condition. We can thus ascertain which of them stand or fall with which others. We may then wish to retain one tendency statement because to abandon it would require us also to abandon other tendency statements standing at a higher level in the quasi-deductive system. And we may wish not to abandon these higher-level tendency statements if other lower-level tendency statements which were equally consequences in the quasi-deductive system of these higher-level statements had been confirmed and not prima facie refuted.

The moral of this for sciences like psychology and the social sciences which make great use of tendency statements is that to assert an isolated tendency statement is to say very little. What is frequently profitable is to put forward a quasi-deductive system of tendency statements all subject to the same, though unknown, condition.* Observations which are prima facie contrary to the lowest-level tendency statements in such a system can then be used to give information about the unknown condition to which the whole system is subject.

For the sake of simplicity only universal tendency statements have been considered. But there can also be statistical tendency statements of the form "Under unspecified conditions such-and-such a percentage of the A's are B's". These statements can be treated on similar lines by interpreting the calculus which would be representing what would be the statistical deductive system were the statements to be ordinary statistical hypotheses as instead representing quasi-propositions about things in a universe of discourse limited in an unspecified way.

As before, when we were considering the status of theoretical concepts in scientific deductive systems, so now a consideration of scientific tendency statements requires us to become self-conscious about our way of scientific thinking, and to think explicitly about the way in which these statements are given a meaning. To say, as I have done, that what they mean are not propositions but quasi-propositions which resemble propositions in that they can be arranged similarly in logical order is only a roundabout way of saying that systems of such quasi-propositions, like systems of propositions, are representable by calculi. We come back, as always in philosophizing, to our means of expression. The way in which a calculus representing a system of tendency statements is used in our scientific thinking is different from, though related to, the way in which a calculus representing an ordinary scientific deductive system is used. An explanation of the difference between tendency statements and ordinary hypotheses can only be given in terms of a difference in the ways in which a calculus is to be interpreted.

* To be profitable the system must be representable by a calculus in which formulae are genuinely derived, according to the rules of the calculus, from other formulae. "No calculus without calculation"; the mere translation of tendency statements into mathematical language (as, for example, in Kurt Lewin's *Principles of Topological Psychology*) is not sufficient to make a quasi-deductive system out of them. The essence of mathematics is not its symbolism, but its methods of deduction. The value, for example, of von Neumann and Morgenstern's mathematicizing of the tendency statements of economic behaviour by their Theory of Games lies, not in merely using the language of theory of sets of points but in the fact that calculations can be made in a calculus using such language; for example, on the game interpretation of the calculus, as to when to raise your opponent in a simple form of two-handed poker (*NM*, § 19).

CONCLUDING REMARKS: IS SCIENCE INVENTION OR DISCOVERY?

A calculus is an artefact, and the interpretation of a calculus is a resolution to employ this artificial tool to organize our experience in a particular way—in a way appropriate both to enable us to make predictions as to the future course of our experience and to yield us intellectual satisfaction by providing a systematic explanation. To what extent, then, should an established scientific deductive system be regarded as a free creation of the human mind, and to what extent should it be regarded as giving an objective account of the facts of nature?

This question would be difficult and important if we wished to maintain, with Kant, that some of the features of scientific laws were products of the human mind, so that in knowing such a law we were reading in nature something which we had written in the mental act of knowing. It would then be essential for a philosopher of science to distinguish between what was to be put down to Nature and what to the knowing mind in each scientific law. If we do not take the Kantian standpoint, the question becomes merely that of distinguishing between what is due to Nature and what is due to our powers of representing by means of our statements and formulae an ordered system of scientific laws, the answer to which is easy to give. Nature does not provide separately both facts and laws; our statements of laws are a way of describing observed facts and of predicting facts at present unobserved. The form of a statement of a scientific hypothesis, and its use to express a general proposition, is a human device; what is due to Nature are the observable facts which refute or fail to refute the scientific hypothesis. The meaning of all the symbols we use—those we use for denoting features of our immediate experience as well as the theoretical terms which we use with a sophisticated interpretation—is determined by our decision to use them in the way we do. What is objective and not a matter for our decision is whether the sentences and formulae containing these symbols express propositions which are true or express propositions which are false. A scientific deductive system of hypotheses is a human artefact in that it is a human interpretation of a humanly constructed calculus. But that it is not only a deductive system of hypotheses but also a deductive system of laws, requires that the hypotheses should be true, which is not in our power to decide.

We construct calculi how we like and interpret them how we like. If we interpret a calculus as a deductive system, by that act of interpretation we commit ourselves to taking propositions which are conclusions from premisses in the system as being true if we take the premisses to be true. If we interpret the calculus as being a scientific deductive system, we hand over to Nature the task of deciding whether any of the contingent lowest-level conclusions are false. This objective test of falsity it is which makes the deductive system, in whose construction we have very great freedom, a deductive system of scientific hypotheses. Man proposes a system of hypotheses: Nature disposes of its truth or falsity. Man invents a scientific system, and then discovers whether or not it accords with observed fact.

The system to be invented may have been suggested by observation, but until hypotheses have been propounded there is nothing to be tested by discovery. One of the objects of this book has been to emphasize the freedom of scientists to construct their deductive systems, and to use theoretical terms in the calculi representing their deductive systems, in any way that they think most profitable. The function of mathematics in science has been shown to be, not that of admitting only hypotheses of a pre-ordained form, but that of providing a variety of methods for arranging hypotheses in a system; knowledge of new branches of mathematics opens up new possibilities for the construction of such systems. The business of a philosopher of science is primarily to make clear what is happening in scientific thinking. If this clarification has the secondary effect of encouraging scientists to construct deductive systems and to use theoretical concepts freely, it will, I believe, assist in the progress of science as well as in the better understanding of what science is doing.

INDEX

Heavy type denotes pages where the use of a term is defined or otherwise explained. Terms occurring frequently throughout the book are only indexed at specially important places.

INDEX

Hertz, H. R., 90 f., 96, 111.
'How?' question, 348 f.
Hume, D., x, 10, 15, 293 ff., 300 f.,
    304, 308, 310n., 317, 320n., 344n.
Huntington, E. V., 60 n.
hyperclass, 134
    probability, 135
    omniselectional, 135
hypothesis(es), scientific, 2, 12, 14 n.,
    123, 293
    in applied deductive system, 55
    Campbellian, 99, 100, 101, 104 ff.,
        110 f., 125, 127
    confirmation of, 13 f., 19 n., 255,
        257 ff.
    conjunction of, 18 f.
    contrary instance of, 14
    establishment of, 13, 14, 262
    fundamental, 112
    instance of, 14
    levels of, 12
    probability of, 119 f., 354, 357
    refutation of, 13 f., 19
    rejection of, 117
    simple, 186
    statistical, 115, 116, 131, 151, 160,
        192 f., 196 ff., 354 f.
    statistical, alternative, 196 ff., 354f.
    testing of, 14, 19
    universal, 115
hypothetical,
    general, 295 n.
    indicative, 314
    subjunctive, 295, 314
'hypothetical infinite population',
    126 ff., 152 n.
hypothetico-deductive establishment,
    299
hypothetico-deductive method, 9, 261,
    264, 303, 353
hypothetico-deductive support, 322

identity (of two classes), 32, 33 f., 47,
    50, 57
idle formulae, 60, 72, 75
impotence, postulate of, 109
inclusion (of one class in another), 33
indicative hypothetical, 314
induction, 257 n.
    contrasted with deduction, 257 ff.,
        292
    by elimination, 260 f., 264, 273
    justification of, 199 f., 255 ff; 294,
        310, 342

logic of, 11 f., 21, 196
perfect, 14 n.
'presupposition' of, 275
by simple enumeration, 10 f., 260 f.,
    264, 273 ff., 278, 280, 283 ff., 289,
    303, 355 f.
validity of, 257, 264 f., 271, 357
inductive belief, reasonableness of,
    264, 269 ff.
inductive policy, 261, 294, 366 f.
    predictive reliability of, 264 ff.
inductive principles of inference, 260 f.
inference, 276, 278
    conclusion in, 278
    objectively valid, 281
    premiss in, 278
    statistical, 199, 263
    subjectively valid, 281
    validity of, 278 ff.
    see also deduction, induction
initial formulae, 25
    of First Calculus, 30
    of Huntington Calculus, 60
    of Second Calculus, 37
instincts, 340
intensional action, 311 ff., 324 ff., 328,
    333 n.
interpretation (of a calculus), 27
    of First Calculus, 31, 35
    of Huntington Calculus, 60
    as quasi-deductive system, 363
    of Second Calculus, 40
intersection (meet, logical product) of
    two classes, 32, 33 f.
interval, 139 n., 183 ff.
invariance (of rules of rejection), 177 f.

Jeffreys, H., xii, 117, 183 ff., 196,
    251 ff., 259, 294
Johnson, W. E., 41, 279 n., 293, 305

Kant, I., 367
Kaufmann, F., 262 f., 355 n.
Kendall, M. G., 171 n., 240 n.
Keynes, J. M. (Baron), xii, 117, 196,
    251 ff., 258 f., 275
Keynes, J. N., 295 n.
Kinetic Theory of Gases, 115 f.
Kneale, W. C., 119 n., 120, 124 n.,
    191 n., 251, 258, 272, 301 n.,
    304
knowledge, immediate (direct, indu-
    bitable, incorrigible), 7
Kolmogoroff, A., 128 n., 130 n.

372